RESOURCE ECOLOGY: SPATIAL AND TEMPORAL DYNAMICS OF FORAGING

Wageningen UR Frontis Series

VOLUME 23

Series editor:

R.J. Bogers
Frontis – Wageningen International Nucleus for Strategic Expertise, Wageningen University and Research Centre, The Netherlands

Online version at http://library.wur.nl/frontis/

The titles published in this series are listed at the end of this volume.

RESOURCE ECOLOGY

Spatial and Temporal Dynamics of Foraging

Edited by

HERBERT H.T. PRINS
Resource Ecology Group,
Wageningen University, The Netherlands

and

FRANK VAN LANGEVELDE
Resource Ecology Group,
Wageningen University, The Netherlands

Springer

A C.I.P. Catalogue record for this book is available from the Library of Congress.

ISBN 978-1-4020-6849-2 (PB)
ISBN 978-1-4020-6848-5 (HB)

Published by Springer,
P.O. Box 17, 3300 AA Dordrecht, The Netherlands.

www.springer.com

Printed on acid-free paper

CONTENTS

	Preface	**ix**
1.	**Introduction to resource ecology** *Frank van Langevelde and Herbert H.T. Prins (The Netherlands)*	**1**
2A.	**Mechanisms determining large-herbivore distribution** *Derek W. Bailey and Frederick D. Provenza (USA)*	**7**
2B.	**Comments on "Mechanisms determining large-herbivore distribution"** *Sipke E. van Wieren (The Netherlands), Michael Drescher (Canada) and Willem F. de Boer (The Netherlands)*	**29**
3A.	**Spatial statistics to quantify patterns of herd dispersion in a savanna herbivore community** *Alfred Stein (The Netherlands) and Nicholas Georgiadis (Kenya)*	**33**
3B.	**Comments on "Spatial statistics to quantify patterns of herd dispersion in a savanna herbivore community"** *Herbert H.T. Prins, Sipke E. van Wieren and Arend M.H. Brunsting (The Netherlands)*	**53**
4A.	**Resource distribution and dynamics: mapping herbivore resources** *Andrew K. Skidmore and Jelle G. Ferwerda (The Netherlands)*	**57**
4B.	**Comments on "Resource distribution and dynamics: mapping herbivore resources"** *Herbert H.T. Prins, Willem F. de Boer and Frank van Langevelde (The Netherlands)*	**79**
5A.	**Foraging in a heterogeneous environment: intake and diet choice** *Emilio A. Laca (USA)*	**81**
5B.	**Comments on "Foraging in a heterogeneous environment: intake and diet choice"** *Michael Drescher (Canada), Herbert H.T. Prins (The Netherlands) and Arend M.H. Brunsting (The Netherlands)*	**101**

6A. Predictive modelling of patch use by terrestrial herbivores 105
John M. Fryxell (Canada)

6B. Comments on "Predictive modelling of patch use by terrestrial herbivores" 125
Ignas M.A. Heitkönig (The Netherlands), Michael Drescher (Canada) and Willem F. de Boer (The Netherlands)

7A. Assembling a diet from different places 129
Herbert H.T. Prins and Frank van Langevelde (The Netherlands)

7B. Comments on "Assembling a diet from different places" 157
Iain J. Gordon (Australia)

8A. Effects of temporal variability in resources on foraging behaviour 159
Norman Owen-Smith (South Africa)

8B. Comments on "Effects of temporal variability in resources on foraging behaviour" 183
Willem F. de Boer and Arend M.H. Brunsting (The Netherlands)

9A. Large-scale movements of large herbivores: livestock following changes in seasonal forage supply 187
Randall B. Boone, Shauna B. BurnSilver, Jeffrey S. Worden, Kathleen A. Galvin and N. Thompson Hobbs (USA)

9B. Comments on "Large-scale movements of large herbivores: livestock following changes in seasonal forage supply" 207
Willem F. de Boer, Thomas A. Groen, Ignas M.A. Heitkönig and Koen Kramer (The Netherlands)

10A. Relevance of key resource areas for large-scale movements of livestock 211
Paul Scholte and Joost Brouwer (The Netherlands)

10B. Comments on "Relevance of key resource areas for large-scale movements of livestock" 233
Herbert H.T. Prins, Arend M.H. Brunsting and Frank van Langevelde (The Netherlands)

11A. Structuring herbivore communities: the role of habitat and diet 237
Sipke E. van Wieren and Frank van Langevelde (The Netherlands)

11B. Comments on "Structuring herbivore communities: the role of habitat and diet" **263**
Herbert H.T. Prins, Arend M.H. Brunsting and Willem F. de Boer (The Netherlands)

12. Prospects for further development of resource ecology **267**
Herbert H.T. Prins and Frank van Langevelde (The Netherlands)

List of references **273**

List of contributors **303**

PREFACE

This book focuses on 'resource ecology', which we defined as 'the ecology of trophic interactions between consumers and their resources'. In our endeavour to shape the discipline of 'resource ecology' further, we invited some world-class scientists to think with us. We thus organised a symposium where we made sure that we paid attention to the original Greek meaning of the word: we discussed, ate and drank together and, for health reasons, made sure that we hade enough time for strolling and bicycling along the banks of the River Rhine. We even climbed a beautiful mediaeval church tower dedicated to Saint Cunera dating back to the year that Columbus sailed for America. In ecology, spatial ecology is an up-and-coming domain of science. It addresses the effects of space and scale on the dynamics of individual species and on the structure and dynamics of multi-species assemblages. More than a few recent studies demonstrate the significance of taking into account the spatial structure of resources on the population dynamics and assemblage structure of consumers. We believe that bringing together scientists specialised in foraging theory with those who know much about spatial ecology can create a new context from which new theory will emerge.

Every chapter was written on invitation. We as editors had outlined what we had in mind and sent that outline to the selected authors. We asked them to have their chapter ready before we started the symposium, and then we sent each contribution to two other members of the symposium and asked them to prepare a thoughtful review of the chapter. We were very happy to invite a number of the postgraduate students from our own research group to attend the symposium: we asked them to take extensive notes but especially to join in the discussions. Because everyone who was present had read the chapters prior to the symposium, we only asked the authors to shortly introduce their chapter, but we gave quite some time to the reviewers to give their comments and share with us their positive ideas about the chapter, but also their critical thoughts and remarks. We then started extensive discussions about the positive and negative sides of each chapter and tried to find our way towards a common understanding. Notes of these discussions were extensively taken by a team of a postgraduate student and a senior researcher. After the symposium, these notes and the comments by the reviewers were sent to the authors. We as editors had independently made our own comments on the written texts; these comments were also sent to the authors. Every author then changed his original paper as he thought fit, after which we went through a second round of editorial interactions. In this process of peer-reviewing some chapters did not make it to the end-stage; we sincerely hope that if we prepare a second edition of this book the research reported in those chapters can be incorporated then. Every chapter was thus double-refereed in a very strict sense of the word.

As stated, we have selected the authors and the range of subjects with a special audience in mind, and a special scientific goal. We have edited this volume accordingly. First, the audience. In the sciences nowadays postgraduate students and post-docs form the backbone of most laboratories and research groups. Together with more senior scientists, they test the new ideas and develop new theories. To show that the forefront of science is, by definition, unsettled, uncertain and undecided, we have included in the present volume a series of commentaries. Each chapter is thus followed by a commentary that highlights some of the discussions we had during the symposium. Even though each chapter was carefully refereed and edited, there is still an ongoing debate about pros and cons. We were thus particularly pleased that a group of postgraduate students were with us during the symposium because the present volume is specifically aimed at post-docs, postgraduate students and final-year undergraduate students. The commentaries also serve a second purpose, namely to demonstrate to students that criticism on each other's work is normal in science. Because we believe that new theory is needed to explain the coexistence of species or the numerical abundance of assorted animal species in various assemblages, which is one of the top unsolved riddles of the 21st century. In this book, we will contribute to the advancement of such new theory. In resource ecology, foraging is the central process because it leads to growth, survival and reproduction of the animal. Resource ecology thus deals with foraging and ultimately with fitness of the consumer, and we believe that a deeper understanding of resource use is a key to unlock the door obscuring coexistence and species diversity rules. We have concentrated the contributions that we solicited from scientists dealing with above-ground herbivory only. In future, we would like to include studies on carnivores, parasites and diseases.

Every chapter ends with some new testable hypotheses: we truly hope that testing these hypotheses will bring the exciting science of resource ecology further.

This is a good place to thank some people whose efforts have been very important for us. First, Rob Bogers: he is the drive and visionary behind the Frontis series of which this book forms a part, and he selected our topic as worthy of receiving finances. Rob and Petra van Boetzelaer helped us to run the symposium smoothly. We thank the Royal Academy for Arts and Sciences *(Koninklijke Nederlandse Akademie van Wetenschappen)* in Amsterdam and the Wageningen graduate school 'Production Ecology and Resource Conservation' for additional generous financial support. Much administrative work for the smooth handling of finances and essential paper work was done by Gerda Martin and Willemien Schouten, for which we are immensely grateful. Herman van Oeveren skilfully redesigned and redrew all graphics to ensure uniform and harmonised figures. Herman and Margreet Mulder helped us with finalising the list of cited literature. We especially thank the (former) PhD students Jasja Dekker, Michael Drescher, Jelle Ferwerda, Thomas Groen, Geerten Hengeveld and Nicol Heuermann for their continued input and support.

Wageningen, Melbourne, Sheffield, 2007

The editors: Herbert H.T. Prins and Frank van Langevelde

CHAPTER 1

INTRODUCTION TO RESOURCE ECOLOGY

FRANK VAN LANGEVELDE AND HERBERT H.T. PRINS

Resource Ecology Group, Wageningen University, P.O. Box 47,
6700 AA Wageningen, The Netherlands
E-mail: frank.vanlangevelde@wur.nl; herbert.prins@wur.nl

> "The question is often raised: why not stay with the real world and generalize our experiences in some succinctly descriptive form? The only answer is that such an approach never proves adequate. In evolutionary biology, it produces *inductive* generalizations that are encapsulated in tendencies or 'rules' (e.g. Bergmann's rule). Causal explanations, the heart of any science, are hard to reach and often impossible to prove by means of such concepts. The descriptive, natural-history stage of science is eventually replaced by a *deductive* theoretical stage, basically mathematical in nature, which creates the abstractions and measurements necessary to deepen causal analysis. (…) The purpose of mathematical theory is to deal with "all possible worlds". The purpose of experiments and field [work] is to deal with the real world: To measure the parameters, to search for new parameters, and to improve the theory which is ultimately our most effective way of viewing the real world." (Wilson and Bossert 1971, pp. 40 – 41)

Resource ecology, the ecology of trophic interactions between consumers and their resources, is central in ecology. It addresses fundamental aspects of the interactions between consumers and resources, and includes competition, plant–nutrient relationships, and predator–prey relationships such as herbivory, parasitism and carnivory. Resource ecology also provides the basis for understanding the diversity, structure and dynamics of multi-species assemblages. From the consumer's perspective, resources, such as energy, nutrients and water, are prerequisites of life that have to be acquired. Therefore, in resource ecology, foraging is the central process because it leads to growth, survival and reproduction of the animal. Resource ecology thus deals with foraging and ultimately with fitness of the consumer.

H.H.T. Prins and F. van Langevelde (eds.), Resource Ecology: Spatial and Temporal Dynamics of Foraging, 1-6.

Resource ecology as link between foraging theory and spatial ecology

Since a consumer's resources are often heterogeneously distributed and exposed to changing conditions, the search for food by animals is unavoidably uneven in space and time. Foraging theory has a long tradition that addresses questions how animals search and forage and what they should do so as to maximise their fitness, based on (i) how different possible behaviours affect fitness; and (ii) what the decision variables are to maximise fitness. Foraging theory has considerable success in explaining observations of foraging behaviour. As foraging behaviour is largely determined by the spatial distribution and variability in time of the resources, many studies recently have been done to investigate movements and spatial decisions in foraging at various spatial scales and under variability in time. Accordingly, resource ecology forms a bridge between the well-developed foraging theory and the emerging field of spatial ecology.

Spatial ecology addresses the effect of space on the dynamics of individual species and on the structure, dynamics, diversity and stability of multi-species assemblages. Although it is an emerging field, numerous theoretical and empirical studies showed the importance of considering the spatial structure of resources on the population dynamics and assemblage structure of consumers. The link between foraging theory and spatial ecology sets resource ecology in a new context from which new theory can emerge. We believe that new theory is needed because existing theories and models appear to be insufficient to explain the co-existence of species or the numerical abundance of assorted animal species in various assemblages. In this book, we will contribute to the advancement of such new theory.

LARGE HERBIVORES AS MODEL SPECIES TO DEVELOP THEORY ON RESOURCE ECOLOGY

Relation between foraging behaviour of animals, their population dynamics and the structure of the assemblage

The theme of the book is the relationship between foraging behaviour, population dynamics and assemblage structure of animals on the one hand, and the spatial heterogeneity and variability in time of their resources on the other. Even though there are many trophic interactions, we focus on herbivory, in particular on the interactions between large mammalian herbivores and the vegetation. What is a 'large herbivore'? We, as other researchers have done, defined a 'large herbivore' as a terrestrial mammal heavier than 5 kg in weight, which obtains most of its diet from vegetative plant parts. This definition includes most antelopes, deer and bovids, for instance, but excludes most rodents, geese and hares. The interactions between large herbivores and the vegetation are appealing because foragers, especially, large herbivores:

- have a large impact on the availability and quality of their resources and therefore affect the spatial resource heterogeneity
- use or can use quite accurate spatial memory in their searching for resources so that they can anticipate to the spatial heterogeneity and variability in time of their resources
- spend the majority of the day searching and foraging
- are highly mobile and have the ability to actively select locations to forage high-quality food or avoid depletion of food or lack of other resources such as water.

Moreover, large herbivores often occur in high diversity, for example on African savannas, and this group receives much attention from the point of view of conservation. Last but not least, the behaviour and requirements of large herbivores are relatively well studied (see the numerous references in this book), which makes them useful model organisms to study the relationship between foraging behaviour, population dynamics and assemblage structure of animals and the spatial heterogeneity and variability in time of their resources.

In the book, we deal with these issues in different subjects to quantify and synthesise resource ecology: from the dynamics in the spatial ecology of large herbivores, their foraging behaviour and their large-scale movements, through their population dynamics, to the structure of the assemblages in which these large herbivores occur. Although we focus on large herbivores in a spatially heterogeneous environment where the grazing resources follow environmental change in time as a particular trophic interaction, we are sure that many mechanisms and principles as discussed in this book are of relevance for understanding other trophic interactions.

Many consumers experience patchy distribution of their resources with certain distances between these patches. Patches are defined as localities (areas) that are more or less homogeneous with respect to a measured variable; in this case the variable is 'food' or 'grazing resource'. These patches can be distinguished at different spatial scales: from vegetated patches at fine scale (e.g., tufts of grass) to disjoint grazing areas at coarse scale (e.g., meadows in an otherwise closed forest). The herbivores may have incomplete knowledge about the spatial distribution of their resources and how this changes over time. In the case of this imperfect knowledge, it is impossible for them to predict accurately where to find sufficient food of sufficient quality. Herbivores have, however, adopted search strategies and have spatial memory to reduce this gap between the true distribution of food (and its quality) and their (imperfect) knowledge about this distribution. This book reviews the current state of knowledge on foraging animals and their strategies to cope with spatial heterogeneity and variability in time of their resources.

Focus of the book is large herbivores in spatially heterogeneous environments where grazing resources follow changes in time

Because of the similarity in the response to spatial heterogeneity and change in time of the grazing resources, the book includes wildlife, free-ranging livestock, and livestock of pastoralists. Traditionally, also pastoralists adopted large-scale movement to secure the availability and quality of the resources for their stock. In both wildlife and pastoral systems, the effects of spatial constraints on migratory movements are well studied. The book enables us to integrate this knowledge. Especially under arid and semi-arid conditions, environmental changeability in time is the rule rather than exception. The book, however, does not exclusively focus on these regions. Also in other regions, such as in temperate regions, there is seasonal variation in growth of the grazing resources due to, for example, changes in temperature and snow cover. In many regions of the world, such large-scale climatic oscillations have an effect on large herbivores and have therefore been included in the book.

Large herbivores adopted a range of strategies to deal with spatially heterogeneous resources that experience change over time, such as trait plasticity in physiology, behaviour, morphology and life history. There are, however, both advantages and disadvantages to such strategies. The disadvantages are costs for the individuals, often expressed in energy, that ultimately reduce the lifetime reproductive success. These strategies are trade-offs between costs and benefits subject to natural selection. At present, the relative costs and benefits associated with different strategies of herbivores that deal with spatial heterogeneity and variability in time of the grazing resources are poorly understood. For example, it is hypothesised that by grazing selectively, animals can achieve nutrient intake rates higher than the average from the environment as a whole. This, however, requires searching of the animal which increases the costs of foraging. At present, it is unsolved how this increased nutrient intake should be set off against the increased energy expenditure. Another issue that needs more thought is whether consumers are selected for maximisation of energy intake or whether they strive to minimise foraging time, or, perhaps to maximise instantaneous or daily intake. These and other hypotheses about the foraging behaviour of large herbivores in spatially heterogeneous environments affected by change in time are discussed in the various chapters.

LAY OUT OF THE BOOK

Each chapter reviews recent developments in resource ecology. At the end of each chapter several testable hypotheses are presented that mark out the current frontiers of this science. The chapters are followed by a comment that discusses the chapter. The commentaries are written to stimulate scientific debate about the issues raised in the chapters.

The first three chapters deal with the distribution of animals and their resources. ***Chapter 2*** discusses several mechanisms that determine the distribution of large herbivores at different spatial and temporal scales. The authors propose a new hypothesis to explain observed foraging behaviour, namely the satiety hypothesis. The satiety hypothesis has been used to explain the avoidance of toxins and the acquisition of nutrients in diet selection. The authors suggest that the satiety hypothesis can be used to better account for the variability in feeding site selection, and thus to better explain the distribution of large herbivores. ***Chapter 3*** further expands on the distribution of animals and provides statistical techniques to describe more accurately patterns in animal distribution. The authors argue that the techniques they propose (F-, G- and J-functions) can be used to better explain the spatial distribution of animals. They analyse the spatial distribution of herds of large herbivores in Laikipia, central Kenya, and discuss possible causes of this pattern. ***Chapter 4*** focuses on the description of the resource distribution and dynamics. The authors present examples where remote sensing is used as tool for producing high-spatial-resolution impressions of the variability of the landscape, and in particular land cover. The authors illustrate this use by mapping nitrogen concentration and phenolic-compound levels in grass and trees in the Kruger National Park, with a spatial resolution of 4 meters. These techniques open doors for new lines of research, where the distribution of herbivores can be linked to the actual resource distribution.

The next set of chapters focus on foraging behaviour of individuals and make predictions about their population dynamics. ***Chapter 5*** deals with intake and diet choice of animals. The author argues that heterogeneity and average herbage mass are frequently related, so that measured effects on intake cannot be unequivocally attributed to total herbage mass. The author concludes that coarser resolution of heterogeneity allows a greater selectivity, and he illustrates this with several examples. ***Chapter 6*** proceeds with the selection of patches of resources. The author shows that trade-offs between resource quality and abundance can change traditional models of patch use. Two aspects of patch use decisions are analysed in this chapter: which patches to visit and how long to stay in a patch, once visited? Empirical data for large herbivores often suggest that optimality principles are often useful in explaining which patches are used in a landscape, but are less successful at explaining how long herbivores choose to stay in a particular patch. ***Chapter 7*** continues with the selection of patches, but includes the intake of different resource types. These resources are unequally distributed over the landscapes, and it is only seldom that food of a herbivore at a given spot exactly matches its requirements. The authors introduce a modelling approach to consider the different satisficing requirements of herbivores. This yields new insights into the causality of the differential way that these animals use the same landscape. ***Chapter 8*** includes temporal variability in explaining foraging behaviour of large herbivores. The author considers how large herbivores adjust their foraging behaviour to cope with variability over different temporal frames. He outlines the conceptual foundation for 'adaptive resource ecology', covering changes in diet composition, daily time allocation, foraging movements, metabolic rate, digestive capacity and fat stores.

The next two chapters link large-scale movement of animals with their population dynamics. ***Chapter 9*** starts with looking at large-scale movements of large herbivores and focus on livestock herded by transhumant pastoralists. The authors analyse changes in the mobility of three pastoral groups, the Aymara of the South-American highlands, Mongolians, and the Maasai of Kenya and Tanzania. They show that pastoralists have successfully evolved methods of herding livestock to access adequate forage in areas of variable climate. ***Chapter 10*** continues with analysing large-scale movements of livestock and make a link between movements and so-called key resource areas. The authors discuss the various assumptions and conclusions regarding key resources and key resource areas, using the floodplains of the Sahel, especially those of Waza-Logone in Cameroon, as examples. They conclude by challenging the relevance of key resources and key resource areas for large-scale movements of livestock.

Chapter 11 illustrates that the mechanisms as discussed in the previous chapter have implications for explaining species diversity. This chapter address the question "Why are there so many species?" with a focus on the diversity of the ungulate community in Kruger National Park. The authors review several mechanisms of resource specialisation between herbivore species that allow coexistence, ranging from diet specialisation and habitat selection to spatial heterogeneity in resources. The authors argue that the focus on the constraints on species' exclusive resources governed by spatial heterogeneity is a useful tool for understanding how competitive interactions structure communities and limit species diversity.

Chapter 12 ends with providing prospects for further development of resource ecology. This chapter proposes six new directions for future research in the field of resource ecology based on the chapters, the hypotheses proposed in these chapters and the comments on the chapters.

CHAPTER 2A

MECHANISMS DETERMINING LARGE-HERBIVORE DISTRIBUTION

DEREK W. BAILEY[#] AND FREDERICK D. PROVENZA[##]

[#] *Northern Agricultural Research Centre, Montana State University, 3848 Fort Circle, Havre, MT 59501, USA*
[##] *Department of Forest, Range, and Wildlife Sciences, Utah State University, Logan, UT 84322-5230, USA*
E-mail: dbailey@montana.edu

Abstract. Grazing distribution is an important component of the foraging ecology of large herbivores. Recognising the differences in foraging behaviours that occur along spatial and temporal scales is critical for understanding the mechanisms that result in grazing distribution patterns. Abiotic factors such as topography, water availability and weather and biotic factors such as forage quantity and quality affect the distribution of large herbivores. Numerous empirical studies have shown that large herbivores typically match the time spent in an area with the quantity and quality of forage found there. Although the observed grazing patterns have been documented, the underlying behavioural processes are still being elucidated. Cognitive foraging mechanisms assume that animals can use spatial memory to remember the levels of forage resources in various locations, while non-cognitive mechanisms require that behaviours such as intake rate, movement rate and turning frequency vary in response to forage resource levels. The ability of animals to use spatial memory during foraging has been demonstrated in several species including livestock, which suggests cognitive mechanisms are possible. Optimal-foraging theory can also be used to help explain behavioural processes. Giving-up rules based on marginal-value theorem appear to work well for large herbivores when a patch or feeding site can be noticeably depleted within an appropriate temporal scale such as a grazing bout or when forage availability is limited. However, giving-up rules do not always explain movements among feeding sites when forage is plentiful. The satiety hypothesis has been used to explain the avoidance of toxins and the acquisition of nutrients in diet selection. We suggest the satiety hypothesis can be expanded to account better for the variability in feeding-site selection. Large herbivocres should move among feeding sites when forage availability becomes limiting or when animals become satiated. Satiation with feeding sites may occur because of the presence of toxins or nutrient imbalances or because of aversive external stimuli. Large herbivores may return to sites that were previously considered aversive due to a combination of individual animal variation and social factors. Large herbivores can now be readily tracked using global positioning system (GPS) technology, which will allow us to test predictions of the satiety and other hypotheses and to better understand behavioural processes associated with foraging.
Keywords. diet selection; feeding site; giving up; grazing; patch selection; satiety

H.H.T. Prins and F. van Langevelde (eds.), Resource Ecology: Spatial and Temporal Dynamics of Foraging, 7-28.

INTRODUCTION

Movement patterns of animals during foraging are a critical component of their behavioural repertoire that affects total energy expenditures, exposure to variable thermal conditions, predation, and availability of food items (Stephens and Krebs 1986; Moen et al. 1997; Yearsley et al. 2002). The distribution of individuals reflects their movement patterns and can therefore be studied to gain insight into the behavioural mechanisms that result in these distribution patterns. The distribution patterns of large herbivores have been well studied and empirical approaches such as multiple-regression models have been used to predict grazing patterns (e.g., Low et al. 1981; Allen et al. 1984; Senft et al. 1985a; Brock and Owensby 2000). However, these models are often site-specific (Senft 1989), and in the case of multiple-regression approaches, they are not spatially explicit and do not address the underlying behavioural mechanisms (Coughenour 1991; Bailey et al. 1996). Since large herbivores have a large impact on their resources, the distribution of grazing across the landscape is also critical for maintaining the productivity and biodiversity of rangelands and pastures (Vavra and Ganskopp 1998; Holechek et al. 2001), which is why land managers have been working on methods to increase uniformity of grazing on heterogeneous rangeland for over 45 years (Williams 1954; Savory 1988; Gordon et al. 1990; Bailey 2004).

Behavioural mechanisms determine large-herbivore foraging patterns at intermediate scales (food patches and feeding sites)

Behavioural mechanisms that result in grazing distribution patterns vary at different spatial and temporal scales, but there appear to be some commonalities among the mechanisms (Bailey et al. 1996; Senft et al. 1987a). These scales can be distinguished by the temporal intervals between behaviours and by the spatial dimensions of the choices (Kotliar and Wiens 1990).

At coarser scales, abiotic factors such as topography, distance to water and temperature often act as constraints within which biotic factors such as forage quantity and quality operate (Senft et al. 1987a; Bailey et al. 1996; Duncan and Gordon 1999). In response to climatic conditions, wild herbivores may move up or down elevation zones to take advantage of the variability in plant phenology. Migrations from one region to another may result from lack of forage or water or changes in precipitation patterns (Senft et al. 1987a). Many species migrate or disperse, which allows them to survive in spite of the variability in weather and climatic patterns (see Boone et al., Chapter 9).

Biotic factors such as forage quantity, nutrient and toxin concentrations affect grazing behaviour at fine and coarse scales. Herbivores select food items that are abundant and that are high in nutrients and low in toxins (Baumont et al. 2000; Provenza et al. 2003). Herbivores also select patches and feeding sites where forage is more abundant, nutrients are more concentrated and toxins are at lower levels (Senft et al. 1985a; Scott et al. 1995; Coppedge and Shaw 1998).

The objective of this chapter is to review some behavioural mechanisms that may be important in determining large-herbivore grazing patterns at intermediate scales (food patches and feeding sites) and to discuss new insights about these mechanisms suggested by recent research. Related mechanisms that result in smaller-scale behaviours such as diet selection (see Laca, Chapter 5 and Fryxell, Chapter 6) and larger-scale behaviours such as migration and transhumance (see Boone et al., Chapter 9) are covered in other chapters.

Many of the studies used to develop and evaluate the proposed behavioural mechanisms were conducted with domesticated livestock. We rely on these studies to describe the mechanisms and often attempt to apply the concepts to other large herbivores including wild ungulates. Our focus on livestock is intentional, and is a result of our interest in wild herbivores, not a lack thereof (Box 2.1). We contend that the behavioural mechanisms and principles developed from experimental studies of livestock may have great value in explaining the foraging behaviours of wild large herbivores.

SPATIAL AND TEMPORAL SCALES FOR UNDERSTANDING FORAGING MECHANISMS

The six spatial scales described by Bailey et al. (1996) are functionally defined and relate to foraging decisions that occur at distinct temporal scales (Table 2.1). The scales vary from a bite that occurs every 1 or 2 seconds to home ranges that change much less frequently (1 month to 2 years), if they do at all. A grazing bout is a period of concentrated grazing typically lasting 1 to 4 hours with large herbivores, which is preceded and followed by non-grazing behaviours such as resting and ruminating (Vallentine 2001). Multiple patches can be selected within a bout, while feeding sites reflect a coarser spatial scale and encompass the entire area grazed within a bout. Herbivore decisions that occur at even coarser scales such as daily, seasonal and lifetime ranges occur much less frequently and often are not relevant to domestic livestock in intensive grazing systems. In this chapter, we target movements and behaviours that occur temporally within and especially between grazing bouts (hours and days). We focus on the spatial selection of patches and feeding sites.

Behavioural mechanisms resulting in distribution patterns vary at different spatial and temporal scales; commonalities among mechanisms exist

ABIOTIC FACTORS INFLUENCE FORAGING PATTERNS

Abiotic factors influencing foraging patterns include slope, distance to water, distance to shade or thermal cover, temperature, wind, fences, barriers, and attractants such as salt or supplement (Bailey et al. 1996). Livestock generally prefer to graze gentle terrain (Mueggler 1965). For example, Gillen et al. (1984) reported cattle avoided grazing in areas with slopes greater than 20%. Areas far from water often receive less use (Valentine 1947). Vertical distance may be more important than horizontal distance from water in mountainous terrain (Roath and Krueger 1982). In the Himalayas, free-roaming livestock use higher elevations as the summer progresses because water availability is usually limited to snowmelt (Mishra et al. 2001; Mishra et al. 2003).

Table 2.1. *Temporal and spatial scales useful for describing and evaluating foraging behaviour of large herbivores (adapted from Bailey et al. 1996; Owen-Smith 2002a). Spatial levels reflect units that large herbivores may select among*

Spatial level	Spatial resolution of selected unit[1]	Temporal interval between decisions	Defining behaviours or characteristics	Response variable	Vegetation entity
Bite	0.0001 – 0.01 m^2	1 – 2 s	Jaw, tongue and neck movements	Bite size	Plant part
Feeding station	0.1 – 1 m^2	2 s – 2 min	Front-feet placement	Bite rate	Plant (grass tuft, shrub)
Food patch	1 m^2 – 1 ha	1 – 30 min	Animal reorientation to a new location. A break in the foraging sequence	Feeding duration	Clump of plants
Feeding site	1 – 10 ha	1 – 4 h	Grazing bout	Foraging movements	Plant species association
Daily range	10 – 100 ha	12 – 24 h	Area where animals drink and rest between grazing bouts	Daily time allocation	Landscape unit
Seasonal range	100 – 1000 ha	3 – 12 months	Migration	Metabolic allocation	Landscape type
Lifetime range	> 1000 ha	Several years	Dispersal or migration	Life history schedule	Geographical region

[1] The spatial resolution of each level will vary among species of large herbivores. These approximate ranges are given to help the reader visualise differences between levels. The temporal intervals between decisions and animal behaviour are used to define the units of selection.

During cold or hot weather herbivores may select areas with more favourable conditions (see Owen-Smith, Chapter 8). During cold and windy weather, animals often seek areas protected from the wind (Houseal and Olson 1995), but cold weather may have little effect on small-scale movements if they can maintain thermoneutral conditions (Duncan et al. 2001). Preferences for certain elevations and aspects may be partially explained by thermoregulation (Harris et al. 2002). Higher elevations are often warmer than lower areas at night. Cows often prefer ridges and avoid valleys at night when temperatures are cooler (Harris et al. 2002). Conversely, during hot weather, animals seek shade (McIlvain and Shoop 1971).

Abiotic factors influencing foraging patterns include slope, distance to water, shade or thermal cover, temperature, wind, fences, barriers and attractants

Preferred sites for non-foraging activities such as resting, ruminating and watering are usually influenced by abiotic factors (Senft et al. 1985b; Bailey 2004). In addition to natural barriers such as cliffs, grazing can be constrained by snow depth (Johnson et al. 2001; Mishra et al. 2003), and domestic livestock are often constrained by fences. In arid and semi-arid areas, sources of water and shade are often limited, which also affects selection of feeding sites. After watering and resting, animals must decide where to begin the next grazing bout. Distances (horizontal and vertical) and routes to feeding sites are determined by the availability and location of water and shade. Attractants such as salt or supplement can modify grazing patterns (Bailey and Welling 1999), as herbivores travel to the attractant and then graze nearby areas later (Bailey et al. 2001c).

BIOTIC FACTORS ASSOCIATED WITH FORAGE RESOURCE AFFECT GRAZING PATTERNS

Forage quality and quantity affect herbivore distribution. The amount of time large herbivores spend in a plant community is proportional to the quality and quantity of forage available (Senft 1989). Senft et al. (1987a) referred to these patterns of grazing observed at landscape scales as 'matching'. Animals match the time spent in a plant community or feeding site with the level of resources found there. Numerous studies have shown that different species of large herbivores spend more time in areas of the landscape or pasture that are more productive and have higher levels of forage quantity and/or quality, and they spend less time in areas with less food (Hunter 1962; Coppock et al. 1983; Duncan 1983; Taylor 1984; Owens et al. 1991).

Though most studies have shown forage availability and quality influence grazing patterns and habitat selection, other currencies have been used to explain and predict where large herbivores will graze. Some studies have found the abundance of certain forage species can explain selection of patches or feeding sites (Marell et al. 2002; Fortin et al. 2003). With snow cover, forage abundance and accessibility may be more important than nutrient concentration (Johnson et al.

Box 2.1. *Use of livestock for studying behavioural mechanisms of large herbivores*

Any discussion of large herbivores should consider livestock, given their importance and abundance throughout the world. FAO (2003) estimates there are roughly 1.4 billion cattle, 1 billion sheep and 0.6 billion goats in the world today. Similarities among domestic and wild large herbivores suggest that behavioural processes observed in domestic species may be applicable to wild species and vice versa. Wild herbivores and livestock have similar rumen or hind-gut (cecal) fermentation digestive systems (Van Soest 1982). Some wild ungulates are closely related to livestock species. For example, North-American bison can mate with cattle and the resultant offspring are fertile (Burditt et al. 2000). Social behaviour of similar-sized domestic and wild herbivores is often similar. In feral conditions, cattle can form highly stable social groups that are similar to wild bovine species such as African buffalo (Reinhardt and Reinhardt 1981; Lazo 1994; Prins 1996).

In studies of rats and ducks, researchers identified some differences in behaviour between domesticated animals and their wild counterparts (Boice 1972; Desforges and Wood Gush 1976; Price 1978). Such studies suggest that the frequency and intensity of behaviour patterns, not the kinds of behaviours, are affected by domestication (Boice 1972; Price 1984). In a study comparing domestic pigs to wild boar hybrids (Gustafsson et al. 1999), domestic pigs used a slightly less costly foraging strategy. Both domestic and wild boar hybrids responded to patch depletion and spent shorter times in a patch on successive visits.

An additional argument for the similarity of domestic and wild large herbivores is the disproportionate contribution of a single order of mammals, the *Artiodactyla*, to successful domestic animals (Stricklin 2001). Large social groups, promiscuous sexual behaviour and adaptability to a wide range of environmental conditions may have contributed to the domestication of large-herbivore species.

Scientific studies of foraging behaviour are often conducted with livestock because of their availability, docility and the ease of manipulating initial conditions for an experiment. Ancestry of livestock can often be determined (e.g., pedigrees). Using techniques such as artificial insemination and embryo transfer, the genotype and rearing conditions of experimental animals can be manipulated. For example, researchers of Montana State University are currently comparing foraging behaviour of offspring from cows that spend much more time grazing high rugged terrain (hill climbers) with that of cows that spend much more time on gentle slopes near water (bottom dwellers). To control for the genetic effects of the male parent, all offspring were sired by the same bull using artificial insemination. By using embryo transfer, all offspring were placed in unrelated cows eight days after conception. Previous grazing patterns of the recipient cows (foster mothers) were established prior to embryo transfer. The result is a 2 x 2 experimental design where the effects of genotype (hill climber and bottom dweller donor cows, 'biological mothers') and early environment (learning early in life from hill climber and bottom dweller recipient cows, 'foster mother') can be quantitatively compared. Such manipulations are virtually impossible with wild species. Thus, we attempt to understand the behavioural mechanisms of foraging from experiments with livestock. The applicability of these proposed behavioural mechanisms to wild species must be evaluated using correlative approaches.

2001). Other researchers have found large herbivores choose areas based on forage quality rather than quantity (Wallis de Vries and Schippers 1994; Coppedge and Shaw 1998; Biondini et al. 1999). Some studies have used indices that combine attributes of forage quantity and quality, such as standing N (kg N/ha), to determine where animals will graze (Senft et al. 1985a; Pinchak et al. 1991). In any case, areas that contain more high-quality forages are often preferred by large herbivores, and their preference is illustrated by the attractiveness of areas that have been burned or fertilised (Hooper et al. 1969; Ball et al. 2000).

Other biotic factors can reduce the attractiveness of forages, patches and feeding sites. Recent work with post-ingestive feedback has shown that animals dynamically select various food items to match nutrient needs and to avoid over-ingesting toxins

(Provenza 1995). The effect of these processes on the selection of patches and feeding sites is an interesting area for further research (Duncan and Gordon 1999). For instance, Scott et al. (1995) found that food preferences affected where a lamb ate, and lambs did not forage in locations of foods to which they were averted.

Human disturbance and predation can affect grazing patterns of wild herbivores. Human settlement and cover affect selection of feeding sites by European roe deer (Mysterud et al. 1999). Predators can also influence feeding-site selection (Brown 1999). Caribou may sacrifice high-quality forage to avoid areas where the risk of predation is high (Ferguson et al. 1988). In Yellowstone National Park, the increase in preferred browse species in meadows after the reintroduction of wolves suggests that elk are avoiding these open areas because of the higher predation risk (Ripple and Betschta 2003). Conversely, an analysis of tracking data of woodland caribou suggested that predation had little effect of movements within large patches and feeding sites (Johnson et al. 2002a, 2002b). Prins (1996) argues that African buffalo ignore the risk of predation when determining where to forage, especially when living as part of a herd.

Animals dynamically select various food items to match nutrient needs and to avoid over-ingesting toxins

BEHAVIOURAL PROCESSES CAUSE OBSERVED FORAGING PATTERNS

As discussed previously, large herbivores spend more time in nutrient-rich than in nutrient-poor areas of the landscape. Although this aggregate 'matching pattern' is well defined, its causes (mechanisms) are still being studied (Bailey et al. 1996). We must understand the behavioural mechanisms (Box 2.2) that result in grazing patterns before we can efficiently manage habitat and manipulate grazing patterns. In what follows, alternative behavioural mechanisms are presented and evaluated based on recent research.

Non-cognitive foraging mechanisms

Bailey et al. (1996) proposed several behavioural mechanisms that could explain observed patterns of matching without invoking cognitive processes. For example, intake rate may vary in response to available forage. If patches become sufficiently depleted so that intake rate drops, animals will leave the patch (Jiang and Hudson 1993). This mechanism is similar to the predictions of the marginal-value theorem for patch residence time (Charnov 1976), where animals should leave a patch when the instantaneous intake rate drops to the average intake in the available habitat. However, short-term intake rate is usually controlled by bite size and is not always related to forage biomass (Gross et al. 1993a, 1993b). In situations where forage is abundant, intake is probably constrained by digestive processes, but if forage quantity is limited, intake may be constrained by the cropping process (Wilmshurst et al. 1999a).

If animals travelled slower in nutrient-rich patches and faster in nutrient-poor patches, they would correspondingly match the time spent in patches with the associated resource level. Rate of travel in some large-herbivore species may vary in different habitat types. For example, foraging velocities of goats increased with a greater abundance of inedible shrubs. However, white-tailed deer increased foraging velocity when the abundance of highly palatable browse increased (Etzenhouser et al. 1998). Goats and deer appeared to use different strategies. Goats moved more quickly in nutrient-poor than nutrient-rich patches, while deer moved faster in nutrient-rich patches because they could harvest the preferred shrubs more efficiently. Although this mechanism at first appears parsimonious, variation in travel rate is not necessarily a valid explanation of time spent in different patches. It is open to circularity.

Foraging mechanisms can explain grazing patterns without assuming large herbivores possess cognitive abilities

Box 2.2. *Affective and cognitive processes in foraging*

Animals process information about foods and foraging sites through two interrelated systems: affective (non-cognitive) and cognitive (Garcia 1989). Taste plays a prominent role in both systems. The affective system integrates the taste of food with post-ingestive feedback (Provenza 1995). This system causes changes in the intake of food items that depend on whether the post-ingestive feedback is positive or aversive. The net result is incentive modification. On the other hand, the cognitive system integrates the odour and sight of food with its taste. Animals use the senses of smell and sight to differentiate among foods, and to select or avoid foods whose post-ingestive feedback is either positive or aversive. The net result is behaviour modification. Cognitive experiences can be further divided to include use of the senses of sight and smell to learn from mother, learn from conspecifics, and learn through trial and error about foods and foraging sites. Together, affective and cognitive processes provide flexibility for animals to maintain homeostasis as their nutritional needs and environmental conditions change.

The anatomical and physiological mechanisms underlying affective and cognitive systems have been fairly well established (Provenza 1995). Taste afferents converge with visceral afferents in the solitary nucleus of the brain stem. Taste and visceral afferents proceed to the limbic system, where the hypothalamus and related structures maintain homeostasis in the internal environment through the endocrine system, the autonomic nervous system, and the neural system concerned with motivation and drive (i.e., incentive modification). Higher cortical centres interact with the hypothalamus through the limbic system, and regulate the internal environment primarily by indirect action on the external environment (i.e., behaviour modification). These alternative means of regulating the internal environment generally function in parallel. For example, the taste of food is adjusted according to the effect of those foods on the internal environment; on this basis, animals use thalamic and cortical mechanisms to select foods and foraging sites that are beneficial and avoid those that are not.

Animals could remain in nutrient-rich areas longer by turning more frequently than in nutrient-poor areas (Bailey et al. 1996). Differences in the tortuousity in movement patterns may occur among species. Observed foraging paths of white-tailed deer were straighter than those of goats in the same landscape in Texas (Etzenhouser *et al.* 1998). Goats and white-tailed deer preferred different browse species, and differences in spatial arrangement of species apparently resulted in differences in tortuousness of foraging paths. In a Montana study, cattle movements during morning grazing bouts were generally linear (Bailey et al. 2004), which suggests that this mechanism is not appropriate for cattle.

Changes in observed foraging velocities and turning frequencies within a patch or feeding site are likely the result of localised differences in availability and arrangement of forage. Fortin (2003) found that bison used area-concentrated searches to locate high-quality patches under the snow. Bison apparently used short-term sampling to avoid digging through snow in areas of low profitability. If forage was abundant, snow craters were congregated, and if forage was sparse the distance between snow craters was greater. Animals also use visual cues to locate food patches (Howery et al. 2000). This ability allows herbivores to move directly to the nearest patch if it is distinguishable (Gross et al. 1995).

Cognitive foraging mechanisms

Using radial-arm mazes, spatial memory has been demonstrated in rats, pigeons and cattle (Olton 1978; Roberts and Van Veldhuizen 1985; Bailey et al. 1989a). Maze studies demonstrate that these animals can learn the locations of food sources and avoid locations that have been depleted. Cattle can remember the quantity and quality of the food at various sites (Bailey et al. 1989b; Bailey and Sims 1998), which enables them to forage more efficiently in arena studies (Edwards et al. 1996; Dumont and Petit 1998; Laca 1998). Anecdotal observations suggest that large herbivores remember the locations of important features of their environment, such as water, shade and thermal or hiding cover. It is likely that all large herbivores have accurate spatial memories and use cognitive processes during foraging.

Bailey (1995) found cattle returned to areas of higher forage quality or quantity more frequently and returned to areas with lower quality or quantity less frequently. Over time, these behaviours should result in typically observed patterns of foraging-site selection where animals match the time spent at various sites with the availability of nutrients found there. Initially animals travel to nutrient-rich feeding sites to graze. Later, as nutrient-rich sites become depleted, animals may shift to other areas of lower forage quality or quantity. In an unpublished GPS (global positioning system) collar-tracking study from the Montana laboratory, cows grazing foothill rangeland followed this type of feeding-site selection pattern. Initially, cows used lower elevations near water where forage quantity and quality were higher. Later, cows used steeper slopes, higher elevations and areas further from water as the more nutrient-rich sites near riparian areas and coulee bottoms became depleted (Bailey and VanWagoner 2004).

Some foraging mechanisms assume that large herbivores have cognitive abilities

MARGINAL-VALUE THEOREM AND BEHAVIOURAL FORAGING MECHANISMS

The marginal-value theorem (Charnov 1976) has been used successfully for explaining patch use in optimal-foraging theory (Stephens and Krebs 1986). The patch model is based on a net energy gain function that varies with time spent in a patch. The primary prediction of the patch model is that foragers should move to another patch when the instantaneous intake rate in any patch drops to the average rate of the entire habitat. The original solution to the problem of allocating time spent in various patches implicitly assumed foragers knew how much time to spend in each patch (McNair 1982). A more reasonable assumption is that foragers compare the current intake rate with intake rates obtained in the past. The time spent in a patch would then depend on a rule-of-thumb decision concerning when to give up and leave a patch (Krebs et al. 1974; McNair 1982). For large herbivores, giving-up rules could explain why animals move as feeding sites become depleted. The standing crop of forage may be reduced sufficiently that short-term (theoretically instantaneous) intake rate decreases, and the animal would then move. Prins (1996) observed that dense herds of African buffalo readily shift from one feeding site to another and on average revisit feeding sites once every 4 to 5 days. In this study, feeding sites were 4 to 50 ha in size and were heavily grazed by buffalo and other competing herbivores. With herds of hundreds of buffalo, the quantity of forage could be reduced within hours and short-term intake rate would decline.

Marginal-value theorem of optimal-foraging theory can explain foraging patterns when resources are limited

At the finer scale of patch selection, the marginal-value theorem and giving-up rules should work even better than at the coarser scale of the feeding site. Large herbivores, especially groups and herds, could noticeably deplete a patch (1 m^2 to 1 ha) within a period of minutes to hours. Wapiti moved to another patch when biting rate began to decline, which is in agreement with the marginal-value theorem (Jiang and Hudson 1993). The time wapiti spent in a patch was usually less than 6 minutes, thus the temporal and correspondingly spatial scale in this study was much finer than feeding-site selection. The marginal-value theorem also successfully predicted patch selection of cattle with patch sizes of 0.6 m^2 (Laca et al. 1993).

These studies suggest that giving-up rules based on optimal-foraging theory work well for large herbivores when a patch or feeding site can be noticeably depleted within an appropriate temporal scale such as a grazing bout or when forage availability is limited. However, at least for cattle, and perhaps for other large herbivores (Box 2.3), giving-up rules based on the marginal-value theorem do not appear to work well for explaining movements among feeding sites when resources are plentiful or when grazing in an area has recently begun and forage is not limiting. In such cases, animals typically under-match – they over-use poorer sites and under-use richer sites (Kennedy and Gray 1993). Wapiti, for example, alternate between high- and lower-quality patches and spend more time foraging in patches of lower quality than predicted by optimality models (Wilmshurst et al. 1995). In gentle terrain with relatively homogeneous vegetation, cattle rarely graze in the same section of a pasture for more than 2 consecutive days (Bailey et al. 1990; Bailey 1995). It is unlikely that changes in forage availability explain this alternation among feeding sites, as less than 5% of the available forage was harvested before cattle switched feeding sites. With GPS technology, researchers can readily observe feeding-site selection. When a 337-ha foothill pasture in Montana was divided into 9 zones based on topographical features, cattle did not graze in the same zone for more than 3 consecutive days (Bailey unpublished data). Indeed, the cows often moved to a zone on the opposite side of the pasture.

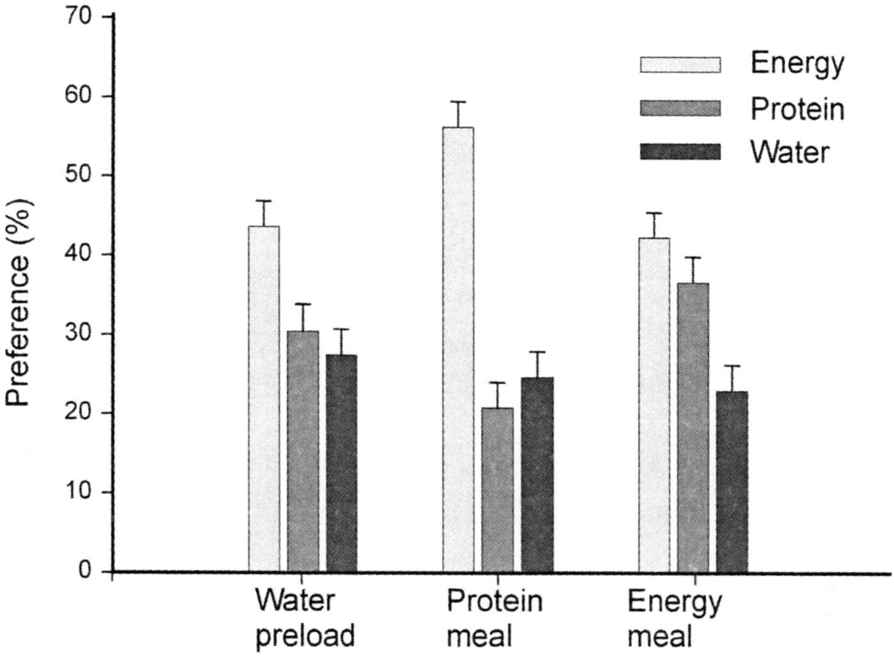

Figure 2.1. *Nutrient-specific satiety. In the study of Villalba and Provenza (1999), lambs were given flavoured straw followed by oral gavage of energy, protein or water; a different flavour was paired with the different nutrient sources during conditioning. Following conditioning, lambs were given a pre-load meal of either water, energy or protein, and offered a choice among straw in the three flavours. Lambs preferred energy > protein = water when fed a basal diet of alfalfa pellets high in protein. When offered straw in the three flavours immediately after a meal high in protein, lambs strongly preferred energy > water > protein. Conversely, when offered straw in the three flavours immediately after a preload of energy, lambs preferred energy = protein > water. Note the strong degree to which preferences for energy and protein changed as the trials progressed from water to protein to energy.*

Why do cattle and other large herbivores alternate among feeding sites when preferred forage is still abundant where they are foraging? Bailey et al. (1996) developed a conceptual model to explain alternation in feeding sites with similar resources. This model relied on memory decay to explain why animals return to areas they previously avoided. However, recent studies with diet selection and anecdotal observations suggest that an animal's memory of aversive events remains for long periods (Provenza 1995, 1996). Conceptual models used to explain diet selection and acquired food aversions of large herbivores may also be applicable for understanding movement patterns.

Box 2.3. *Nutrient-specific satieties*

Animals learn to discriminate among foods based on feedback from nutrients (Provenza 1995, 1996). Energy- or protein-deficient sheep learn to prefer poorly nutritious foods such as straw when their intake is accompanied by intraruminal infusions of energy (Villalba and Provenza 1996; 1997a; 1997c) or protein (Villalba and Provenza 1997b), both of which can condition strong food preferences. Sheep maintain ratios of energy to protein that meet nutritional needs (Egan 1980; Provenza et al. 1996; Wang and Provenza 1996) by discriminating between feedback from energy and protein following food ingestion (Villalba and Provenza 1999, Figure 2.1).

Preferences for foods high in protein or energy are governed by the nutritional state of insects (Simpson and Raubenheimer 1993, 1999), rodents (Gibson and Booth 1986, 1989; Perez et al. 1996; Ramirez 1997; Gietzen 2000; Sclafani 2000), and ruminants (Cooper et al. 1993; Kyriazakis and Oldham 1993, 1997; Kyriazakis et al. 1994; Berteaux et al. 1998; Villalba and Provenza 1999). Animals prefer a food high in energy after a meal high in protein and vice-versa (Figure 2.1). Elk optimise macronutrient intake by selecting appropriate patches of grass, though there has been debate over whether energy (Wilmshurst et al. 1995) or protein (Langvatn and Hanley 1993) is more important (see Fryxell et al., Chapter 6). Modelling efforts and experimental analyses show that both are critical, and that the preferred ratio of protein to energy depends on needs for growth, gestation and lactation (Wilmshurst and Fryxell 1995; Fisher 1997; Villalba and Provenza 1999).

The synchrony of nutrient fermentation also affects food intake, which increases with appropriate ratios of energy and protein and decreases with an excess of either (Kyriazakis and Oldham 1997; Villalba and Provenza 1997b; Early and Provenza 1998). The balance of energy and protein affects the rates of production of end products of microbial fermentation such as organic acids and ammonia. The rate at which energy and protein are released cannot exceed the rate at which they can be processed without causing excesses of organic acids or ammonia and decreases in intake. Hence, rates of fermentation of energy and protein influence intake (Cooper et al. 1995; Francis 2002).

SATIETY HYPOTHESIS IN DIET SELECTION

Ruminants eat an array of plant species, varying in nutrients and toxins. This selection makes intuitive sense, but no theories adequately explain this diversity. Some maintain that this variety in diet selection reduces the likelihood of overingesting toxins (Freeland and Janzen 1974), whereas others contend it meets nutritional needs (Westoby 1978). Nevertheless, herbivores seek variety even when toxins are not a concern and nutritional needs are met. The satiety hypothesis provides an explanation that encompasses both the avoidance of toxins and the acquisition of nutrients (Provenza 1995, 1996).

A key concept in the satiety hypothesis is aversion, the decrease in preference for food just eaten. Aversions are due to interactions between sensory (taste, odour, texture – i.e., flavour) and post-ingestive (effects of nutrients and toxins on chemo-, osmo-, and distension-receptors) effects unique to each food. Flavours result when sensory receptors in the mouth and nose respond to gustatory (sweet, salty, sour, bitter), olfactory (an array of odours), and tactile (astringency, pain) stimuli. These receptors interact with visceral receptors that respond to nutrients and toxins (chemo-receptors), osmolality (osmo-receptors), and distension (mechano-receptors). Collectively, these neurally-mediated flavour-feedback interactions enable animals to discriminate among foods, each of which possesses a distinct utility, and they encourage animals to eat a variety of foods and to forage in a variety of locations (Pfister et al. 1997; Early and Provenza 1998; Villalba and Provenza 1999; Scott and Provenza 1998, 2000; Atwood et al. 2001a, 2001b).

Satiety hypothesis may explain variability in diet selection

The satiety hypothesis makes three testable predictions. First, the hypothesis attributes varied diets to transient food aversions due to flavours, nutrients and toxins interacting along concentration gradients (Provenza 1995, 1996; Provenza et al. 2003). Gustatory, olfactory and visual neurons stop responding to the taste, odour and sight of a particular food eaten to satiety, yet they continue to respond to other foods (Critchley and Rolls 1996). Second, aversions should become pronounced when foods contain too high levels of toxins or nutrients or nutrient imbalances. Aversions also result when foods are deficient in nutrients or when amounts of nutrients required for detoxification are inadequate. Aversions occur even when a food is nutritionally adequate because satiety and surfeit are on a continuum. Finally, cyclic patterns of intake of different foods are due to eating any food too often or in too large an amount (Provenza 1995, 1996), and the less adequate a food is relative to an animal's needs, the greater and more persistent the aversion (Early and Provenza 1998; Atwood et al. 2001a, 2001b).

The satiety hypothesis helps to explain why sheep prefer to eat clover in the morning and grass in the afternoon, even though clover is more nutritious than grass (Newman et al. 1992; Parsons et al. 1994). Hungry sheep initially prefer clover because it is more digestible than grass. As they continue to eat clover, however, sheep acquire a mild aversion likely from the aversive effects of byproducts of nutrient fermentation – excess organic acids produced from soluble carbohydrates and ammonia produced from highly digestible proteins (Cooper et al. 1995; Francis 2002) – and from the aversive effects of cyanogenic glycosides (Burritt and Provenza 2000). The mild aversion causes them to eat grass, which is relatively lower in nutrients and toxins than clover, in the afternoon. During the afternoon and evening, the aversion subsides as sheep recuperate from eating clover. By morning, they are ready for more clover.

While most explanations for varied diets stress why animals seek other more nutritious alternatives, they do not account for why animals stop eating highly nutritious foods. Transitory food aversions explain why animals stop eating particular foods, and flavour-feedback interactions provide a mechanism for the response. Aversions yield benefits: obtain a nutritious diet, reduce ingestion of toxic foods, optimise foraging and rumination times, sample foods and maintain a diverse microflora in the rumen. These are often mistaken as the cause of varied diets.

SATIETY HYPOTHESIS IN FEEDING-SITE SELECTION

Though developed to explain the dynamics of diet selection, the satiety hypothesis may be useful for explaining observed shifts in feeding sites at both low and high levels of forage availability. As preferred food items become limited, large herbivores begin to eat less preferred foods. Similarly, large herbivores switch from preferred to less preferred feeding sites as forage availability becomes limited (Figure 2.2). According to the satiety hypothesis, the behavioural mechanisms for switches between feeding sites, often described in optimal-foraging theory as a 'giving-up rule' (Stephens and Krebs 1986), involve satiating on a particular food or foraging location as they become increasingly less adequate (deficient, excessive or imbalanced) relative to needs.

While the satiety hypothesis emphasises flavour-feedback interactions involving so-called gut defences, large herbivores may also switch to alternative feeding sites when forage is still abundant as a result of feedback from both gut- and skin-defence systems (Box 2.4). Feedback from the gut-defence system can cause animals to switch to alternative sites if either a necessary nutrient is deficient or if nutrients or toxins are excessive. If available food items at a feeding site are deficient in a nutrient or cause toxicosis, animals become averse to the food/site, which causes them to switch to a site that contains the needed nutrient. Lambs challenged by an imbalance of energy or protein in their basal diet forage in areas where food items that complement their imbalance occur (Scott and Provenza 2000). Cattle trained to avoid eating a high-quality food by pairing the food with a toxin (LiCl) not only avoided eating the food, they also avoided the area where the food was first fed when it was the only food available in that area (Cibils et al. 2004).

Satiety hypothesis can account for the dynamics of feeding-site selection

Feedback from the skin-defence system also may affect feeding-site selection. Abiotic factors such as steep slopes, distances far from water, and cold or hot temperatures may make a feeding site adverse, whereas the opposite would cause place preferences. Cattle avoid areas where they have received electric shock (Markus et al. 1998; Cibils et al. 2004), and animals avoid areas where the risk of predation is high (Ferguson et al. 1988; Brown 1999; Ripple and Beschta 2003). The converse is true in species where risk of predation is low, for example African buffalo (Prins 1996).

In summary, there is growing evidence that animals satiate on nutrients and toxins, and this causes the dynamics of food selection as discussed above. Less is known about gut defences and feeding sites. Animals may become satiated with a feeding site when it contains food items that result in a mild aversion due to nutrient imbalances (Villalba and Provenza 1999) or excesses of toxins (Provenza 1996). Even less is known about skin defences and locations. In general, researchers have not tested experimentally hypotheses that animals satiate on feeding sites based on gut-defence responses or on locations per se based strictly on skin-defence responses. However, the stereotypic behaviour observed in wild herbivores confined in zoos (Pollard and Littlejohn 1996; Bashaw et al. 2001) or in livestock confined in small stalls (Redbo 1992; Redbo et al. 1998) are undoubtedly examples of a satiation based on location. The apparent aversive nature of cages in zoos and small stalls in livestock operations can occur even though the diets are well balanced and varied. Providing periodic environmental change may reduce stereotypy in general (Hutchins et al. 1984; Bashaw et al. 2001).

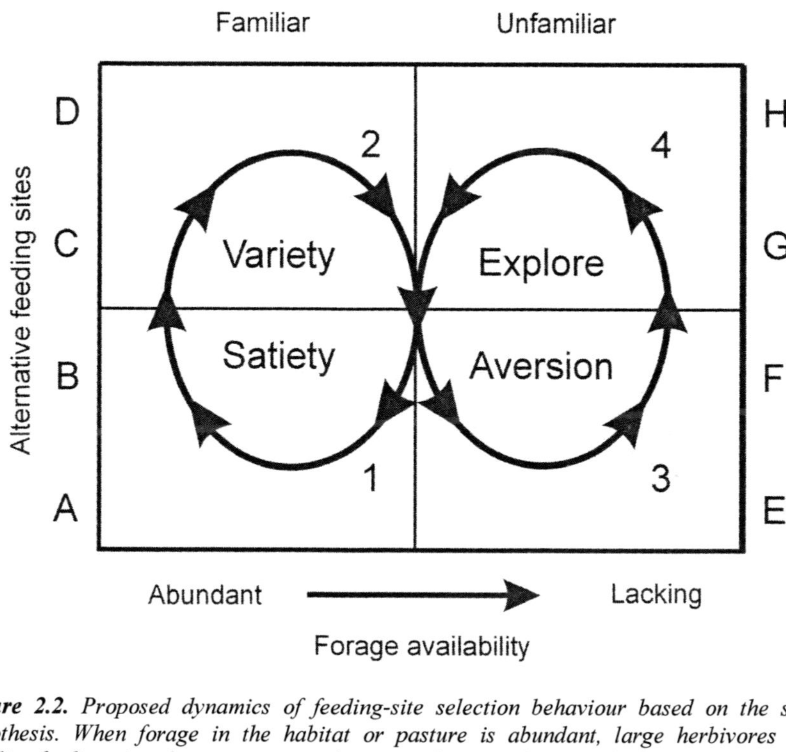

Figure 2.2. *Proposed dynamics of feeding-site selection behaviour based on the satiety hypothesis. When forage in the habitat or pasture is abundant, large herbivores select familiar feeding sites that are more productive and require less travel effort (quadrants 1 and 2). After one or more visits to the same feeding site, animals may become satiated to the site because of mild aversions to the foods (gut-defence system) or the location (skin-defence system) (Box 2.4). Aversions encourage animals to choose alternative feeding sites and increase the variety of sites selected. As forage levels in familiar and more desirable feeding sites begin to decrease, animals become more likely to investigate new options (aversion – explore, quadrants 3 and 4). If environmental conditions are acceptable and the forage resources are at least comparable with current levels at previously visited sites, animals will then include them as familiar alternatives (quadrants 1 and 2).*

Box 2.4. *Skin and gut defences*

Animals learn about the consequences of their behaviours due to feedback from external (exteroceptive) and internal (interoceptive) environments in two fundamentally different ways. They associate what happens in specific places (exteroceptive) with positive and aversive consequences, thereby acquiring place preferences and aversions. They also associate specific foods with positive and aversive post-ingestive consequences (interoceptive), thereby acquiring food preferences and aversions. Place/food preferences and aversions are dynamic and transitory, as discussed for the satiety hypothesis.

Animals learn about foods and places in different ways. As the pre-eminent psychologist John Garcia points out, "All organisms have evolved coping mechanisms for obtaining nutrients and protective mechanisms to keep from becoming nutrients" (Garcia 1989). In many birds and most mammals, auditory and visual stimuli and sensations of pain and satisfaction are associated with the so-called skin-defence system, evolved in response to predation. The taste of food and sensations of nausea and satiety are part of the so-called gut-defence system evolved in response to toxins and nutrients in foods. Odours are associated with skin- or gut-defence systems, depending on the behaviour. The odour of predators forewarns the skin-defence system, while the odour of food serves as a cue for the gut-defence system.

The way skin- and gut-defence systems work is illustrated in trials with hawks fed distinctively coloured or flavoured mice (Garcia y Robertson and Garcia 1987). When hawks normally fed white mice are given a black mouse, followed by an injection of a toxin, the hawks eat neither black nor white mice. They do not discriminate between mice as a food item based on colour. However, when a distinct taste is added to black mice, hawks learn to avoid black mice on sight after a single black-mouse toxicosis event. The hawks discriminate between food sources based on taste. These and other experiments show that not all cues are associated readily with all consequences (Garcia and Koelling 1966). Animals made ill following exposure to audiovisual and taste cues show much stronger aversions to the taste than to the audiovisual cue. In contrast, if they receive foot-shock following the same cues, they show much stronger aversions to the audiovisual than to the taste cues.

The same kind of response has been demonstrated for food and place aversions (Garcia et al. 1985). Toxins decrease palatability, but they do not necessarily cause animals to avoid the place where they ate a particular food; this is the essence of the hawk–mice toxicosis experiment. Conversely, an attack by a predator may cause animals to avoid the place where they were eating, but it does not decrease the palatability of the food. While place aversions are specific to the site, food aversions depend on the food and are generally independent of the location where the food was eaten.

Two kinds of memory underlie both skin and gut defences. Declarative (cognitive) memory is generally thought of as the process by which earlier conscious experiences are recalled (LeDoux 1992). The original learning and remembering are both conscious events. Such declarative memory is mediated by the hippocampus and the cortex. In contrast, emotional (non-cognitive) memory is mediated by the thalamus, amygdala and cortex, and in all likelihood operates independently of conscious awareness. Nevertheless, emotional information may be stored within declarative memory and retrieved in parallel, "their activities joined seamlessly in conscious experience" (LeDoux 1994). Thus, for example, acquired preferences and aversions are part of emotional memory, and in the case of specific foods, post-ingestive feedback and memory of the feedback event occur whether or not animals are conscious (Provenza et al. 1994). The emotional memory (satiating feedback from nutrients or nauseating feedback from excesses of nutrients or toxins) is stored within declarative memory of specific foods and the context in which they were eaten.

PROPOSED MECHANISMS BY THE SATIETY HYPOTHESIS

Memory and sampling

When forages are abundant, for example after a pasture change or during late spring while forage is growing rapidly, animals will first select preferred areas they know (Figure 2.3). Although forage is still plentiful, they may periodically change feeding sites as they satiate with characteristics (foods/location) of the site. One might expect they would first use areas that they remembered had higher levels of resources in the past. As forage in nutrient-rich or environmentally favourable feeding sites becomes depleted, animals will explore other areas of a pasture or habitat. As more feeding sites are visited, animals may become averted to some sites because of undesirable forage or environmental conditions. Animals should alternate among more novel sites with more favourable conditions along with previously preferred sites.

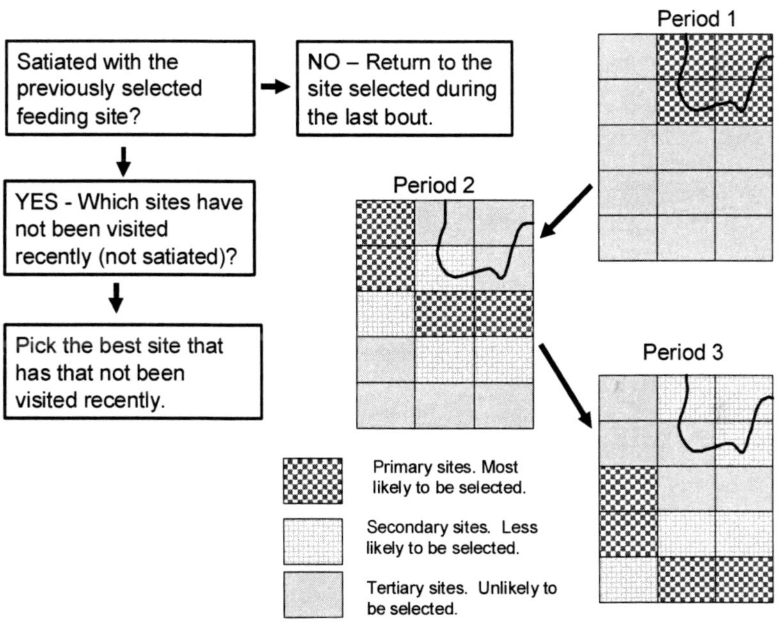

Figure 2.3. *Description of the decisions herbivores may make for selecting feeding sites based on the satiety hypothesis. If animals are not satiated with a feeding site, they will likely return to the same area. If animals are satiated, they will select another site. The pattern of changes in feeding-site selections is presented. Initially (period 1), animals alternate among the best feeding sites near the stream. As the animals become satiated with the streamside sites, they select sites further from water (period 2) (stream given by black line). As animals become satiated with those sites, they will select sites even further from water (period 3) and may return to the streamside sites, which could contain high-quality forage regrowth.*

Over time, this proposed mechanism should result in more time spent in nutrient-rich sites and less time spent in less productive sites. The alternation among sites should result in animals spending similar amounts of time in locations with similar resource availabilities. Nevertheless, animals should also sample less productive sites as they become satiated with the sites containing higher and intermediate levels of resources. This is an important prediction of the satiety hypothesis because large herbivores such as cattle periodically visit virtually all areas of a pasture or habitat. Gillen et al. (1984) found cattle sign (tracks and faecal pats) in virtually every part of a mountain pasture in Oregon, even in upland areas where forage utilization was less than 10%.

Satiation and alternation among feeding sites could have long-term benefits for large herbivores. In environments where change is the only constant, satiating on the most familiar (that visited most recently) causes animals to explore continually the less familiar (that visited in the past) and the novel (the unknown). Eating a variety of foods produces many health benefits related to nutrition, parasite loads and ingestion of compounds that in low concentrations might promote health (Engle 2002). Indeed, it is likely the most important thing an animal can do to maintain health is to eat a variety of foods that vary in concentrations of nutrients and toxins. Thus, periodic sampling of alternative foods and feeding sites may ensure an adequate food supply in rangeland environments where forage conditions can change dramatically in both time and space due to variable climatic conditions, thereby contributing to fitness.

As animals become satiated, alternative feeding sites may become more attractive. However, it is unlikely herbivores would readily revisit sites that were perceived to be very poor because of forage conditions, predation or difficulty in reaching, for example long distances from water. Animals likely remember sites that were unusually attractive or adverse for long periods. Bailey (1995) observed that steers avoided an area with lower forage quality for 21 consecutive days after first sampling the site. The satiety hypothesis predicts that the more averse the consequence, the longer the period before an animal will again engage in the behaviour.

Variation among individual animals

Differences among individuals in food intake and preference depend in part on variations in how animals are built morphologically and how they function physiologically, and marked differences are common even among uniform groups of animals in needs for nutrients (Scott and Provenza 1999) and abilities to cope with toxins (Provenza et al. 1992). Differences in foraging behaviour among individuals result from variability in genotype and experiences early in life (Provenza et al. 2003). Individual herbivores also have very different grazing patterns. Cattle breeds developed in mountainous terrain use rugged topography much more uniformly than breeds developed in more gentle terrain (Bailey et al. 2001b). Bailey et al. (2004) tracked cows in Montana and found that some individuals use steep, high terrain, while others prefer gentle terrain (Figure 2.4). Differences in grazing patterns

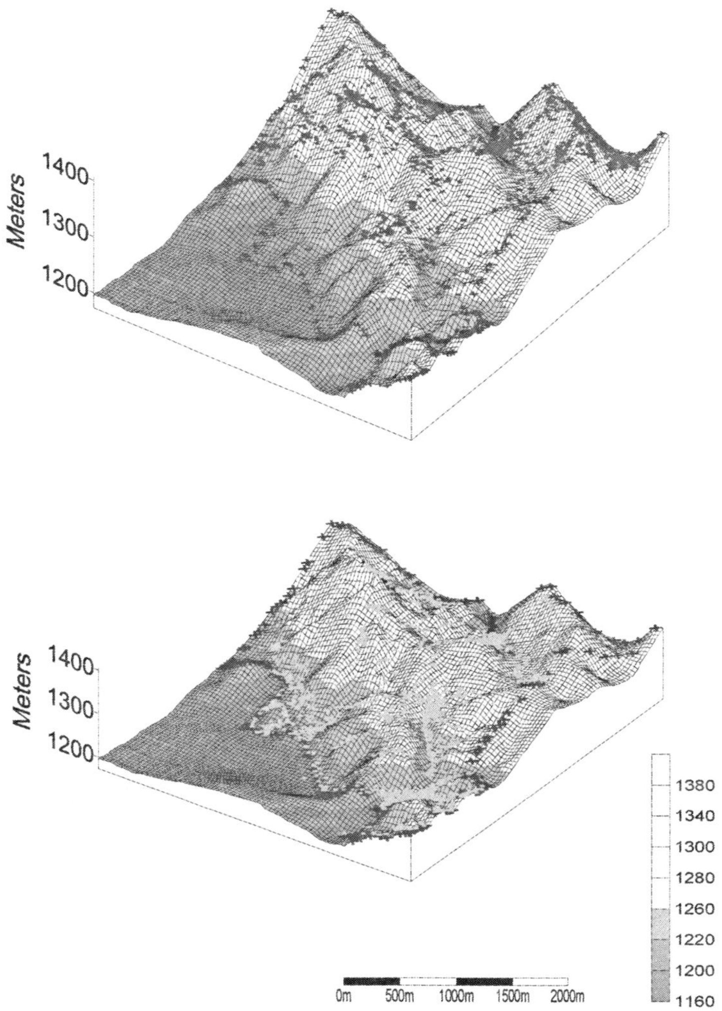

***Figure 2.4**. Distribution patterns of two cows recorded by GPS (global positioning system) tracking collars in a 337-ha rangeland foothill pasture in northern Montana during a 3-week period in late August and early September. Cow locations were recorded every 10 minutes during the day and every 20 minutes at night. Both cows were 7 years of age and were lactating. During the previous year cow 1179 (upper figure) was observed more frequently on steep slopes and high elevations, while cow 1080 (lower figure) was observed much less frequently in these areas. These cows were purposely selected as extremes from the previous year's observations collected by observers on horseback. Cow 1080 was Hereford, a breed developed in England. Cow 1179 was Tarentaise, a breed developed in the French Alps. In addition to these 2 cows, 178 other lactating cows were grazing in the pasture during the period of observation. Water was available in streams*

observed among cattle breeds (Herbel and Nelson 1966; Bailey et al. 2001a; 2001b) suggest feeding-site selection may be at least partially heritable.

Experiences early in life affect feeding-site selection in sheep and cattle. Key and MacIver (1980) evaluated habitat selection of 2 breeds of sheep. The Welsh mountain breed that preferred native rangeland was smaller in size (36 kg), and had lower reproductive rates (90%). The Clun forest breed was larger (54 kg), had higher reproductive rates (150%), and preferred more productive seeded habitats. These researchers conducted a cross-fostering experiment where Welsh dams reared Clun lambs and Clun dams reared Welsh lambs. After weaning when lambs were 6 to 7 months of age, Welsh-reared Clun lambs preferred native rangeland, and Clun-reared Welsh lambs preferred seed habitats. Howery et al. (1996) determined that different cows grazed in different areas within extensive mountain pastures during summer and they used the same areas during the next four summers; they then studied the habitat preferences of offspring from these cows. When evaluated after weaning at 2 and 3 years of age, daughters preferred the areas that their mothers preferred. This study also included a cross-fostering experiment where female calves from cows that preferred one area of the habitat (drainage) were reared by unrelated cows (foster mothers) that preferred a different habitat (adjacent drainage). Cross-fostered offspring preferred the areas where they were reared by their foster mother. We suggest that nature (genome) and nurture (learning) interact to influence habitat selection behaviours including not only where individual animals go but also their propensity to roam (Moore 2002).

Social influences

When ingesting a novel food is followed by toxicosis, herbivores acquire a strong aversion to the food that can last at least 3 years. Nevertheless, the aversion will diminish if animals with the aversion observe peers consuming the food (Ralphs and Provenza 1999). A similar response may occur with selection of feeding sites.

Social interactions among herbivores influence grazing patterns (Mendl and Held 2001). In sheep and cattle, social interactions within groups encourage animals to eat a broader array of foods (Scott et al. 1995), and to forage in a greater variety of locations (Howery et al. 1998), as individuals maintain the cohesiveness of the group (Dumont and Boissy 2000) and respond to ever-changing preferences of individuals within the group. Social interactions may allow animals within the herd to transfer information about the environment as experienced animals help naïve animals find food (Bailey et al. 2000; Ksiksi and Laca 2000). Where herds of African buffalo choose to forage may be a communal decision (Prins 1996). Individuals appear to orientate toward various feeding sites, and the direction the herd takes is a result of a consensus of herd members.

SYNTHESIS

It is likely that satiation, variation in individuals within the herd, and social interactions all interact to cause alternation among similar feeding sites and periodic sampling of all sites. For example, cattle may become satiated with feeding sites near water and travel to a new feeding site further from water. If forage conditions at the new site are not as good as the conditions near water, they probably will not visit there again soon, especially if the longer travel distance was perceived as adverse. However, some individuals within the herd might not perceive the travel as adverse as others and visit sites further from water again. As these individuals travel to the further sites, others in the herd who had avoided the sites may now visit them with their peers. Cattle often follow individual animals with 'purposeful movement' (Greenwood and Rittenhouse 1997), and in a Montana study, cattle followed animals that walked away from the group at a faster pace in a specific direction (Bailey et al. 2000). This behaviour appeared to facilitate the transfer of knowledge of feeding locations from experienced to naive animals. Social interactions may also encourage animals to resample areas previously visited and perceived as less desirable.

Forage depletion, satiation, individual variation and social interactions interact and result in alternation among feeding sites

Thus, the combination of forage depletion, satiation, individual variation and social interactions can result in a dynamic process where animals periodically sample locations within their home range. This periodic sampling of poor sites could be due to memory decay, as is incorporated in the cognitive foraging model proposed by Bailey et al. (1996). However, while memory decay probably occurs at some level, forage depletion, satiation, variation in individual animals, and social interaction may better account for the variability in many observed behaviours.

Recent technologies such as GPS tracking collars and geographical information software (GIS) enable researchers to evaluate movements of large herbivores at temporal scales from minutes to months. Previously, it was impractical and cost prohibitive to observe bout-to-bout or day-to-day movements of large herbivores over weeks to months. These difficulties limited the number of empirical studies of feeding-site selection of large herbivores. Much of optimal-foraging theory was developed for diet and patch selection, which could be observed during periods of minutes to hours and in laboratories and small field areas. The three hypotheses given in Box 2.5, developed from the conceptual behavioural mechanisms of feeding-site selection described in this chapter, could not have been tested 10 years ago, but they can today.

Box 2.5. *Testable hypotheses for future research*

Hypothesis 1. Abiotic factors such as slope, horizontal and vertical distance to water can increase travel effort to reach feeding sites, and in mountainous terrain these factors interact to influence use of feeding sites. For example, steep slopes may be more aversive if they are further from water (Mueggler 1965). We expect that indices that combine the effects of slope and distance to water (horizontal and vertical) will be more useful in predicting large herbivore grazing patterns than if terrain attributes are considered independently. Similarly, indices of effective temperature that combine ambient temperature and wind speed or ambient temperature and solar radiation may explain terrain use of large herbivores more accurately than climatic factors considered independently. In rugged topography, large herbivores can move and thereby modify elevation, aspect and slope to seek or avoid wind and direct sunlight. Ongoing research suggests spatially explicit models that include terrain and environmental factors can explain movement patterns of livestock and likely other large herbivores (Harris et al. 2002).

Hypothesis 2. Based on the satiety hypothesis, large herbivores should alternate among feeding sites not only when forage resources are sparse but when they are plentiful as well. When forage availability is limited, animals should move to an alternative feeding site as short-term intake begins to decline. When forage availability is plentiful, animals should move to alternative feeding sites without a measurable change in short-term intake rate or forage abundance due to satiating on nutrients and toxins in the forages on offer. Studies of cattle suggest that this might occur (Bailey et al. 1990; Bailey 1995; Laca et al. 1993), but more rigorous examinations with livestock and other large herbivores are needed.

Hypothesis 3. When forage is plentiful, large herbivores should satiate more quickly to a feeding site when the terrain and/or forage in a pasture or habitat is homogeneous than when it is heterogeneous. For example, livestock should alternate among feeding sites more frequently in pastures seeded with monocultures than in pastures with mixtures or on rangeland with mixtures of native vegetation. Large herbivores should become satiated and alternate among feeding sites more frequently in gentle terrain than in mountainous terrain. We have a limited amount of data to support this hypothesis. Cattle were tracked in two similar-sized pastures (approximately 350 ha) in Montana for 30-day periods using GPS collars. The first pasture contained rugged terrain and was grazed in late summer. This pasture was stratified into four sections based on topography and was considered more heterogeneous due to topography and the variability in forage quality typically observed in late summer. The second pasture contained more gentle terrain and was grazed during autumn. This pasture was stratified into three sections based on topography and was considered more homogeneous. Cattle in the first (heterogeneous) pasture were observed in the same section of the pasture for 4 or more consecutive days during 87% of the total days of observation. In contrast, cattle in the second (homogeneous) pasture were in the same section of the pasture for 4 or more consecutive days during only 20% of the total days of observation. Although these data support the satiety hypothesis, more extensive studies are needed with livestock and with other large herbivores under a variety of conditions designed specifically to test predictions of the satiety hypothesis.

CHAPTER 2B

COMMENTS ON "MECHANISMS DETERMINING LARGE-HERBIVORE DISTRIBUTION"

SIPKE E. VAN WIEREN[#], MICHAEL DRESCHER[##] AND WILLEM F. DE BOER[#]

[#] *Resource Ecology Group, Wageningen University, P.O. Box 47, 6700 AA Wageningen, The Netherlands*
[##] *Ontario Forest Research Institute, 1235 Queen St. East, Sault Ste. Marie, Ontario, P6A 2E5, Canada*
E-mail: sip.vanwieren@wur.nl

How many prey to take of a certain type or how long to stay in a patch are key questions of a foraging animal according to the optimal foraging theory (OFT) (Krebs and Davis 1986). Within the OFT, the goal for herbivores generally is some form of energy maximisation within the limits of certain constraints. Although the application of energy as single currency has had some success, it is widely recognised that focusing on energy alone is not sufficient to explain the foraging behaviour of herbivores. Especially the complex, and ever changing, nature of their diet, together with the many constraints to be taken into account, poses problems (Krebs and Davies 1986; Simpson et al. 2004; Illius et al. 2002; Bailey and Provenza, Chapter 2). Essential here is that herbivores tend not to stay in a patch as long as predicted, and/or do not select a diet which provides maximal energy gain (Van Wieren 1996; Bailey and Provenza, Chapter 2). Because of this, alternative models have been developed, among them the sufficing principle (defined by Ward (1992) as choosing between different options when information-processing limits the ability of an animal to make optimal decisions), and the satiety hypothesis. The question here is if and/or how the satiety hypothesis fits into the OFT. Interestingly, the satiety hypothesis is not formally stated in the chapter. The closest we can get to a definition is that, according to the satiety hypothesis, "the behavioural mechanisms for switching between feeding sites involve satiating on a particular food or foraging location as they become increasingly less adequate (deficient, excessive, or

H.H.T. Prins and F. van Langevelde (eds.), Resource Ecology: Spatial and Temporal Dynamics of Foraging, 29-31.

imbalanced) relative to needs". A key concept in the satiety hypothesis is 'aversion' due to flavours, toxins and nutrients, leading to a decrease in preference of food just eaten and to satiation, after which the animal stops eating a particular food. Despite the lack of a clear definition, the satiety hypothesis leads to three testable predictions. All tests require some measurement of aversion in relation to some 'nutrient' level. How this exactly needs to be done does not become clear. In some experiments, with only a few forage species, the satiety hypothesis did 'help' to explain the preference patterns found. It may, however, be very difficult to estimate the relative effects of individual forage species when herbivores take a mixed diet, a problem in common with the analysis using predictions derived from the OFT. More problematic is that the satiety hypothesis is not directed at a specific goal. Basically it deals with *a posteriori* effects after food has been ingested and, as such, is more related to the 'giving-up rules' or even 'constraints' which are part of the OFT. For the satiety hypothesis really to be(come) an alternative for the OFT, it should formulate clear goals for the foraging animal. Moreover, according to the satiety hypothesis animals can also choose among different forage types, and maximisation principles (not only aversion) are expected also to operate when deciding what to eat. It is, however, not clear if the authors have the intention to formulate an alternative theory or that they feel that the satiety hypothesis should in one way or the other be incorporated in OFT.

The ultimate goal of foraging animals is to maximise Darwinian fitness, and, as this is still the ruling paradigm, any foraging theory should at least in principle be embedded in this paradigm. Although energy maximisation has frequently been used as a proxy for fitness, it is clearly a special case, since fitness maximisation and energy maximisation subject to constraints are in general not equivalent (Illius et al. 2002; Simpson et al. 2004). Although still far away from a new foraging theory, some concepts of the OFT are presently being rethought. Is there, for instance, an alternative for energy maximisation? The complex nature of both the food base and the requirements for a number of nutrients that constitute the herbivore's world, calls for the inclusion of more nutrients in the 'goal' than energy alone. There is growing evidence that some herbivores regulate the intake of multiple nutrients independently and, instead of maximising intake, avoid ingesting surpluses and deficits relative to regulated points (Simpson et al. 2004). The goal, then, becomes the regulation of a multidimensional 'intake target'. Regulation implies that the animal strives to a certain state and it is only a small step to relate this state to the concept of 'homoeostasis', another fundamental paradigm in biology (Bradshaw 2003). There are approaches such as multiple criteria or multiple objective optimisation, or approaches that include conflicting demand (Schmitz et al. 1997). We suggest that the satiety hypothesis could, perhaps, find a place within the homoeostasis concept, because it, too, deals with balancing the intake of nutrients and tolerance levels, while requirement levels could be included. If it is possible to predict the requirements for homoeostasis properly, then it will become possible to predict at least part of the optimal behaviour of animals. In testing the predictions, indicators of performance like body weight could be used as a common currency.

Although striving for homoeostasis may be used as a convenient substitute for fitness maximisation, it is likely not equivalent to it? How does the struggle for life works out if all the members of a population (only) strive towards homoeostasis? It is possible that competitive and evolutionary processes have shaped the 'regulation points' to a higher level than that strictly required for maintaining homoeostasis? In that case some maximisation principle, again, needs to be invoked, and included in the models. Whatever that may be, the homoeostasis concept enables us to understand much of the behaviour of animals over a relatively short time span. It also makes animal performance the currency to evaluate, and this is a much more encompassing and integrative evaluation criterion than energy intake only.

If the regulation of a certain state is the goal rather than energy maximisation, the expected behaviours of animals are somewhat relaxed. The so frequently observed 'non-optimal' behaviour of herbivores can then be more realistically understood from the viewpoint of sufficing, than from being 'suboptimal' because either not all the constraints have apparently been included or because some basic assumptions (e.g., complete knowledge of the home range) have not been met. As the greater part of the regulation mechanism deals with evaluations across relatively short time periods, it is expected that much insight can be gained from studying foraging herbivores relative to these short periods. Selection of feeding sites (1-10 ha) within a daily range of 10-100 ha seems to be appropriate.

CHAPTER 3A

SPATIAL STATISTICS TO QUANTIFY PATTERNS OF HERD DISPERSION IN A SAVANNA HERBIVORE COMMUNITY

ALFRED STEIN[#] AND NICHOLAS GEORGIADIS[##]

[#] *International Institute for Geo-Information Science and Earth Observation (ITC), P.O. Box 6, 7500 AA Enschede, The Netherlands*
[##] *Mpala Research Centre, P.O. Box 555, Nanyuki, Kenya*
E-mail: stein@itc.nl

Abstract. Understanding the spatial distribution of species is a fundamental issue in ecology, yet quantitative descriptions of animal species' distributions are rare. In this chapter, we use a spatial-statistics approach to describe the spatial distribution of herds of large herbivores in Laikipia, central Kenya. We used Global Positioning System technology and spatial point pattern analysis (*F*-, *G*- and *J*-functions) to characterise herd distributions of the 9 most abundant species comprising large herbivore communities in African savannas. *F*-function analysis is based on estimating the probability of a herd occurring within radius *r* of randomly selected focal points. *G*-function analysis is similar, but based on randomly selected focal herds. The *J*-function is derived from the ratio of *G*- and *F*-functions. Comparing results from the different functions was instructive about possible causes of spatial patterning at the landscape level. All species displayed consistently aggregated distributions under *F*- and *J*-function analyses, partly because wildlife has been displaced by humans and livestock from sections of the study area. By contrast, the *G*-function provides a description of dispersion under more natural conditions because areas lacking herds are excluded from the analysis. *G*-function results showed 5 species to display random or nearly random dispersion patterns (zebra, impala, Grant's gazelle, eland and hartebeest), while the remainder were aggregated (African elephant, giraffe, African buffalo and Thomson's gazelle). When data for all species were pooled, *G*-function results revealed an emergent property of this community: wild herbivore herds were arrayed across the landscape in a significantly regular fashion. Two possible causes of this pattern, invoking interspecific complementarity in habitat preference, or disaggregation by prey herds to counter predators, could not be distinguished. Both mechanisms may have been operating in savannas over such long evolutionary time that their effects cannot be separated without experimentation.

Keywords. nearest-neighbour distances; savanna; spatial point pattern; herbivores; Laikipia

H.H.T. Prins and F. van Langevelde (eds.), Resource Ecology: Spatial and Temporal Dynamics of Foraging, 33-51.

INTRODUCTION

Animals often aggregate in groups or herds, and usually do not randomly distribute. These herds, in turn, display spatial distribution patterns. Variation in animal grouping and distribution (Krause and Ruxton 2002) has been explained on physiological grounds, invoking metabolic requirements, on ecological grounds, invoking habitat preference (Lamprey 1963), feeding style, competition, facilitation (Prins and Olff 1998; Arsenault and Owen-Smith 2002) and food distribution (Voeten 1999), and on climatic grounds (Walker 1990). A key difference exists between habitat preference and prey preference. Habitat preference largely determines where animals have a good hiding, living and mating place, whereas prey preference determines the areas where prey animals tend to live. These two concepts are to some degree complementary, as prey animals may have another habitat preference than the predators.

Spatial statistics can quantify patterns in animals to explain wildlife diversity

Buckland and Elston (1993) modelled wildlife distributions in space, Augustin et al. (1996) applied an autologistic approach to modelling spatial distributions, and Li et al. (1997) used regression to model the spatial distribution of the red crowned crane. None of these studies, however, employs a point pattern spatial-statistics approach. Such an approach may be beneficial to discover and quantify patterns in herds, which in turn may lead to an ecological explanation and hence to a better understanding of wildlife diversity. Since spatial heterogeneity influences resource use and, thus, competitive interactions between herbivore species, analysing the distribution of animal groups may be helpful to understand these underlying mechanisms. It is important in that sense to test whether the distribution of the groups deviates from random. For a random distribution, the distribution of the groups is assumed to have no underlying mechanism. Resource ecology may benefit from such explanations since they may help to explain consumer distribution and resource use.

Distribution patterns have an important effect on sampling and analysis (Borchers et al. 2002). For example, preferential sampling can be carried out if herds are known to disperse according to preferential habitats, or adaptive sampling may be useful if herds tend to cluster, both leading to estimates with a lower variance. Detailed distributional data, however, are rare for large vertebrate species comprising a community or guild. After more than 50 years of research on wild herbivores in Africa, for example, we could find no statistically supported descriptions of herd distribution at the ecosystem or landscape level. Systematic sample surveys (Norton-Griffiths 1978; Grunblatt et al. 1995; Khaemba and Stein 2000) have been routinely used to monitor wildlife distribution and dynamics because this method is efficient in terms of effort and repeatability (Caughley 1977; Krebs 1989). They provide impressions of spatial

Analysing the distribution of animals provides understanding how spatial heterogeneity influences competitive interactions

variation in density (e.g., Sinclair and Arcese 1995a), but the data do not allow formal descriptions of dispersion, rather, the method itself is based on the generally untested assumption that focal species are randomly dispersed.

Advances in Global Positioning System technology (GPS; see Wint 1998) and spatial point pattern analysis (Ripley 1981; Diggle 2003), permitted us to characterise spatial distributions of the 9 most abundant large herbivore species in the Laikipia ecosystem of central Kenya (in order of decreasing abundance: plains zebra, *Equus burchelli*; impala, *Aepyceros melampus*; Grant's gazelle, *Gazella granti*; Thomson's gazelle, *Gazella thomsoni*; eland, *Taurotragus oryx*; African buffalo, *Syncerus caffer*; African elephant, *Loxodonta africana*; hartebeest, *Alcelaphus buselaphus*; and giraffe, *Giraffa camelopardalis*).

The data, representing point measurements of herds, were collected during a total count of wildlife in an area of 7,100 km^2, using 10 aircraft equipped with GPS receivers (Georgiadis 1997). Three methods of spatial point pattern analysis were used to characterise the distributions of wild herbivore herds for each species separately, and for all species combined. Possible causes of observed patterns of dispersion within and among species are discussed.

GEOSTATISTICAL METHODS

Distribution data

Data in this study were collected during a total count within Laikipia District over three days in September 1996 (Georgiadis 1997; the region is also described in Heath 2000 and Georgiadis et al. 2003). The area was divided into three sections, each sub-divided into daily counting blocks of approximately 200 to 300 km^2, and each block was allocated to one aircraft per day. Ten high-winged aircraft were used simultaneously to search each block systematically. Each aircraft flew at heights between 70 and 130 m above ground level, following transects spaced 1 km apart. Each aircraft carried a crew of pilot, front-seat observer and two rear-seat observers. Whenever an animal or a group of animals was spotted, the aircraft deviated from its flight line to circle the observed animals until their number was counted. Observers in the aircraft estimated group size. This may be different from ordinary harem sizes, as these aggregate and disaggregate on a daily basis. Geographical coordinates of their positions were recorded using a Trimble GPS receiver. Overlaps and double counts at the boundaries of the blocks were identified and subtracted from the total wildlife numbers as a correction for count overlaps. This resulted in a data set of 1828 locations where at least one animal was observed (Figure 3.1).

For these purposes the observation of one or more animals at a given location is termed a herd. We make the basic assumption that each location is equally likely to host a herd. Deviations from randomness are then of interest, both for individual herds, and in their mutual relationships. Also of interest in this study is the data

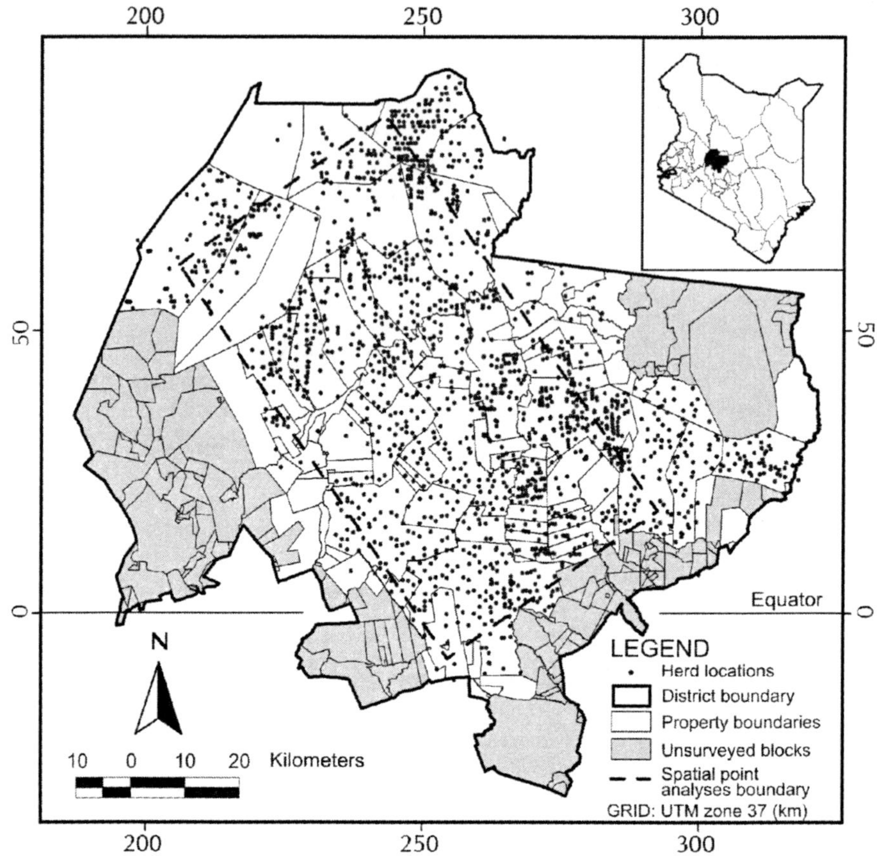

Figure 3.1. *Map of the study region in Kenya showing the set of all locations where a herd (at least one animal) was observed, i.e., the process X•. The rectangular box delineates the area used for point pattern analysis*

quality issue 'positional accuracy' (Goodchild and Jeansoulin 1998). Estimates of dispersion used here were affected (but to the same degree) by subjective variation among observers in their assignment of individuals to a herd. The scale at which these decisions were manifested as errors should be limited to distances no greater than the counting-strip width (0.5 km). Similarly, errors in the position of herds, due to the aircraft not passing directly overhead when the GPS location was recorded, should also amount to less than 0.5 km. Visibility for all species declines with distance from the aircraft, affecting dispersion patterns at distances up to the interval between transects (approximately 1 km; see also Ottichilo et al. 2000a; 2001). As is shown in this study, departures from complete spatial randomness generally occurred at values of $r > 1$ km; we assume conclusions drawn from these results will not be qualitatively affected by these errors.

Measures of dispersion

Spatial aggregation is the most frequent dispersion pattern encountered in nature, due to the prevalence of potent aggregating forces such as habitat specificity, social structure and organisation, philopatry, predator avoidance, and limited dispersal. Analysis of spatial patterns as spatial processes has now found a wide acceptance (see Lawson et al. (1999) for a recent overview in disease mapping, Manly (1999) for applications in biology, and Dale (1999) for applications in plant ecology). Spatial processes yielding observed animal counts are characterised by a simple stochastic model applied to a region A. Herds are represented by the coordinates of their centre of gravity marked by observed animal species. As a result, A is summarised by a mapped point pattern, consisting of the presence of at least one animal out of the nine species. The density of the processes is denoted by λ. Density in this study is similar to the number of herds per unit area. It is estimated by the ratio of the number of herds, divided by the size of the area.

To describe the spatial point pattern generated by the distribution of the nine ungulates, we let $Y = (X_1,\ldots,X_9)$ be a nine-variate point process in A with jointly stationary components. On the one hand we have the total pattern Y, on the other hand the 9 different patterns of the herds of individual species. Stationarity means that the position of the herds is independent of the location, although differences in densities may emerge due to stochastic influences. The process consisting of all points regardless of type is denoted by

$$X_{\bullet} = \bigcup_{i=1}^{9} X_i$$

In this paper, statistical inference for Y is based on distances, either those between a fixed reference point $a \in A$ and the points of the process $X_{\bullet}$, or those between the points of $X_{\bullet}$ themselves. We take stationary processes as the starting point for our research; for non-homogeneous populations we refer to Diggle and Chetwynd (1991).

Thus, for each $a \in A$, let $\rho(a, X_{\bullet})$ denote the distance from a to the nearest herd. Then the empty-space function of $X_{\bullet}$ for $r \geq 0$ equals

$$F_{\bullet}(r) = \Pr(\rho(a, X_{\bullet}) \leq r),$$

the probability of observing at least one herd closer than r to the arbitrary point a in the area. Under the assumption of stationarity, $F_{\bullet}(r)$ does not depend upon a. The heuristic explanation of 1 - $F_{\bullet}(r)$ is the probability that a circle with radius r placed at random in the area does not contain a herd, thus explaining the terminology. The empty-space function of X_i, $i \in \{1,\ldots,9\}$, is denoted by $F_i(r)$. By the stationarity assumption, neither $F_{\bullet}(r)$, $F_i(r)$, $i = 1,\ldots,9$ depend on the choice of the reference point a. A completely spatially random (CSR) pattern of herds with density λ shows an F-function equal to $F(r) = 1 - \exp(-\pi\lambda r^2)$. A clumped distribution has an F-function below this function, as on short distances fewer herds are encountered than for a random pattern, whereas a regular pattern has an F-function above it.

Turning to inter-herd distances, the nearest-neighbour distance function $G_\bullet(r)$ is the distribution function of the distance from a typical herd to its nearest neighbour,

$$G_\bullet(r) = \mathrm{Pr}^{!a}\left(\rho(a, X_\bullet) \le r\right)$$

for $r \geq 0$. Here, $\mathrm{Pr}^{!a}$ denotes the distribution of $X_\bullet$ at $a \in A$, *i.e.*, the conditional probability distribution that the distance from the point a to an arbitrary herd is less than or equal to r, given the location a (Stoyan et al. 1995). The function $G_\bullet(r)$ can be interpreted as the conditional distribution that a herd occurs within a distance r from a herd located at location a. Formally, it equals $\mathrm{Pr}\left(\rho(a, X_\bullet \setminus a) \le r \mid y \in X_\bullet\right)$, where $a, X_\bullet \setminus a$ denotes the full process $X_\bullet$ from which the herd at location a is excluded. A heuristic description of 1 - $G_\bullet(r)$ is the probability that within a circle with radius r centred on a randomly selected herd no further herd occurs. Again, $G_\bullet(r)$ does not depend on a because of stationarity. The univariate nearest-neighbour distance functions are denoted by $G_i(r)$, $i \in \{1,\ldots,9\}$. The empirical distribution function (EDF) for the G-function is obtained for each distance r by counting the number of herds at distances less than or equal to r from each of the herds. For example, for the ith species with n_i occurrences in A, it is

$$\hat{G}_i(r) = \frac{1}{n_i} \sum_{r_i \le r} 1$$

A random pattern of herds with density λ shows a G-function equal to $G(r) = 1 - \exp(-\pi\lambda r^2)$. A clumped distribution has a G-function higher than this function, as on short distances more herds are encountered than for a random pattern, whereas a regular pattern has a G-function below it.

Comparison of inter-herd distances to distances with respect to a reference point, say $a = 0$, yields the $J_\bullet(r)$ -function, defined by

$$J_\bullet(r) = \frac{1 - G_\bullet(r)}{1 - F_\bullet(r)}$$

for all $r \geq 0$ satisfying $F_\bullet(r) < 1$. For Poisson processes (*i.e.*, fully random processes without any aggregation or regularity), $J_\bullet(r) = 1$; $J_\bullet(r) > 1$ indicates inhibition between the points, aggregated patterns generally result in $J_\bullet(r)$-function values smaller than one. Moreover, the $J_\bullet(r)$ function is constant beyond the effective range of interaction (Van Lieshout and Baddeley 1999). The $J_\bullet(r)$ function is a useful index for the type and strength of spatial interaction (Stein et al. 2001).

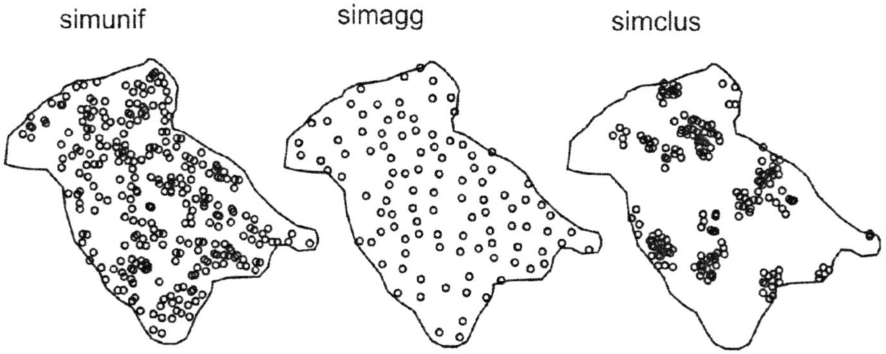

Figure 3.2. *Simulated random (simunif), aggegated (simagg) and clustered (simclus) process in the Laikipia area. The patterns have 300, 100 and 191 herds, respectively. See Box 3.1 for further explanation*

Box 3.1. *F-, G-, and J-functions to describe spatial point patterns*

To show the functioning of the different functions, three processes were simulated: a uniform pattern (denoted as simunif), an aggregated pattern (denoted as simagg) and a clustered pattern (simclus) (Figure 3.2). For the aggregated pattern, a simple sequential inhibition process was used with an inhibition parameter equal to 5 km, and a Matèrn Cluster process was applied for the clustered process. Parameters were set such that the area corresponded with Laikipia area, yielding 300, 100 and 191 herds, respectively. The F-function for simagg was well inside the confidence bounds, the simagg was above the confidence bounds, the simclus was below the confidence bounds for randomness (Figure 3.3). Deviations from randomness were much clearer when a G-function was estimated (Figure 3.4). First, the simagg pattern is below the confidence envelope for randomness. It further showed very nicely the inhibition in the simagg data, i.e., a 0-valued G-function for distances up to 5 km. The simclus pattern is now above the simulation envelope for randomness. Finally, the J-function estimates (Figure 3.5) show that up to distances of 2 to 2.5 km simunif yields a horizontal function, which then drops, whereas the simagg pattern yields a J-function above the simulation envelope, and the simclus yields a J-function that falls well below the confidence envelope.

From the above it follows that the *F*- and the *G*-function show an opposite behaviour. A pattern with a *G*-function above that for CSR means aggregation, because as measured from a herd there are more short distances to other herds than expected on the assumption of spatial randomness. A *G*-function below that for CSR means regularity, because small distances to herds occur less often than expected under the assumption of spatial randomness. For the *F*-function the opposite applies. In case of uniformity the expected distance from an arbitrary point in the area to the nearest herd is smaller than what would be expected on the basis of CSR. Therefore, in this case the *F*-function is above that of CSR. In case of aggregation, however, small distances are under-represented and the expected distance to the closest herd would be larger. Therefore, the estimated *F*-function falls below that for CSR. Opposite behaviour of the *F*- and *G*-functions therefore leads to the same conclusion when considering deviation from CSR. The *J*-function leads to similar conclusions as both the *F*- and the *G*-function, but it does not rely in any sense on the Poisson model. In Box 3.1, the *F*-, *G*-, and *J*-functions are demonstrated using simulated point patterns.

To facilitate calculation of nearest-neighbour distances during spatial point pattern analysis using the S-Plus software package (Kaluzny et al. 1998), we rotated the pattern by an angle of 55°, so that the area is close to a rectangle of size 60 by 110 km (Figure 3.1).

SPATIAL DISTRIBUTION OF HERDS OF LARGE HERBIVORES

Population size and density

In total, 57,928 animals of the nine species were observed in this study, distributed over 3,034 herds. The maximum herd size equals 473 animals, whereas 322 solitary animals were observed, i.e., herds of size 1.

Abundance varied among the nine species 23-fold, the largest herds occurring for plains zebra and the smallest for giraffe (Table 3.1). Mean herd size varied 4.5-fold, with plains zebra having the highest mean herd size (30.5 individuals per herd) and giraffe the lowest mean herd size. The impala, although less abundant than the

Table 3.1. *The nine animal species under study, including their mean and median herd size, standard deviation and total count*

Process	Species	No. of herds	Total count	Group size			Area (km^2)	Density ($n\ km^{-2}$)
				mean	median	sd		
X_1	plains zebra	1,034	31,517	30.5	18	39.7	7,103	0.1456
X_2	impala	431	5,707	13.2	10	13.5	6,567	0.0656
X_3	Thomson's gazelle	211	4,255	20.2	12	31.0	5,487	0.0385
X_4	Grant's gazelle	436	3,507	8.0	6	8.0	5,735	0.0760
X_5	eland	258	3,164	12.3	6	20.9	5,681	0.0454
X_6	elephant	162	2,287	14.1	9	22.2	5,319	0.0305
X_7	hartebeest	206	1,681	8.2	7	9.5	6,048	0.0341
X_8	buffalo	69	1,563	22.7	15	25.7	5,370	0.0128
X_9	giraffe	218	1,340	6.2	4	8.0	5,292	0.0412

Thomson's gazelle, aggregates in smaller herds and hence has a higher number of herds. The buffalo forms relatively large herds (22.7 individuals per herd), whereas all other species form much smaller herds (8.0 - 14.1 individuals per herd) (Table 3.1; compare Voeten 1999). Densities, i.e., number of herds km^{-2}, vary 11-fold in the area, the highest density occurring for plains zebra, the lowest density for buffalo (Table 3.1).

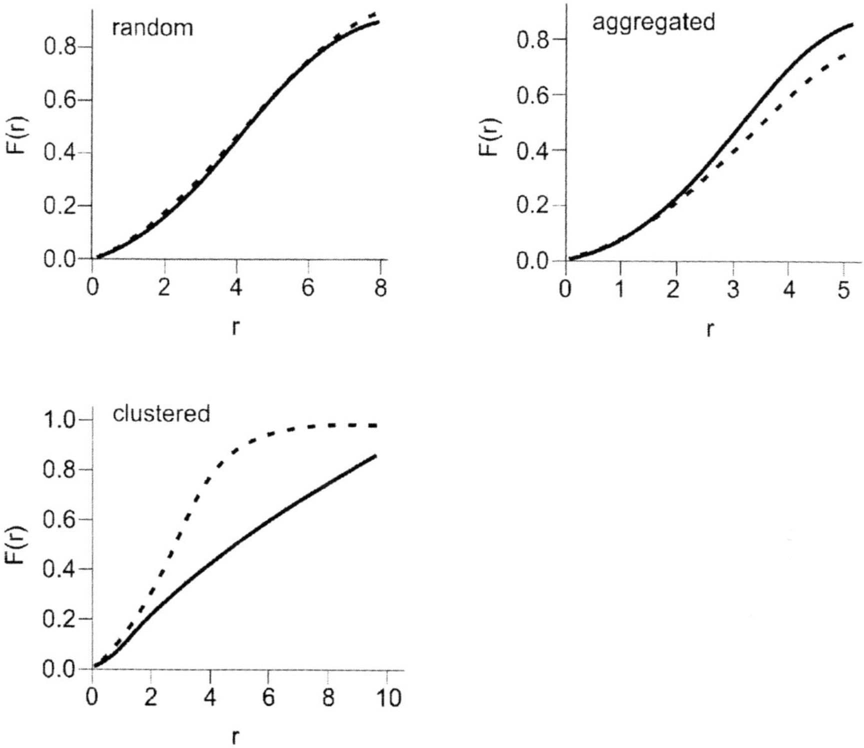

Figure 3.3. *F-Functions for the random, aggregated and clustered processes. See Figure 3.2 for the corresponding patterns*

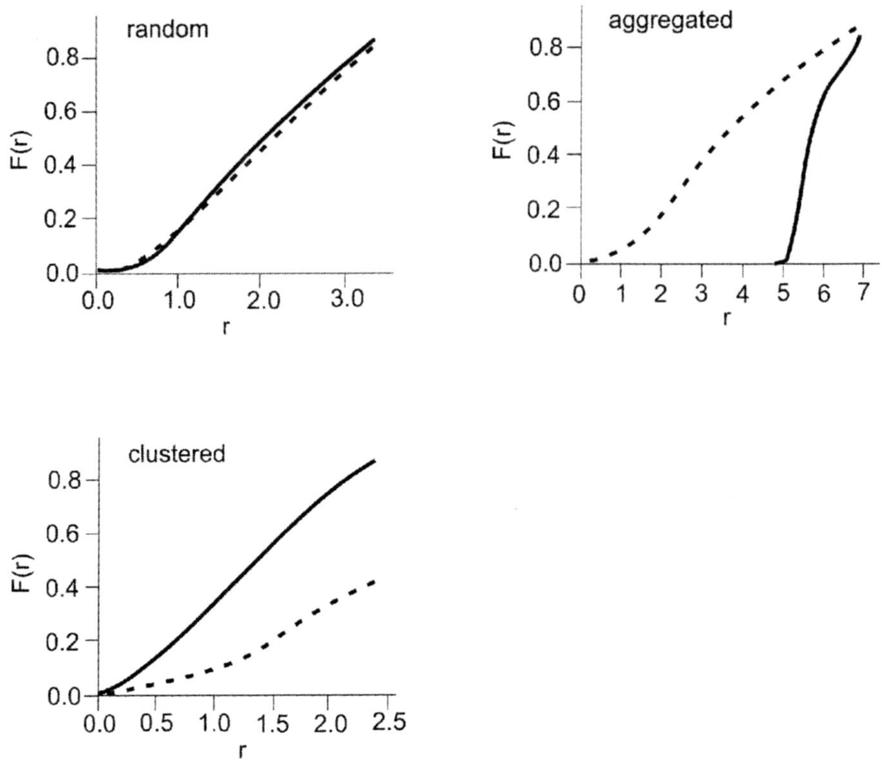

Figure 3.4. *G-functions for the random, aggregated and clustered processes. See Figure 3.2 for the corresponding patterns*

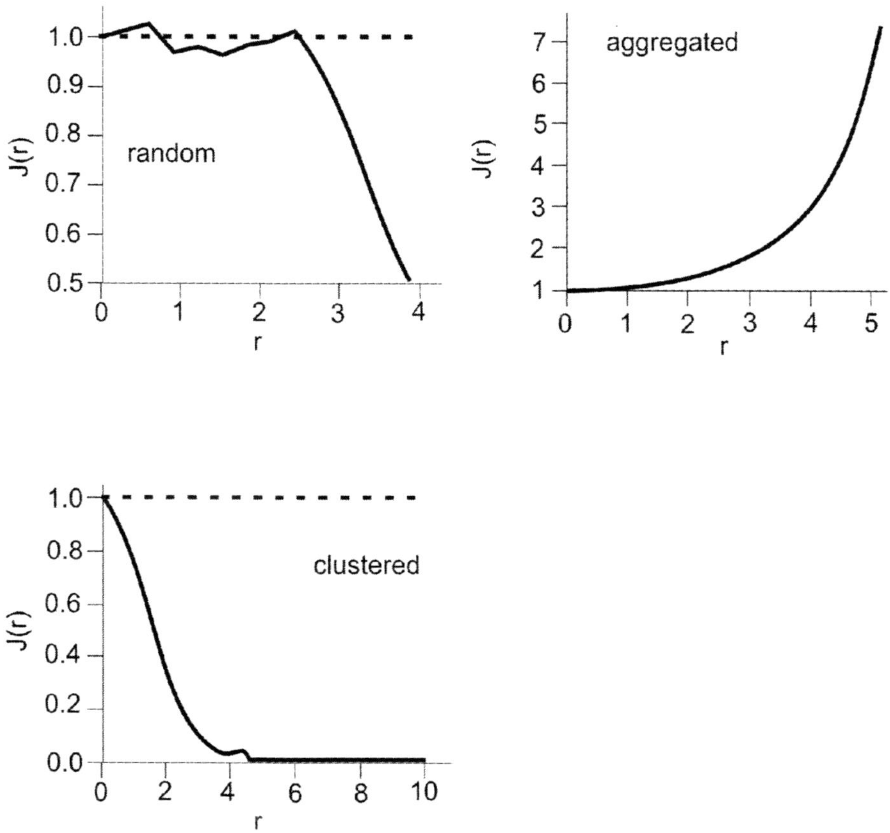

Figure 3.5. *J-functions for the random, aggregated and clustered processes. See Figure 3.2 for the corresponding patterns*

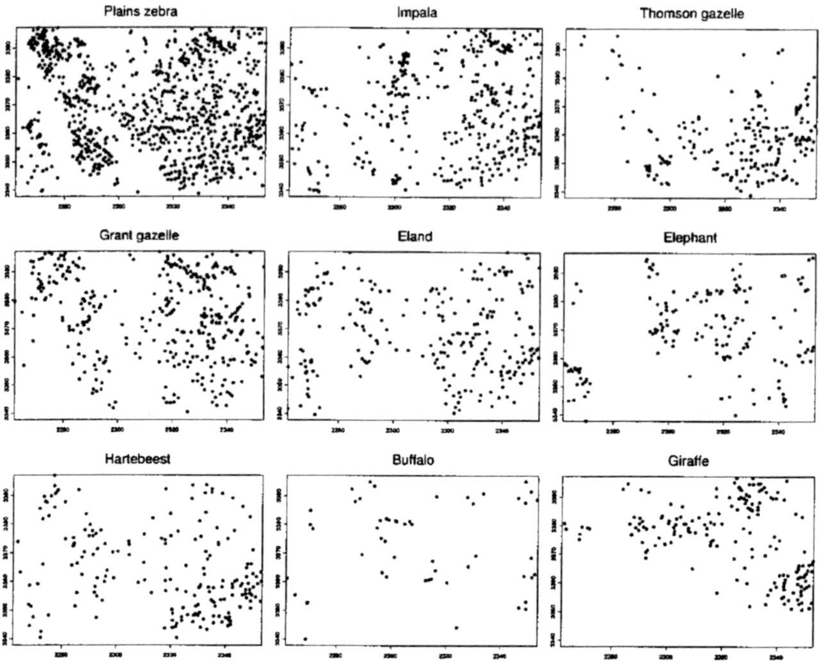

Figure 3.6. *Spatial point patterns for the nine animal species in the study area. Coordinates are in km along a 55° rotated image*

Spatial point patterns

Spatial point patterns for individual species cover different areal extents, with convex hulls ranging from 7,103 km^2 for the plains zebra to 5,300 km^2 for the elephant and giraffe, suggesting a restricted use of the habitat for these two species (Table 3.1).

The plot of the combined spatial pattern is fairly dense with no apparent spatial pattern (Figure 3.1). A section with almost no herds occurs in the northern part of the area. Evidence of aggregation emerges from individual species point patterns (Figure 3.6). The plains zebra has both a high density and a highly aggregated spatial point pattern. The impala occurs more often in the southern part of the region, the Thomson's gazelle in the south-western part, the giraffe in the eastern part of the area. Buffalo is typically a somewhat rare species, at least its herds are found at a lower frequency than those of the other species. Each species appears to exhibit some aggregation. Such visual assessments are to be tested using distance measures for confirmation.

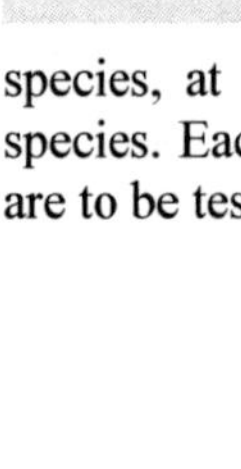

Evidence of aggregation emerges from individual species point patterns

The estimated $\hat{F}_i(r)$-function plots for individual species (i = 1,…,9) display significantly aggregated patterns at radii > 0.5 km (plains zebra) to 3 km (buffalo) (Figure 3.7). The $\hat{F}_i(r)$-function for plains zebra, which had by far the highest density, shows a value equal to 0.6 at distances of 3.5 km. Similar results were obtained for Grant's gazelle and impala. Despite their differing densities, therefore, herd spatial distributions for these species are comparable. For the remaining species, $\hat{F}_i(r)$-functions display lower steepness, but so too were expectations of functions observed under CSR conditions. Accordingly, the $\hat{F}_\bullet(r)$-function plot for all species combined (Figure 3.7) shows the observed $\hat{F}_\bullet(r)$-function to be more aggregated than random at all radii.

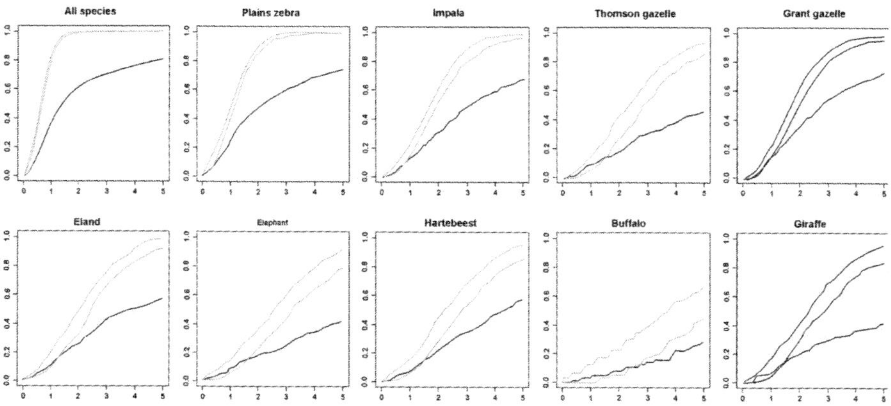

Figure 3.7. *The estimated* $\hat{F}_\bullet(r)$ *function for* $X_\bullet$ *the combined data, and the* $\hat{F}_i(r)$*-functions (solid lines) of the nine animal species with envelopes (dotted lines) for testing for spatial randomness. Coordinates along the horizontal axis are in km. The figure seems to imply regularity, as the estimated* $\hat{F}(r)$ *functions are below the confidence bounds, both for all species and for each individual species. This is hard to justify, as a probable cause may be the empty sub-area in the northern part of the study area*

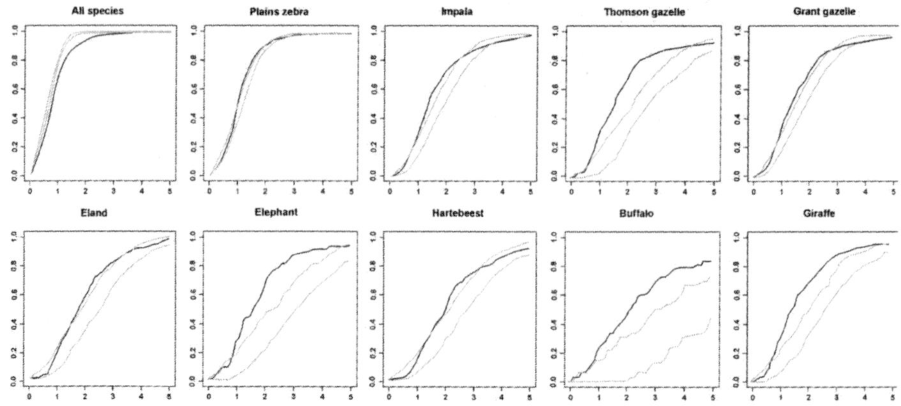

Figure 3.8. *The estimated* $\hat{G}_{\bullet}(r)$ *-function for* $X_{\bullet}$, *the combined data, and the* $\hat{G}_i(r)$ *-functions (solid lines) for the nine animal species with envelopes (dotted lines) for testing for spatial randomness. Coordinates along the horizontal axis are in km. Figure numbering is from a (top left) to j (bottom right). All species show an aggregated pattern, with the exception of the plains zebra, which displays a random distribution. The pattern of all species, on the contrary, is more regularly distributed*

$G_i(r)$-function results for each species treated individually also show a tendency for aggregation in all cases, except the most abundant species, plains zebra, but only at intermediate values for r (Figure 3.8). For impala, Grant's gazelle, eland and hartebeest, departure from CSR was never marked, while Thomson's gazelle, elephant, buffalo and giraffe display clearly aggregated distributions. In contrast to all results presented above, the $\hat{G}_{\bullet}(r)$ -function plot for the combined data for all 9 species shows a significantly more regular pattern than CSR (Figure 3.8) at radii between 0.5 and 5 km.

The $\hat{F}_{\bullet}(r)$ -function plot for $X_{\bullet}$, the combined population of species drops below the lower bounds of CSR at a distance of approximately 0.5 km, showing that the pattern of herds is significantly aggregated beyond this distance (Figure 3.9). The individual $\hat{J}_i(r)$ -functions for the plains zebra and the giraffe (Figure 3.9) become significantly aggregated at r = 0.7 km, whereas for other species this occurs at greater values of r (r = 0.8 – 1 km). For most species the trend is relatively smooth but the pattern for elephant and to a lesser extent buffalo appears stepped at r = 1. This may reflect the tendency for these species to occur in stable groups that are spatially aggregated for social reasons.

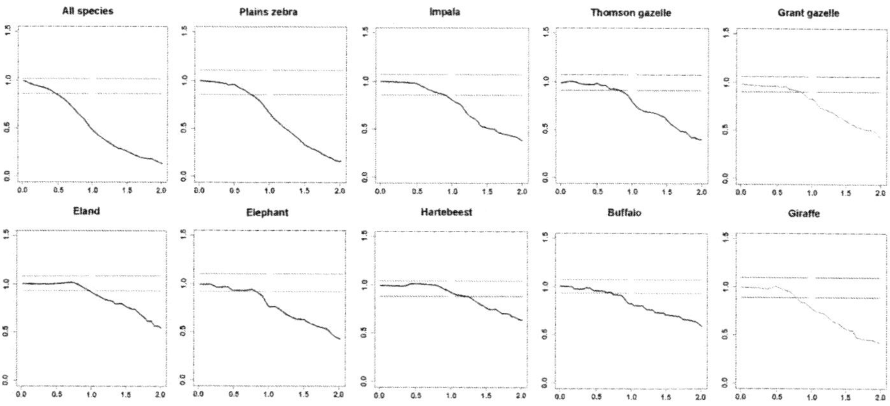

Figure 3.9. *The estimated* $\hat{J}_{\bullet}(r)$ *-function for* $X_{\bullet}$ *the combined data and the* $\hat{J}_i(r)$ *-functions (solid lines) for the 9 animal species with envelopes (dotted lines) for testing for spatial randomness. Coordinates along the horizontal axis are in km. Figure numbering is from a (top left) to j (bottom right). These J-functions indicate randomness for distances up to 0.4 km (for all species) and to distances between 0.7 km and 1 km for individual species. Beyond that distance, herds are distributed in a more aggregated way*

DISCUSSION

The nine species featured in this study are subject to all forces that influence spatial aggregation: habitat specificity, social structure and organisation, philopatry, predator avoidance, and limited dispersal. Based on *F*- and *J*- function results, all displayed aggregated distributions. By contrast, some of the same species displayed random distributions under the *G*-function. Comparison of results from the different functions used to estimate dispersion is instructive about spatial patterning at the landscape level. The area also has some fencing, but although this may lead to lower densities, it is not considered to be very important for the spatial-pattern study.

Results from the *F*-functions were derived from points selected randomly over the entire study area, and are therefore influenced not only by natural forces affecting herd dispersion (such as those listed above), but also by 'unnatural' factors, such as displacement of wildlife by humans, cultivation and livestock. By excluding wildlife from large sectors of the study area, these factors accentuate observed degrees of aggregation and may partly account for the consistently aggregated dispersion patterns displayed in *F*-function plots for all species, relative to CSR. In ecosystems such as this, therefore, where wildlife are displaced from some areas by humans and livestock, all species are likely to violate the assumption of random dispersion patterns typically made when sample counting.

Comparison of results from the F-, G- and J-functions used to estimate dispersion is instructive about spatial patterning

A way to reduce the influence of these 'unnatural' factors on the analysis of dispersion would be to excise human-occupied sectors from the study area, and repeat the *F*-function analysis. However, the *G*-function provides a more efficient way of reducing 'unnatural' influences because randomly chosen herds comprise foci for analysis, and areas lacking herds are automatically excluded from the analysis. By this measure, herds of plains zebra, impala, Grant's gazelle, eland and hartebeest were dispersed in a random, or only slightly aggregated fashion, whereas herds of elephant, buffalo, giraffe and Thomson's gazelle were distinctly aggregated (at least at values of r between 1 and 4 km). These results support the widely held view that herds of elephant and buffalo are likely to be too aggregated to be effectively censused by sample counting with regularly spaced transects, even in a 'natural' context.

Comparison of *G*-function results among species yielded no clear association between dispersion pattern and feeding preference (grazer, mixed feeder or browser), dependence on drinking water, mating system or tendency to migrate. Factors that have a potentially organising influence on herds within species, such as territoriality (impala, Thomson's gazelle, Grant's gazelle, hartebeest), or intraspecific competition, may have been operating, but were evidently not strong enough to cause regular patterns of dispersion within species in this landscape. Strong social bonds within and between herds are known to influence elephant herd associations (McComb et al. 2001), possibly accounting for the extreme aggregation displayed by this species. By contrast, plains-zebra harems are known to associate and disassociate on a daily and seasonal basis (Rubenstein pers. comm.), but this evidently does not result in significant aggregation at the landscape level. At least in this woodland-dominated habitat, 'exogenous' forces such as patchiness of preferred habitat are more likely to account for aggregations of Thomson's-gazelle herds, which prefer open, short grassland.

G-function results for each species treated separately, showing random or aggregated patterns, contrasted strikingly with results from the pooled data for all species, which showed that wild herbivore herds were arrayed across the landscape in a significantly regular fashion. This result implies that factors causing individual species to be aggregated or randomly dispersed either (1) complement each other when combined across space, or (2) are organised by factors that exert their influence on the entire community. As an example of the first possible cause, which we refer to as 'habitat-preference complementarity', we expect that animals that are abundant are more randomly dispersed (often called habitat generalists) than rare species (habitat specialists). While the two most abundant species (plains zebra, impala) showed random *G*-function distributions, there was no overall association between rank of relative abundance and dispersion pattern ($P > 0.065$). By contrast, we expect habitat specialists to be found within preferred habitat patches, to display lower herd densities in transitional

Species aggregation is caused by factors that complement each other or that exert their influence on the entire community

habitats, and to be absent from unsuitable habitats. When the data of all species are pooled, the net effect is for herds to become regularly spaced across the landscape.

Clear differences in the distribution of groups emerged. For example, habitat preference results in grouping that is closely linked to the location of available habitat, whereas this relationship between consumer presence and resources is affected by the presence of competitors. Further details concerning the ecological explanation of the observed patterns have to be worked out in the future, as this extends the scope of this chapter (see Box 3.2).

Box 3.2 *Testable hypotheses for future research*

Hypothesis 1. Based on 'habitat preference complementarity', we expect that abundant species are more randomly dispersed (i.e., habitat generalists) than rare species (habitat specialists).
Hypothesis 2. We expect habitat specialists to be aggregated within preferred habitat patches (in fact, one then tests whether these habitat patches are aggregated in the landscape), to display lower herd densities in transitional habitats, and to be absent from unsuitable habitats.
Hypothesis 3. As is shown in this chapter, herds become regularly spaced across the landscape when the distribution data of all species are pooled. This brings us to the following hypothesis: abundant animals tend to be dispersed randomly in contrast to rare species. The rare species mutually exclude each other because they are habitat specialists and, thus, do not occur in the same habitat. The over-layering of these patterns results in a 'regular' pattern.

As an example of the second possible factor referred to above, which we refer to as 'prey preference complementarity', predators are hypothesised to have a disaggregating effect on dispersion of preferred prey herds (although it is hypothesised that predators lead to increased herd size, i.e., 'safety in numbers', see Krause and Ruxton 2002), which, when the data of the prey species are pooled, is manifested as an organising effect by predators on the dispersion of preferred prey. Herds of preferred prey species, which could be aggregated in the absence of predation, react to the presence of predators by moving apart, becoming less aggregated, and alleviating pressure exerted by predators. Since the presence of predators is cued to multiple prey species, the net effect on combined prey herds is to cause a more regular pattern of dispersion.

G-function results for individual species in this study appear to conform to expectations of both habitat and prey preference complementarity in ways that are hard to separate. In the former, we observe large-sized species (elephants, giraffe and buffalo) preferring habitats featuring high vegetation biomass, small-sized species (Thomson's gazelle) requiring open habitats with low biomass, and mid-sized species (eland, plains zebra, hartebeest, impala and Grant's gazelle) distributed across a variety of savanna habitats featuring grasslands associated with a range of tree densities. Because extreme habitat types, featuring either high or low vegetation biomass, are likely to be rarer and more patchy than intermediate habitat types, herbivore species preferring extreme habitat types are likely to display more aggregated distributions than are species preferring intermediate habitat types.

Similarly, for prey-preference complementarity, herds of mid-sized prey species (eland, plains zebra, hartebeest, impala and Grant's gazelle), which are more likely to be preferred by the dominant predators in this ecosystem (hyenas, lions and leopards), are expected to display less aggregated (even random) distributions. By contrast, herds of large-sized species (elephant, giraffe and buffalo) are expected to experience lower predation pressure (Sinclair et al. 2003, although buffalo has quite a high predation rate, see Prins and Iason 1989), and are thus expected to be more aggregated. Herds of the smallest-sized species (Thomson's gazelle) conform to the expectation of an aggregated distribution, although they experience high predation pressure.

Attempting to discriminate between these alternatives, further *G*-function analyses were performed on the combined data for large- and small-sized (habitat specialist or 'non-preferred' prey) species, and on the combined data for the mid-sized (habitat generalist or 'preferred' prey) species. If habitat preference complementarity is operating, the result should be a tendency by both groups to shift towards a more regular dispersion pattern. If prey preference complementarity is operating, the result for the non-preferred prey group should be to remain aggregated, while the preferred prey group should become regularly spaced. Results (Figure 3.10) tend to support habitat preference complementarity, in that the 'non-preferred' prey group is randomly dispersed, and while the 'preferred' prey group tends towards a regular dispersion pattern, this is far less marked than the result for all species combined (Figure 3.8).

These results do not rule out the possibility that both mechanisms – habitat and prey preference complementarity – have been operating in this landscape over such a long evolutionary time that experimentation is required to detect their separate effects. Recent evidence of the effects of reintroduced wolves on prey in North America shows that predators can greatly affect movements, local densities and sizes of prey herds (Hebblewhite et al. 2002; Hebblewhite and Pletscher 2002; Kunkel and Pletscher 2001; Ripple et al. 2001). However, explicit evidence of disaggregating effects by predators on prey herd dispersion is lacking. There is also growing evidence that predators affect the presence of other predator species (e.g., Durant 1998; Creel 2001). Spatially explicit models of predator–prey dynamics have tended to oversimplify responses of predators and prey to each other (Lima 2002), but are beginning to examine the effect of predators more realistically (Cosner et al. 1999; Abrams 2000; Alonso et al. 2002; Connel 2000; Forrester and Steele 2000; Křivan and Vrkoč 2000; Pitt and Ritchie 2002; Vucetich et al. 2002).

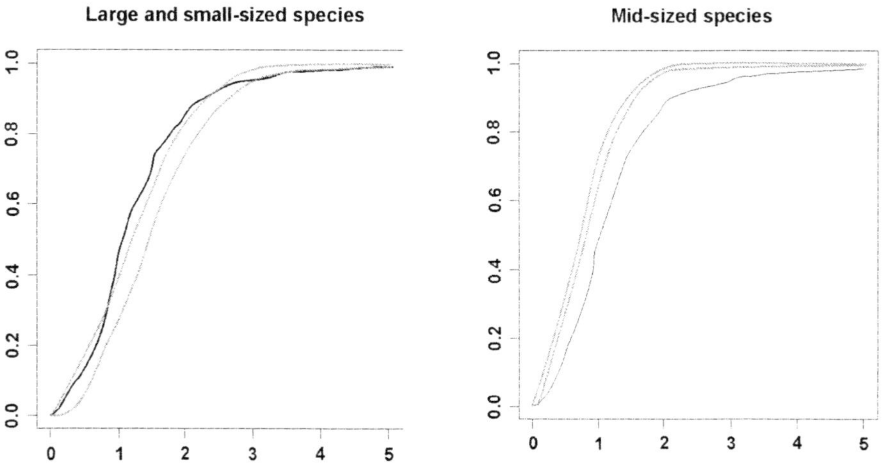

Figure 3.10. *The estimated* $\hat{G}(r)$ *function for large- and small-sized species (left-hand side) and mid-sized species (right-hand side). Coordinates along the horizontal axis are in km*

SYNTHESIS

Combination of GPS technology with spatially explicit statistical techniques, such as plots of the *F*-, *G*- and related *J*-functions, yields novel ways of characterising wild-herbivore dispersion patterns. In particular, the striking contrast between aggregated or random patterns displayed by species analysed separately and the regular pattern observed with pooled data from all species, reveals an emergent and unexpected property of the herbivore community that demands further explanation. This issue and some other hypotheses for future research are formulated in Box 3.2.

Further, the applied value of the spatial analysis of these total count data is, for example, the potential to correct systematic bias in the sample survey methodology. Further studies should reflect the degree to which violation of the assumption of random distribution affect the estimate and which species are mostly affected by this.

ACKNOWLEDGEMENTS

We are grateful to Dr. Jan de Leeuw and Dr. Wilbur Ottichillo at the ITC International Institute for Geo-Information Science and Earth Observation, The Netherlands, and to Mohammed Said at the Department of Resource Surveys and Remote Sensing (DRSRS), Kenya, for their contributions to the ecological component of this study. We are also grateful to Mr. Nasser Olwero for creating Figure 3.1. Finally, we are very grateful to Dr. M.N.M. van Lieshout and Dr. Leo Barendregt for giving many valuable suggestions on the statistical procedures.

CHAPTER 3B

COMMENTS ON "SPATIAL STATISTICS TO QUANTIFY PATTERNS OF HERD DISPERSION IN A SAVANNA HERBIVORE COMMUNITY"

HERBERT H.T. PRINS, SIPKE E. VAN WIEREN AND AREND M.H. BRUNSTING

Resource Ecology Group, Wageningen University, P.O. Box 47, 6700 AA Wageningen, The Netherlands
E-mail: herbert.prins@wur.nl

The famous statement "Lies, damned lies, and statistics" is attributed to Benjamin Disraeli (1804 – 1881), British Prime Minister for the Conservative Party. When he made that pronouncement, he possibly referred to one of the original meanings of statistics, namely, the (quantitative) description of nation states. In the 19th century, the three developmental lines merged of what we now call 'statistics', that is to say, the quantitative description of societies, the study of sets of objects and the analysis of probabilities. Statistics became the quantitative investigation of equal elements or objects belonging to one set. At that time, two major ways to test hypotheses emerged; one was the experimental way (which became dominant in physics and chemistry) and the other was the statistical way, where a theory or hypothesis was confronted with observations (typical for biology and medicine). Finally, statistics developed into a powerful tool to discover underlying mechanisms that explain variation in patterns or processes.

Alfred Stein and Nicholas Georgiadis (Chapter 3) were apparently motivated by two of these meanings of 'statistics'. First is the quantitative description of equal elements (in their case 'herds' of animals) belonging to one 'set' (all herds in Laikipia District, Kenya; for more information on that area, see, e.g., Heath 2000). The second is the discovery of possible underlying mechanisms that could explain the observed distribution of 'herds' in a landscape.

H.H.T. Prins and F. van Langevelde (eds.), Resource Ecology: Spatial and Temporal Dynamics of Foraging, 53-56.

Their quantitative description of the 'herds' of Laikipia is an exciting one, even though it has some obvious problems. The first problem is that they use the term 'herd' very loosely; not a "group of individuals of the same population that form a socially coherent unit over time and space" but merely as "one or more animals at a given location". Note that 'location' is not defined in the text, but we may infer perhaps that it could be a cell with a radius of 250 m. Because Stein and Georgiadis do not use a biological definition of 'herd', their analysis is, by definition, not able to make a statement about these biological social units. Their analysis, however, may make statements about what normally is just called a 'group' (in their case this would be about "groups of individuals of the same species that are observed within a certain (here not reported) distance of each other". Because they chose to use the term 'herd', their conclusion may be suspect. Indeed, harem size of plains zebra is normally about 7 individuals (Voeten 1999); it is unclear to us what biological unit the mean herd size of 30-odd zebra (their Table 4.1) could represent. Likewise, while a 'herd' size of buffalo of 23 (*loc. cit.)* can occur in nature (e.g., Prins 1996), one wonders whether lone bulls and herds of buffalo cows were all taken together under the term 'herd'. We would not advocate that, since 'bachelor bulls' often select different terrains and have a different spatial use than herd animals (Sinclair 1977; Prins 1989a; 1989b; 1996).

A second problem is about the contrast between easily observable animals and ones that are difficult to spot from the air. The chapter of Stein and Georgiadis assumes that all individuals and all species in all terrain types are equally well-spotted; indeed, they state that their work is based on a 'total count'. Yet, we know from other studies that this is not realistic: small animals are more difficult to spot than big ones, greyish animals more difficult than reddish-brown, and it is easier to get a realistic count of animals on grasslands than of those occurring in thickets (e.g., De Leeuw et al. 1998; Said 2003). Conclusions about patterns of spatial use of animals have to take into account the distribution of the substrate in which they occur: in other words, the spatial patterns of randomness or aggregation may have been caused by vegetation patterns, but the authors do not provide information on these.

Problems aside, the true contribution of this chapter is the development of tools to describe spatial patterns of elements or objects. Stein and Georgiadis show how the F, G and J functions can be used to describe and analyse how animals are distributed in a spatial context. This is useful, because a full description of the exact locations where each individual occurs does not yield insight: one would not see the wood for the trees. Of course, these functions can also be used to describe and analyse other spatial patterns, e.g., the occurrence of large trees in a savanna landscape, water points in an arid environment, camp sites of nomadic people, etc., etc. Stein and Georgiadis point the way, and we think it is a way that should be followed.

In their analysis of the patterns of spatial occupancy of the different species, they find, by-and-large, that groups and singletons of plains zebra, impala, Grant's gazelle, eland and hartebeest were randomly dispersed, but that groups and singletons of elephant, buffalo, giraffe and Thomson's gazelle were distinctly aggregated. They then started searching for explanations for these patterns. Here they use the second important function of statistics, namely, the unearthing of possible underlying mechanisms that explain an observed distribution. Thus, statistics can function as a heuristic tool for the development of testable hypotheses. In their search for testable hypotheses that potentially explain their observed result, they tested whether feeding style, dependence on drinking water, mating system or tendency to migrate could explain the dichotomy between the two groups of ungulates. They did not find such a simple factor that could explain why, for example, in September 1996 (when their data were collected) Thomson's gazelle occurred aggregately and Grant's gazelle randomly. We think this lays bare two things. First, explaining an observed (or 'discovered' pattern) without a preconceived hypothesis, is not simple and can rapidly develop into a hunt for the notorious needle in a haystack. Second, because there is no clear underlying hypothesis data 'happened' to be collected in September (which is the middle of the dry season) and sampling at the end of the dry season (or for that matter at the end of the wet season) could have yielded two different sets of species of which individuals are dispersed randomly across the landscape versus species of which the individuals aggregate at a particular scale of observation. The work of, for example, Voeten (1999), Voeten and Prins (1999), Ottichilo (2000), Ottichilo et al. (2001), Oindo (2001), Oindo et al. (2001) or Said (2003), shows how dynamic wild herbivores use these landscapes in time and space. In other words, Stein and Georgiadis should not have made deductions on an analysis of data from a single time of the year: it is much more likely that repeated sampling should have yielded the desired 'explanatory factor'. We hope that Nick Georgiadis will be able to collect simultaneous GPS data (using 10 aircraft!) of the wild herbivores in Laikipia over different seasons.

Finally, Stein and Georgiadis make another tantalising observation for which they have no good explanation. If they group the nine different species, and they run their F, J and G tests on the amalgamated observations of the spatial occurrence of groups and singletons of all species in Laikipia District, they find that these "wild herbivores are arrayed across the landscape in a significantly regular fashion". We call this 'tantalising', because we have our doubts about their explanation that predation could explain this. Repeated sampling is crucial for the confirmation of this pattern that was observed only once. Their idea that predation could explain this, is not shared by us. There are several reasons to be hesitant.

First, the 'safety in numbers hypothesis' does not predict regular spacing but predicts (unpredictable) aggregation in space. *Second*, studies of the association between species in the face of predation in East Africa do not suggest regularity across the landscape. Parts of Laikipia look like parts of the Serengeti, and Sinclair and Norton-Griffiths (1982) observed significant associations between species, while in Manyara, De Boer and Prins (1990) found randomness in the association between species on the scale of tens of hectares. Detailed observations of Voeten (1999) in Tarangire of wildebeest and plains zebra show very similar habitat requirements of these species at certain times of the year, leading then to strong spatial association, but dissociation and dissimilar requirements at other times of the year. *Third*, Stein and Georgiadis are cushy when ascribing the patterns they find to predation. They state "for prey preference complementarity, herds of mid-sized prey species (eland, plains zebra, hartebeest, impala and Grant's gazelle), which are more likely to be preferred by the dominant predators in this ecosystem (hyena, lion and leopard), are expected to display less aggregated (even random) distributions. By contrast, herds of large-sized species (elephant, giraffe and buffalo) are expected to experience lower predation pressure". They concede that buffalo can experience very high predation pressure (Prins and Iason 1989), and acknowledge that the small Thomson's gazelle (about 25 kg; see for body mass of different ungulates Kingdon 1982) does not fit their expectation either, so two out of nine species do not agree with their hypothesis. But why classify eland (about 470 kg) as mid-sized together with, e.g., Grant's gazelle (55 kg), but buffalo (630 kg) or giraffe (600 kg) together with elephant (3550 kg)? Further, are predation data known for Laikipia? In Timbavati (South Africa), for example, 55% of the annual mortality was caused by predation in case of impala, but for giraffe this was 34% (Hirst 1969): is this difference really big enough to support the dichotomy as envisioned by Stein and Georgiadis? In the Serengeti, Sinclair and Norton-Griffiths (1982) found that between 13 and 21% of the annual mortality of wildebeest was caused by predation, while Sinclair (1985) reported 30%, but Hirst (1969) found it to be 96% in Timbavati, and Mills (1984) found a figure of 93% for the Kalahari. Perhaps Laikipia has more of the appearance of the Serengeti than of the Kalahari, but does it function like the Serengeti? Without a study of predation pressures in Laikipia itself, how well supported are Stein and Georgiadis' thoughts about the relation between body mass, predation and level of aggregation of the individuals in their study? *Fourth*, where Stein and Georgiadis expect aggregation in the case of giraffe, a study on individuals in the Masai Ecosystem of northern Tanzania by Van der Jeugd and Prins (2000) reveals much randomness. *Fifth*, from a natural-selection point of view one would expect also spatial patterns to reflect fitness maximization strategies of individuals of the nine different populations of herbivores. It would make sense to test a hypothesis based on the ideal free distribution of these individuals, but a hypothesis based on predation by several predator species resulting in more regularity across the landscape of groups and individuals of several species of prey, could, to our taste, easily lead to group-selectionist type arguments (*sensu* Wynne-Edwards 1962), which makes us very weary of accepting predation by several predators as a potential explanatory factor for the regular distribution of individuals of several ungulate species.

In conclusion, we think that statistics is very important to describe patterns concisely and precisely. Stein and Georgiadis' statement how useful certain statistical techniques are, is amply demonstrated by their paper. On the other hand, we are less enthusiastic about statistics as a heuristic tool for finding explanatory factors, although we fully acknowledge its use. We think that potential explanatory factors should be firmly grounded in existing theories and linked-up with hypotheses that have been tested. Finally, we think that one needs more observations on the (ir)regularity of herbivore distributions, of single species or of pools of species, at different times of the year, and with a better understanding of the associated resource distribution patterns before one jumps to conclusions.

CHAPTER 4A

RESOURCE DISTRIBUTION AND DYNAMICS

Mapping herbivore resources

ANDREW K. SKIDMORE AND JELLE G. FERWERDA

International Institute for Geo-Information Science and Earth Observation (ITC), P.O. Box 6, 7500 AA Enschede, The Netherlands
E-mail: skidmore@itc.nl

Abstract. The distribution of food is an important predictor for the distribution and density of herbivores in an ecosystem. Determining the distribution and densities of resource quantity and quality in space and time is therefore a crucial step towards understanding the spatial arrangement of herbivores. In recent years remote sensing has become the tool of choice for producing high spatial-resolution impressions of the variability of the landscape, and in particular land cover. Remote sensing is slowly moving away from mapping the surface into discrete land-cover classes. More and more, it is now used to produce highly accurate probability maps of presence, depicting the percentage of individual pixels covered with a certain surface element. This more closely represents the continuous nature of natural phenomena. Recent studies have indicated that it is possible to measure the chemical composition of foliage too. Recently a case study in Kruger National Park confirmed that it is possible to measure nitrogen concentration and phenolic compound levels in grass and trees accurately, with a spatial resolution of 4 meters. This opens doors for new lines of research, where the distribution of herbivores can be linked to the actual resource distribution.
Keywords. remote sensing; resource mapping; resource quality; resource quantity; spatial analysis

INTRODUCTION

A herbivore's search for food (vegetation) and its consumption is driven by the spatial and temporal variation in the vegetation resource, and so vegetation is the central resource considered in this chapter. We use the term 'resource' as defined by Morrison and Hall (2002): it is "any biotic or abiotic factor directly used by an organism, and includes food, nutrients, water, atmospheric gas concentrations, light, soil, weather (i.e., precipitation, temperature, evapotranspiration, etc.), shelter, terrain, and so on. For large herbivores, foraging is a central activity and food

H.H.T. Prins and F. van Langevelde (eds.), Resource Ecology: Spatial and Temporal Dynamics of Foraging, 57-77.

(vegetation) is a key resource. Vegetation resource abundance is the absolute amount of a food item in an explicitly defined area. Vegetation resource availability is a measure of the amount of a resource actually available to the animal. Finally, Morrison and Hall (2002) define resource preference as the likelihood that a resource will be used if offered on an equal basis with others. In the context of herbivory, and specifically vegetation resources, resource preference refers to likelihood that a particular (part of a) plant will be chosen. In this chapter, resource quantity is synonymous with the definition of resource abundance.

The focus of this chapter is describing how vegetation resources can be mapped for herbivores at different scales

Quality tells us something about how well a resource meets the needs of a herbivore relative to the amount of the resource to be consumed. Resource quality is linked to the concept of resource preference, but also implies that vegetation will have components that both attract and deter an animal. Components of vegetation that are 'attractive' to herbivores include factors such as protein, sugar, starch, etc., while 'deterrents' include factors such as tannin, polyphenols or physical attributes such as thorns. Both the quantity (biomass) and quality (the foliar concentration of nitrogen, phosphorous, calcium, magnesium, potassium and sodium) of grass are important factors influencing the feeding patterns and distribution of wildlife and livestock in savanna rangelands (Drent and Prins 1987; Prins 1989a; 1989b; 1996; McNaughton 1990; 1995). Therefore, mapping the quantity and quality of the vegetation resource is a prerequisite to understanding wildlife distribution patterns.

There is great variation in herbivore resources, and the variation depends on the scale considered. The concept of scale is loosely defined. Traditionally, mapmakers use scale as the ratio of the distance (unit) on a map to the same distance (unit) on the ground. In other words, cartographers define 'large scale' as a small portion of the Earth's surface studied in detail (e.g., a scale of 1:500) while a small scale is large area viewed more generally (e.g., a scale of 1:100,000). With digital maps, the cartographic concept of scale is blurring, as software varies the detail appropriate for the areal extent being viewed. Thus, detail and areal extent have become independent as cartographers turn to digital media instead of paper media. To confuse matters, the common usage of a 'large scale' implies an approach that covers regional or even wider areas. In other words, in common usage, scale gives a sense of extent, while cartographers define it as a ratio. In the Oxford English Dictionary (OED), this confusion is apparent: scale is defined as "… represent in dimensions proportional to the actual ones …", which fits the cartographer's definition, while the OED goes on that "… scale up and scale down, make larger, smaller, in due proportions …", which matches the 'common' use of scale. There have been attempts to match cartographic with ecological scale,

The concept of scale in ecology and cartography is differently defined

but such definitions are arbitrary and subject to change depending on whim, because the terms have no clear definition and cannot therefore be used comparatively (for example, see Table 4.1 derived from Estes and Mooneyhan 1994).

Other terms have been proposed to define scale in ecology, such as 'fine' and 'broad' (Levin 1992), but again such terms are inherently imprecise, and their use has not become widespread. Recent research indicates that scale can be precisely defined in ecology and may therefore be used as a tool for comparison as well as understanding of ecological pattern and process.

Table 4.1. *Comparative definition of cartographic and ecological scale (Estes and Mooneyhan 1994)*

Cartographic scale	Ecological scale
1:10,000 or larger	Site
1:50,000 to 1:250,000	Regional
1:250,000 to 1:1,000,000	Continental
1:1,000,000 or smaller	Global

Another interesting concept related to scale, is that different properties emerge when viewed at different scales. For example, a toadstool may be seen as an individual object in a forest, but may form part of a 'fairy ring' when viewed at a distance. Similarly, a 'patch' of grassland which forms part of a heterogeneous habitat for an elephant, may be an unsuitable habitat for a mouse as it is too homogeneous. Thus scale can be defined, scale parameters can define herbivore resources, and through scale, emerging properties may be discovered. Scale, and its influence on grazing resources, will be referred to throughout this chapter. Box 4.1 explains semivariograms as a tool to describe spatial scale.

'Emergent properties' is an important concept when considering scale

Box 4.1. *Semivariogram*

Curran (1988) defines a semivariogram as a function that relates semivariance to the separation distance of points in space and provides an unbiased description of spatial scale; he goes on to describe the concept of semivariograms in detail. Wavelets partition the variance of a data function, such as a satellite image, on a scale-by-scale basis (Lindsay et al. 1996) and have been described by Ogden (1997) and Bruce and Hong-Ye (1996). Recently, Murwira and Skidmore (2003) developed two new methods to define scale: intensity is the maximum variance exhibited when a spatially distributed landscape variable (such as vegetation cover) is measured with a successively increasing window size or scale, while dominant scale is the scale at which the intensity is displayed. Wavelets and variograms were adapted in order to calculate dominant scale and intensity. Both techniques have been described mathematically and conceptually.

The concept of dominant scale and intensity, as derived from variograms and wavelets, has been proven for the regular landscape of the Netherlands, as well as (semi-)natural areas in northwest Zimbabwe by Murwira and Skidmore (2003). They further showed that 80% of the variation in herbivore (in this case elephant) presence could be explained using intensity and dominant scale. In other words, the patchiness or heterogeneity of the landscape varies according to the object (species) of study (for example, the patchiness relevant to a mouse is clearly different to that of an elephant). The heterogeneity of habitat is clearly relevant to species, and scale in ecology can be treated rigorously using the concepts of dominant scale and intensity.

TOOLS TO MAP ENVIRONMENTAL VARIABLES

Until a decade ago it was virtually impossible to display more than one environmental factor on a single map. The habitat type, defined as a mappable unit of land 'homogeneous' with respect to vegetation and environmental factors, circumvented this problem and was the basis of land-systems (or landscape-guided) maps developed in the 1980s (Walker et al. 1986; Zonneveld 1988). Land-systems mapping is based on the assumption that environmental factors show an interdependent change throughout the landscape, and that the environmental factors are constant within the 'homogeneous' area (see an example delineating forest types in Figure 4.1). Thus, the term habitat arose as a way to overcome operational difficulties in species distribution mapping. However, the variation of one environmental factor affecting the distribution of a species tends to be independent of the other environmental factors (Corsi 2000), with the conclusion that 'homogeneity' is seldom the case. If 'homogeneity' cannot be assumed, then the relationship between species and habitat types implied by the land-systems approach requires refinement as we cannot assume that habitat types are homogeneous with respect to the environmental variables that affect the species distribution.

Mapping traditionally uses the concept of land systems. This is being replaced by the concept of landscape continuum

Remotely-sensed imagery has been traditionally classified using computer methods synonymous with the land-systems approach. Each pixel is assigned to a class. The end product is a thematic map of a limited number of classes (Figure 4.1).

Figure 4.1 details land cover and land use classes important for herbivores in the Narok District of Kenya. In these figures, it is possible to detect change in land cover, and relate changes in herbivore abundance and diversity to these changes in land cover.

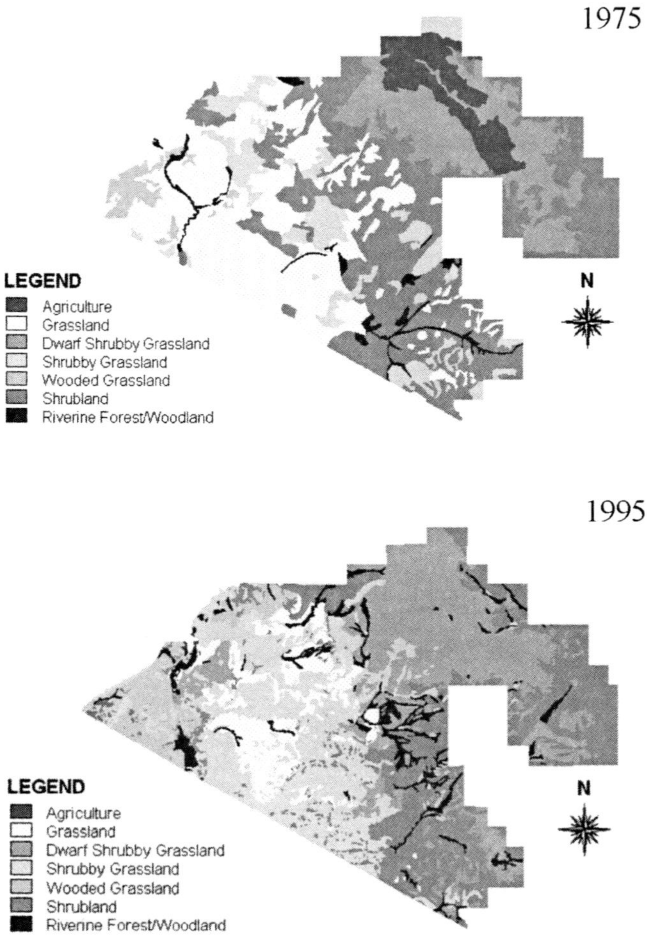

Figure 4.1. *Land cover/land use map of the Narok District, Kenya, prepared by computer classification of Landsat imagery from 1975 and 1995*

A large number of environmental factors act in concert to determine a habitat. GIS makes it possible to integrate the variation of environmental factors as independent layers of information, and to integrate these independent environmental surfaces into a map displaying the suitability of land as a habitat for a specific species. It is a reductionist approach, where the landscape is viewed as a series of separate data layers, which are combined by computer models, with features often being managed and displayed as a continuum (Skidmore 1989). Models utilising environmental factors in order to define an 'envelope' of suitable conditions for a species are now standard GIS tools in environmental impact assessment and habitat modeling (e.g., Skidmore et al. 1996).

A new tool to be briefly introduced here is hyperspectral remote sensing. This type of imagery appears to be most promising for mapping vegetation quality and quantity. As pigment concentrations, leaf characteristics, canopy structure, and biochemical content vary between different vegetation types, so does absorption and reflectance (Knipling 1970; Asner 1998; Martin et al. 1998). Schmidt and Skidmore (2003) summarise the wavelength positions of plant spectral features as calculated by a number of different authors, and conclude that spectral characteristics of vegetation and their biochemical constituents differ because of different experimental set-ups. However, it is also clear from the summary of (Schmidt and Skidmore 2003) that it is possible to identify common absorption features in vegetation, and that specific biochemicals and physical plant characteristics contribute to each absorption feature.

Hyperspectral imagery is an important new tool for mapping herbivore forage quality and quantity

SPATIAL VARIATION IN HERBIVORE RESOURCES

Quantity of herbivore resource

The production of forage is a key determinant of the suitability of habitat for herbivores. The growth of plants is dependent on their ability to assimilate carbon dioxide as organic compounds using energy absorbed from light during photosynthesis. Factors affecting the rate of photosynthesis include available moisture, temperature, light intensity, nutrients and carbon-dioxide concentration. Other factors affecting plant growth include atmospheric pollutants, defoliation, seasonality, different photosynthetic pathways, i.e., C_3 or C_4, as well as leaf amount, type and duration.

These factors influence plant growth at different scales. At a continental scale, available moisture (derived as a function of rainfall and temperature) and soil fertility have been identified as critical environmental factors explaining resource production (Coe et al. 1976), while geographers classify ecosystems using climatic factors (such as rainfall, temperature, seasonality, latitude and degree of continental influences) (Köppen 1931; Holdridge 1947; Bailey et al. 1996). At a regional scale (a few hundred km), the influence of topography, geology and soil emerge as key explanatory variables of primary production (Walter 1971; Woodward 1987;

At different scales, various environmental factors determine plant biomass production

Skidmore et al. 1998). For example, Walter (1971) shows that the vertical profile of soil water depends on soil texture, which is important in determining the distribution of savannas within regions of tropical seasonally (summer) wet climate. At a local scale (i.e., a few kilometres), soil physical and chemical conditions, microclimate (explained by topographic variables such as slope, aspect, elevation and position in the terrain) (Skidmore 1989; Schmidt et al. 2003), as well as indices that integrated microclimatic variables such as total solar-radiation influence (Kumar and Skidmore 2000) influence vegetation type and biomass production.

A number of approaches for measuring productivity have been explored. An important variable explaining canopy primary production is leaf area index (LAI), which represents the size of the interface between plant and atmosphere for energy and mass exchanges (Kumar et al. 2001). LAI is the ratio of the area of leaf (measured on one side only) to the area of ground beneath (Hutchings 1986). The estimation of LAI from remote-sensing measurements has received much attention. For example, a simplified semi-empirical reflectance model for estimating LAI of a green canopy was introduced by Clevers (1988, 1989).

LAI and NDVI, measured by remote sensing, can estimate the productivity of ecosystems

A second measure of ecosystem productivity is based on the idea that productivity is the rate at which energy flows through an ecosystem (Rosenzweig and Abramsky 1993). Ecologists usually use a proxy index to measure productivity over larger areas, such as rainfall in semi-arid landscapes (Rosenzweig and Abramsky 1993). Evapotranspiration can be a suitable index of productivity for terrestrial systems where there is a wide range of temperature and precipitation regimes (Rosenzweig 1968). The normalised difference of the red and near-infrared bands (so-called normalised difference vegetation index or NDVI) estimated from satellite remote-sensing systems is closely related to net primary productivity (NPP) as well as actual evapotranspiration for many vegetation types (Box et al. 1989). The relationship between NDVI and net primary productivity can be established through deductive logic.

Net primary production (NPP) is the net amount of carbon captured by land plants through photosynthesis each year. In a series of modelling experiments, Haxeltine (1996) proved theoretically that NPP is roughly proportional to FPAR on a seasonal and annual basis, thereby validating the empirical observations of Monteith (1972, 1977). FPAR (the fraction of photosynthetically active radiation that is intercepted by green vegetation) is a fundamental variable for the prediction of NPP and hence biomass production. It is possible to estimate FPAR using NDVI (Sellers et al. 1994, 1996) and to estimate biome-averaged global FPAR values against observed monthly maximum FPAR, with an explained variance of 76% (Haxeltine 1996).

However, a problem limiting the application of remote sensing to map the quantity of tropical grasses is that NDVI saturates at higher canopy density: the saturation level is usually reached at about 0.3 g cm^{-2} (Mutanga and Skidmore 2004b). In other words, the widely used vegetation indices (such as NDVI) asymptotically approach a saturation level after a certain biomass density or LAI (Tucker 1977; Sellers 1985; Clevers 1994; Gao et al. 2000), thus yielding poor estimates of biomass during the productivity peak of seasons. Results from Said (2003) used regional data from Africa to demonstrate that the NDVI response is linear in areas of intermediate rainfall but shows little variation at high and low rainfall (Figure 4.2). In other words, the correlation is low for semi-arid areas as there is little vegetation, and also low for forests as the large amount of vegetation saturates the NDVI response. For intermediate-rainfall zones (such as grasslands and savannas) NDVI can yield reasonable estimates of LAI or biomass.

Only at intermediate biomass quantities, traditional remote sensing can accurately map biomass

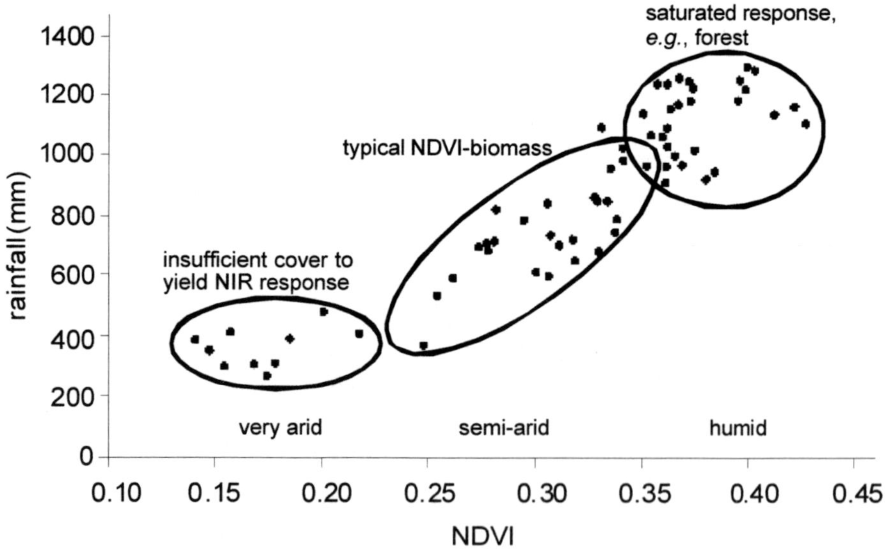

Figure 4.2. *Relationship between rainfall and NDVI; results indicate a poor correlation in the very arid and humid regions of East Africa (Said 2003)*

In order to improve the estimation of biomass from remotely sensed imagery, Guyot et al. (1988) and Baret et al. (1992) showed that canopy LAI, leaf chlorophyll content and leaf inclination angle may be estimated from the 'red edge' wavelength (see Box 4.2). For grassland in the Kruger National Park, South Africa, Mutanga and Skidmore (2004b) tested the utility of the widely used vegetation indices for estimating biomass (in particular, NDVIs involving all possible two band combinations between 350 nm and 2500 nm were tested, in other words also ranging across the red edge). The narrow-band hyperspectral data contained in 647 discrete channels allowed the computation of 418,609 narrow-band NDVIs for biomass estimation. Figure 4.3 shows that a modified NDVI (R755-R746/R755+R746) involving narrow bands located in the red edge, yielded a higher correlation coefficient with biomass as compared to the standard NDVI (R833-R680/R833+R680).

Using the red edge of hyperspectral imagery, biomass can be more accurately estimated

Box 4.2. *The use of the 'red edge' wavelength in remote sensing*

The red edge is the region of transition in low reflection in the red part of the electromagnetic spectrum, to a region of high reflectance found in the near-infrared region. This transition is a typical element of green-vegetation reflectance spectra. It is suggested that the wavelength shift in the inclination point of the red edge is determined by the brightness of the reflectance in the red and near-infrared, and the variation in this measure is dominated by LAI. Clevers (1994) showed that the red edge seems to be independent of soil reflectance for areas of low LAI and that solar zenith angle appears to have only a minor effect on the position of the red edge. Elvidge et al. (1993) and Clevers and Jongschaap (2001) also demonstrated that the red edge was able to detect changes in the amount of vegetation in areas of sparser vegetation; in contrast, NDVI does not vary in value at low LAI. In other words, evidence from the laboratory and crop fields pointed to the use of red edge for better estimating the biomass of sparse vegetation.

The practical difficulty in utilising the red edge is that traditional broadband satellite sensors such as Landsat or SPOT do not utilise this information. Baret et al. (1992) argued that the small amount of independent information in the broadband two-channel sensors cannot reliably estimate the numerous variables controlling canopy reflectance. In particular, for arid and semi-arid environments, hyperspectral imagery is required to resolve subtle variations in canopy reflectance, in order to estimate biomass-related variables (Asner 1998). In other words, the information contained around the red edge slope appeared to be required for the more accurate estimation of biomass, especially in semi-arid natural landscapes such as found in Southern Africa, which may change from sparse to dense grass biomass depending on precipitation.

The techniques for remotely estimating biomass (or resource quantity) have been steadily improved by using the full information content of hyperspectral imagery, next-generation image pre-processing and calibration image processing algorithms, as well as improved field sampling techniques. In other words, the early results with vegetation indices have been refined and developed to a point where biomass may be reliably estimated, also in areas of sparser and denser grass vegetation.

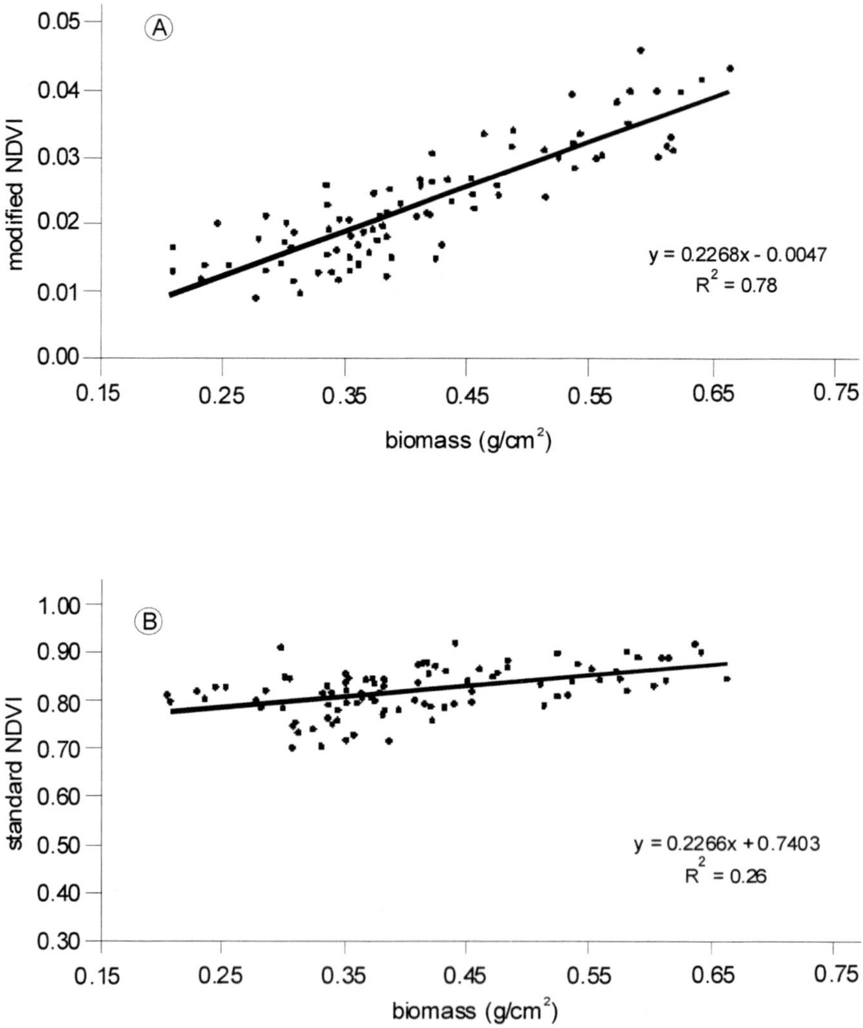

Figure 4.3. *Relationship between biomass and the best-modified NDVI (A) as well as the standard NDVI calculated from a near-infrared (833 nm) and red band (680 nm) (B). Note the R^2 improves from 0.26 for the standard NDVI to 0.78 for the modified NDVI. Taken from Mutanga and Skidmore (2004b)*

QUALITY OF HERBIVORE RESOURCE

The circulation of nitrogen within ecosystems is a classic example of a nutrient cycle: nitrogen is a nutrient essential for plant and animal survival. By understanding and mapping an important herbivore resource variable such as foliar nitrogen, ecologists may be able to model the distribution of herbivores based on their resource preference, as well as understand and predict the diversity of species across a landscape.

There is generally a strong positive correlation between leaf nitrogen concentration and photosynthesis (as long as other factors such as water availability or light are not limiting) (Field and Mooney 1986; Mooney 1986). Of the nitrogen found in a leaf, a large fraction (over 50%) is contained in the carbon-fixing enzyme ribulose biphosphate carboxylase (Mooney 1986) which is a critical component in photosynthesis. It is therefore not surprising that there is a strong positive correlation between photosynthetic capacity and leaf nitrogen content. The relationships between leaf chlorophyll concentration and leaf nitrogen concentration in petiole sap are strong, and linear (Vos and Bom 1993).

The enhancing effect of increased nitrogen supply on dry-matter production as well as protein (including vitamin-B compounds) concentration is well established in the agricultural literature; see Marschner (1995) for a review. Using artificial media as well as field experiments, insect herbivore growth is positively related to nitrogen content (Lincoln et al. 1982), while significantly higher abundance of large herbivores occurs on nutrient-enriched sites in southern Africa (East 1984; Scholes and Walker 1993; Owen-Smith and Danckwerts 1997; Prins 1996). Foliar nitrogen concentration has been shown to be an important environmental factor (Coe 1983).

By understanding and mapping resources such as foliar nitrogen, ecologists may be able to model and explain the distribution of herbivores

Turning to a food resource that large herbivores in Africa more typically consume (i.e., native grass in Kruger National Park, South Africa), obtaining adequate protein from vegetation is a critical parameter determining the success of herbivores. For example, Dublin (1995) demonstrated that the elephant shifts from a grass diet during the wet season to a woody-species diet during the dry season, as the latter maintains a higher percentage of crude protein (13-17%). In contrast, the crude protein of long grasses declines from about 11% to 3% over the course of the dry season (Dublin 1995). Mutanga and Skidmore (2004b) measured the reflectance of a native grass species *Cenchrus ciliaris* grown under three different nitrogen treatments (i.e., low, moderate and high nitrogen fertilization). They demonstrated that higher canopy nitrogen concentration in African native grass is significantly correlated with a shift of the red edge to longer wavelengths, confirming the results of Clevers and Jongschaap (2001) who had earlier demonstrated this for a ruderal (crop) species.

In another experiment, Mutanga et al. (2004b) tested the utility of using four variables derived from continuum-removed absorption features for predicting canopy nitrogen, phosphorus, potassium, calcium and magnesium concentration: (i) continuum-removed derivative reflectance (CRDR), (ii) band depth (BD), (iii) band depth ratio (BDR), and (iv) normalised band depth index (NBDI). It was shown that a stepwise regression using normalised bands (calculated from continuum-removed reflectance spectra) could explain the variation of *in situ* grass quality, with R^2 values ranging between 0.43 and 0.80 (Figure 4.4). This is an encouraging result for mapping the distribution of forage quality under natural conditions in Africa.

Recently, Mutanga and Skidmore (2004a) demonstrated that grass foliar chemistry can be successfully mapped. In this case, nitrogen was predicted from hyperspectral imagery (HyMap) flown over a test area in the Kruger National Park, South Africa. The spatial distribution of foliar nitrogen, and *ipso facto* foliar protein, was for the first time mapped as a continuous variable across a savanna landscape. Mutanga and Skidmore (2004a) further showed that a fenced area (Roan Camp), which had been treated with fire, had a noticeable difference in foliar nitrogen concentration: the burnt area had significantly higher foliar nitrogen content. This subset of the image (Roan Camp area) demonstrates the utility of mapping foliar nitrogen.

Grass foliar nitrogen can be accurately mapped using hyperspectral imagery

Having established that foliar quality (nitrogen) can be mapped from hyperspectral images, and that large African herbivore strategy is to obtain adequate protein from vegetation, what evidence is there that other herbivores respond to foliar quality? A study of Australian herbivores by Braithwaite et al. (1983) showed that arboreal mammal density varied in response to nutritional quality indicators, as did goose species in the Netherlands (Ydenberg et al. 1983). These results were obtained at a landscape scale, and did not consider intra-specific variations in leaf chemistry, nor the foraging decisions made by individual animals. But the clear message was that nutritional quality determines herbivore density.

Thus far, the discussion concerns foliar ‘attractants’ such as protein. Can deterrents, plant biochemicals that adversely effect palatability or digestibility, play a role in determining herbivore behaviour? And if so, can deterrents be mapped across a landscape? Studies have highlighted the role of biochemical compounds acting as deterrents for herbivores, and were it not for the widespread and general occurrence of such allelochemical substances, plants would be totally consumed by insect and mammalian herbivores (Boughey 1973). For example, the above results of Braithwaite et al. (1983) were further developed by Lawler et al. (1998), who found that leaf chemicals acting as foraging deterrents (acylphloroglucinol compounds) played a large role in determining which eucalyptus leaves were palatable to herbivores (koalas and ringtail possums), and which leaves they avoided. Those trees with leaves that had higher levels of acylphloroglucinol compounds were virtually ignored by the animals. Interestingly, they hypothesised that acylphloroglucinol compounds are undetectable by the herbivores, but essential leaf oils or terpenes (such as cineole) act as cues to the ultimate deterrent in foliage (Figure 4.5). When viewed from a resource distribution perspective, their results indicate a large variation in secondary metabolite compounds between individual trees within a species.

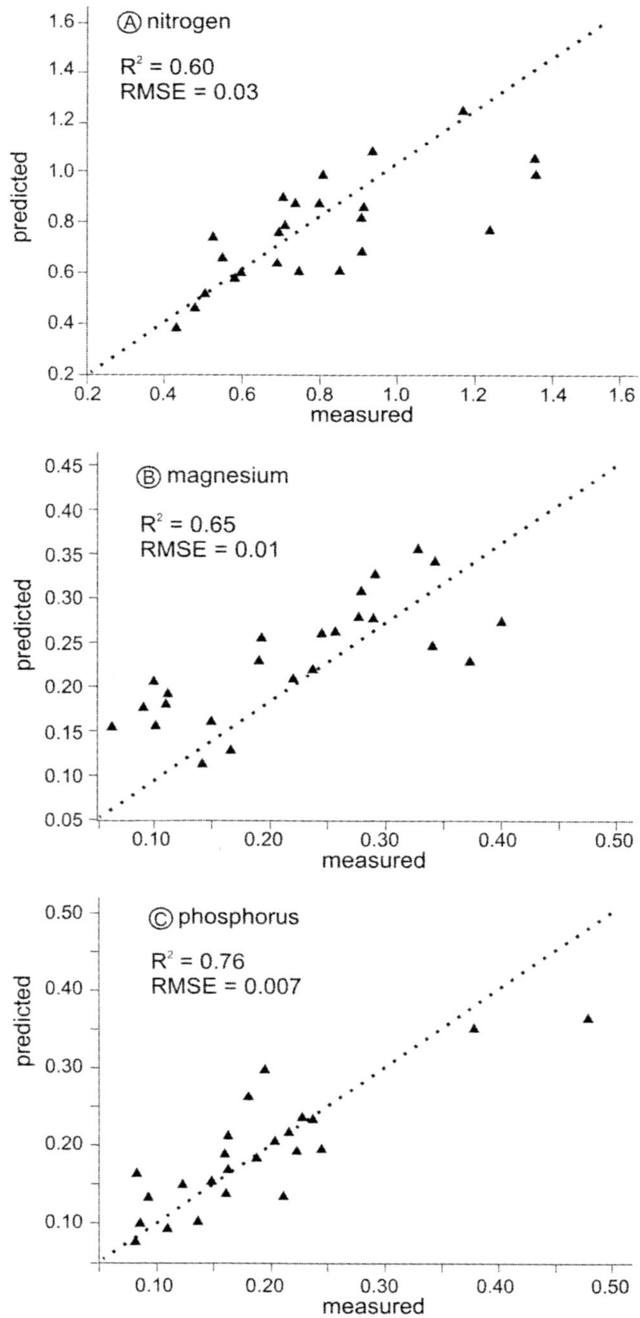
Ⓐ nitrogen
R² = 0.60
RMSE = 0.03
predicted
measured
Ⓑ magnesium
R² = 0.65
RMSE = 0.01
predicted
measured
Ⓒ phosphorus
R² = 0.76
RMSE = 0.007
predicted
measured

(cont.)

Figure 4.4. (cont.)

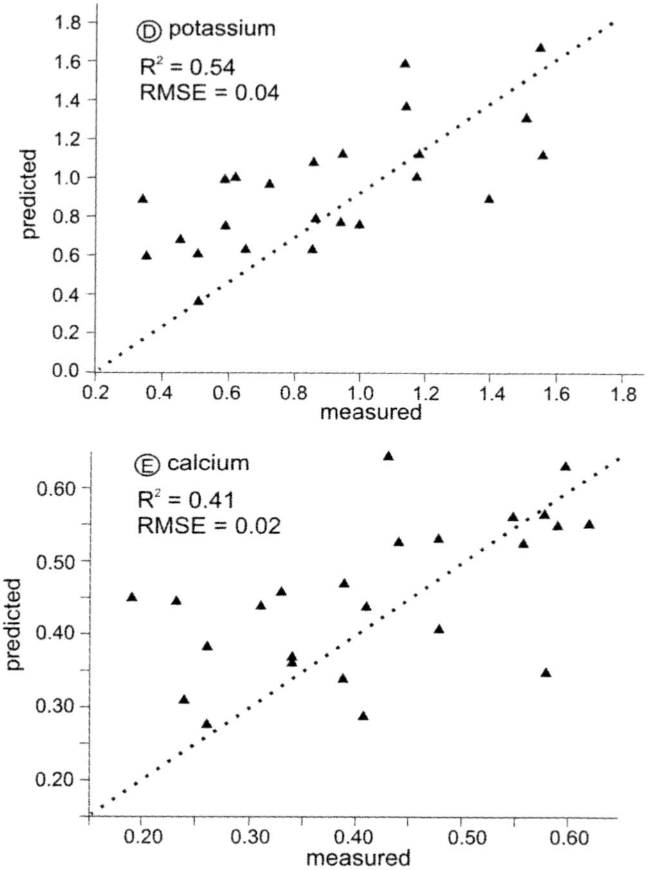

Figure 4.4. *Measured versus predicted biochemicals for a randomly selected test data set (n = 24) using continuum-removed derivative reflectance (CRDR). Regression equations developed from the training data set (n = 72) were used to predict biochemicals on an independent test data set. From Mutanga et al. (2004b)*

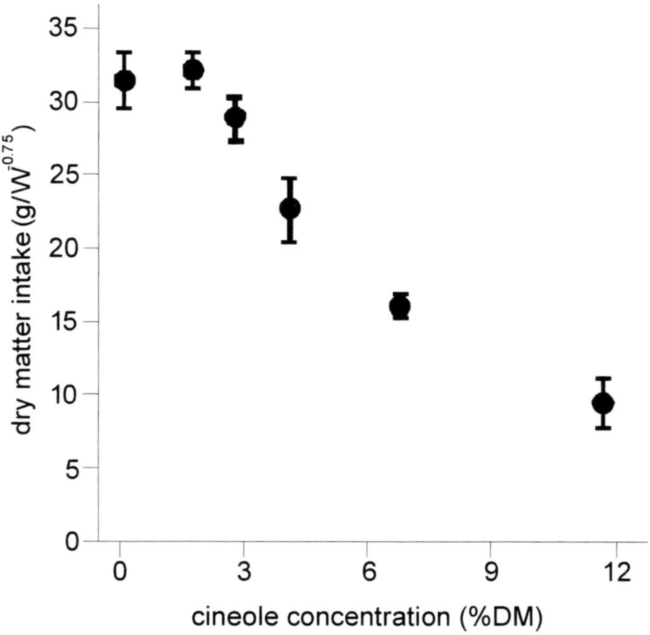

Figure 4.5. *Dry-matter intake by koalas decreases as the concentration of terpene (cineole) increases (from Lawler et al. 1998)*

Continuing this line of research, Dury et al. (2001) used a spectrometer in the laboratory to estimate the correlation between reflectance of dried and fresh eucalypt leaves and sideroxylonal compounds (R^2 of 0.72 and 0.53 for dry-leaf and fresh-leaf spectra, respectively). It is clear that there may be a number of target compounds (either acylphloroglucinol or essential oils) that perhaps can be detected from aircraft sensors and ultimately mapped using hyperspectral imagery.

Turning to the African savanna, Scholes (1997) summarises knowledge concerning plant defences against herbivores. The fine-leaved savanna trees have developed a physical (thorny) defence system, but are still browsed. The broad-leaved savannas are hardly browsed as a result of tannin in two forms: condensed tannins, which are effective against mammals, and hydrolysable tannins, which are effective against insects (Cooper and Owen-Smith 1986; Owen-Smith et al. 1987; Cooper et al. 1988). Tannins are estimated to be the fourth most abundant group of biochemicals produced by vascular plants (Hernes and Hedges 2000); they have been found to play an important role in several ecosystem processes. Recent studies have shown that condensed tannins may reduce nitrogen cycling in forest ecosystems (Northup et al. 1995; Kraus et al. 2004) and that increased concentrations of tannin forage have a negative effect on browsing herbivores. Condensed tannins in plants may reduce nutrient intake through the alteration of gut efficiency (Robbins et al. 1987; Mangan 1988; Mole 1989; Jones et al. 2001; Kondo et al. 2004; Bailey and Provenza, Chapter 2). By learning from previous encounters, forage selection may therefore focus on reducing tannin intake levels (Cooper et al. 1988; Belowsky and Schmidtz 1994; Clauss et al. 2003b; Matson et al. 2004).

Further, the ratio of condensed tannins to leaf nitrogen is a good predictor of acceptability to browsing antelope (Owen-Smith and Cooper 1987), hinting that in contrast to the Australian findings of Lawler et al. (1998), forage quality attractants (in the form of leaf nitrogen) in combination with deterrents may be important in the African savanna. Ongoing studies are considering the concentration of deterrents (specifically tannin and polyphenols) in mopane shrubs and trees, and assessing whether hyperspectral imagery may be used to estimate these leaf biochemicals (Ferwerda et al. 2002).

Recently, Ferwerda et al. (2006a; 2006b) showed that tree chemical composition can be mapped using hyperspectral remote sensing. By combining data on individual absorption points in the infrared part of the light spectrum with a neural-network system, the spatial distribution of condensed tannins and total polyphenol concentration in foliage was mapped with a prediction accuracy (R^2) of 52 % for total polyphenol concentration and 67 % for foliar condensed-tannin concentration. Their calibration routine only considered tannin concentration in mopane (*Colophospermum mopane*) trees and shrubs. By masking the pixels that were not predominately mopane, they were able to show regional differences in condensed-tannin and polyphenol concentration (Figure 4.6). For this study, a comparison of nutrient and deterrent concentrations in vegetation was possible through utilising the same imagery as Mutanga and Skidmore (2004a). It is clear that regional differences in tannin and polyphenol concentration may be observed at a regional scale, and are related to differences in available nitrogen.

Also deterrents can be accurately mapped using hyperspectral imagery

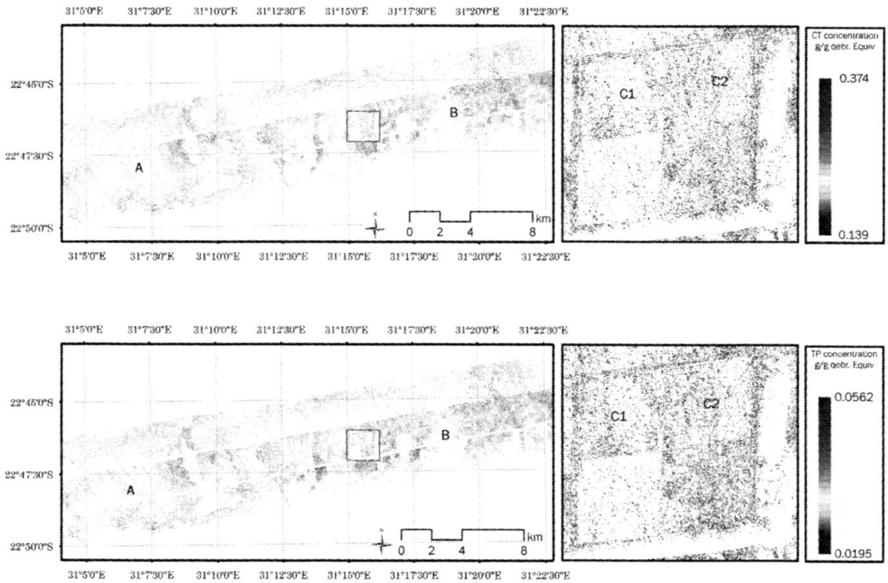

Figure 4.6. *Condensed-tannin and total polyphenol concentration in Mopane trees on the Northern Plains of Kruger National Park, as calculated with a neural network from HyMap derivative reflectance images. Tannin concentration is higher on nutrient-poor shales and sandstones (A) than on the basalt-based soils (B), with lower concentration on sites that recently experienced fire (C1) than on the sites that did not recently experience fire (C2) (Ferwerda 2005; Ferwerda et al. 2006a; 2006b)*

Forage quality is also related to plant species. The preference for particular plant species (that is forage quality) varies between herbivores (Lawler et al. 1998; Bos 2002). In general, short-growing grass species are favoured by grazers (Aerts et al. 1996; Bos 2002). In addition, areas of intermediate biomass have the highest nutritional return (Fryxell 1991). Particular species dominate these broad vegetation types. For example, in the Serengeti, McNaughton (1995) defines short grasslands dominated by species of *Sporobolus*, mid-grasslands by *Themeda triandra*, and tall grasslands by *Loudetia* and *Hyperthelia*. The ability to map species may allow the resource component to be defined, and the suitability for herbivores, based on the plant-species occurrence and abundance, to be modelled. Is it possible to map the occurrence of individual plant species?

Maps of plant species or groups of species are another interesting indicator of forage quality

A number of authors have analysed the spectral difference between plant species and plant communities in the laboratory by visually looking at the shape of the reflectance curves (Elvidge 1990; Vogelmann and Moss 1993) or by statistical analysis of the differences in a more quantitative manner (Gausman et al. 1973; Richardson et al. 1983; Atkinson et al. 1997; Schmidt and Skidmore 2001). The common objective of these studies is to determine whether there is a difference in the spectra between species, as well as whether biochemical or biophysical characteristics contribute to these differences. Some success has been achieved in discriminating between forest species (Franklin 1994; Gong et al. 1997), as well as major physiogomonic categories (i.e., tree, shrub, grass) (Kalliola and Syrjanen 1991). For grasses, there have been few successful mapping exercises using remote-sensing imagery (Lewis 1994; Fayaerts and Van Gool 2001), though recent results prove it is possible to discriminate between African grassland species based on their hyperspectral reflectance spectra. Schmidt and Skidmore (2001) successfully discriminated 10 grass species for a rangeland in the Masai Mara, Kenya. At the herbaceous and grassland level, Schmidt and Skidmore (2002) demonstrated that 27 salt-marsh vegetation types in the northwest of the Netherlands could be discriminated.

A link between the salt-marsh vegetation types and herbivores (Schmidt and Skidmore 2002), is explored by Bos (2002), who showed that geese and hare herbivory modified the vegetation types, and also that grazing intensity was in turn influenced by the vegetation types and species. In fact, there are a number of empirical studies that show that short-growing plants characteristic of early succession are favoured by grazing (Aerts et al. 1996). Ungrazed high marshes are dominated by tall *Elymus* (Bakker 1989; Andresen et al. 1990; Olff et al. 1997). Based on similar published results, interviews with expert ecologists, as well as additional field work, an expert system using knowledge from expert ecologists was developed to map and monitor salt-marsh vegetation (Skidmore et al. 2001; Schmidt and Skidmore 2003), with accurate maps of herbivore resources being produced at the vegetation type level. These expert-system approaches can be modified to generate wildlife habitat maps directly, or wildlife habitat may be deduced from the vegetation map.

Another example of ecological (point-based) research being linked to resource availability affecting herbivores is formed by local hotspots of resident animals existing in the Serengeti in Tanzania (McNaughton 1995), these hotspots being associated with high sodium concentration in the vegetation. Sodium is particularly important to herbivores during late pregnancy and lactation. Mutanga et al. (2004a) demonstrated that vegetation with high sodium concentration may be differentiated from low-sodium-concentration vegetation using hyperspectral remote sensing. Again, ecological knowledge about large-herbivores resources can potentially be mapped over extensive areas and linked to animal behaviour, providing information for management.

TEMPORAL VARIATION IN HERBIVORE RESOURCES

As discussed in the section on "Quantity of herbivore resource", there are a number of methods to estimate resource quantity spatially. For example, the biomass available in Kenya has been estimated by analysing the time series of advanced very high-resolution radiometer (AVHRR) images normalized using the difference vegetation index (NDVI). High positive values of NDVI correspond to dense vegetation cover that is actively growing, whereas negative values are associated with bare soil or sparse vegetation, clouds and snow. In Figure 4.7, the average NDVI for Kenya was calculated over a 12-month period and plotted every year for 4 years. 1982 was a dry year with low average biomass, and 1985 a wet year with much higher mean biomass.

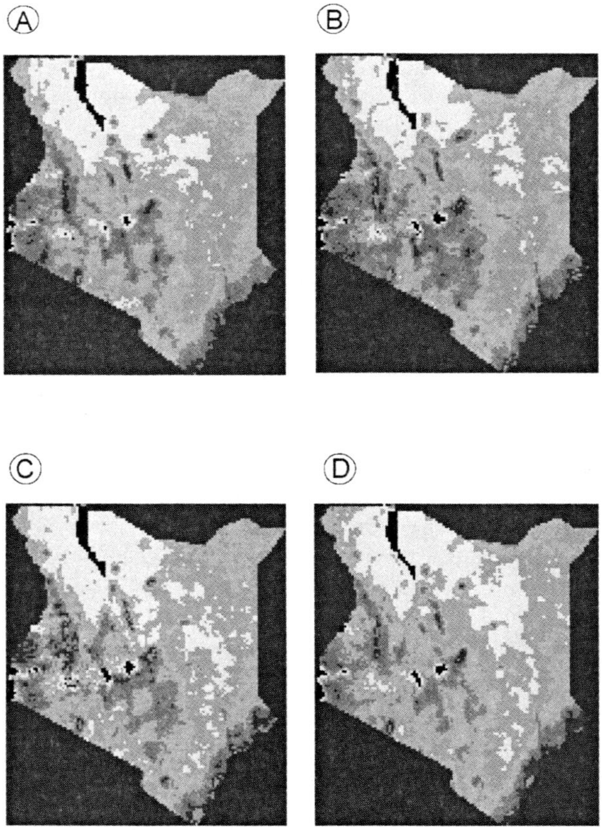

Figure 4.7. *Average Normalised Difference Vegetation Index calculated from NASA AVHRR imagery for Kenya for (A) 1981, (B) 1982 , (C) 1983 and (D) 1984. Dark tones are high biomass, light indicates low biomass. Note that images were prepared for this chapter using a NOAA AVHRR data set for Africa (processing involved calculating the highest NDVI value per 10-day period in order to remove cloud and other artefacts, and then averaging the 10-day periods over the whole year)*

Oindo (2003) used a 20-year time series of NDVI processed by the Global Inventory and Monitoring and Modeling Studies (GIMMS) at NASA, and re-sampled to a pixel size of 7.6 km. The inter-annual maximum NDVI used in this study was considered to represent the NDVI at the height of the growing season (Lewis et al. 1998). The data were corrected for cloud and atmosphere contamination as well as variations in sensors over the period (Los 1998). Morin (2000) suggested that temporal variation in productivity may be a factor that generates diversity. The results produced by Oindo (2003) suggested that a greater number of herbivore species and individuals coexist in more seasonal environments, in other words, environments that have a greater variability in maximum NDVI. In contrast, for bird species in Kenya, Oindo et al. (2001) found that the standard deviation of maximum NDVI represents the amount of woody vegetation, and this factor limits the distribution of birds. High bird species diversity was also shown by this time series analysis to occur in stable and predictable environments in Kenya (Oindo et al. 2001).

Satellite imagery can be used to map the change in biomass over time using NDVI

The study by Said (2003) also used time-series data to estimate the ungulate-species richness in East Africa, and proved that time-series climatic indices (based on rainfall and potential evapotranspiration) are a better predictor of species richness than time-series NDVI indices (69% versus 55% of the variance explained). Ottichilo (2000a) used long-term animal census data from the Department of Resource Surveys and Remote Sensing (DRSRS) in Kenya and showed that the total of all non-migratory wildlife species in the Masai Mara ecosystem in Kenya declined by 58% in 20 years as a result of land use change, drought effects and poaching. Said (2003) confirmed that loss of wildlife wet-season range and habitat fragmentation through land use changes has a much wider impact on large herbivores than do the effects of climatic variations. This was caused by signs of competition between large herbivore species during non-migratory periods mainly associated with intra-specific competition (i.e., wildebeest and buffalo) for food.

The above results, generated using NDVI and climatic surfaces and analysed over extended time periods, show that spatial patterns of variation in resources are clearly impacting on large herbivores. Classical ecology, using point-based observations, has built up a large body of knowledge demonstrating the influence of time on ecological processes. For example, Dublin (1995) showed that there is a change in elephant feeding from grass to trees during the dry season as a response to reduction in the resource quality (foliar nitrogen) of grasses through the dry season. Such results may be combined with spatial-temporal models to better understand the resources available to large herbivores and assist in their management.

Since 1988, the Environment and Natural Resources Service of the UN FAO has been operating the Africa Real Time Environmental Monitoring Information System (ARTEMIS) (see http://metart.fao.org/default.htm). The system acquires and processes routinely hourly estimates of rainfall and vegetation index (NDVI) images, using Meteosat and NOAA data. The system covers the whole of Africa, and the outputs are produced on a 10-day and monthly basis for use in the field of 'early warning for food security' and 'desert locust control'. An example of the type of output that may be generated from these data can be found at the FAO website. A vegetation index was generated by calculating the difference between the 1997 and 1982-1996 average NDVI, and the difference sliced into 5 classes. For each class, a moving average-NDVI difference through a growing season or other period can be calculated, and used to estimate whether the biomass production is less than, or exceeding, the long-term average biomass production.

SYNTHESIS

In this chapter, we advocate that determining the distribution and densities of resource quantity and quality in space and time is a crucial step towards understanding the spatial arrangement of herbivores. In recent years remote sensing has become the tool of choice for producing high-spatial-resolution impressions of the variability of the landscape, and in particular land cover. Remote sensing is slowly moving away from mapping the surface into discrete land cover classes. More and more, it is now used to produce highly accurate probability maps of presence, depicting the percentage of individual pixels covered with a certain surface element. In this chapter, we presented several examples to illustrate the progress in remote sensing supporting resource ecology. These examples showed that biomass and nutrient concentrations in the vegetation may be monitored. The consequences for large herbivores from such approaches are obvious in terms of modelling resource quantity and quality over time. Some hypotheses for future research are formulated in Box 4.3.

Box 4.3. *Testable hypotheses for future research*

Hypothesis 1. Spatial variation of resources as charted by remote sensing can be used to test herbivore-foraging models.

Hypothesis 2. Although regional forage quality patterns can be mapped using remote sensing, local variation cannot accurately be captured with current-day sensor systems.

Hypothesis 3. Nutrient levels in forage as measured with hyperspectral remote sensing can be as accurate as nutrient levels measured using traditional lab-based techniques.

CHAPTER 4B

COMMENTS ON "RESOURCE DISTRIBUTION AND DYNAMICS: MAPPING HERBIVORE RESOURCES"

HERBERT H.T. PRINS, WILLEM. F. DE BOER AND FRANK VAN LANGEVELDE

Resource Ecology Group, Wageningen University, P.O. Box 47, 6700 AA Wageningen, The Netherlands
E-mail: herbert.prins@wur.nl

What is a resource, how are resources distributed, and how do they change over time, together with the possibilities of mapping these resources through remote sensing, are the subjects of this chapter. Skidmore and Ferwerda (Chapter 4) follow Morrison and Hall (2002) in their definition of what a resource is, namely "*a resource is any biotic or abiotic factor directly used by an organism, and includes food, nutrients, water, atmospheric gas concentrations, light, soil, weather (i.e., precipitation, temperature, evapotranspiration, etc.), terrain, and so on*". The central notion of a resource is that it is used. However, Morrison and Hall and also Skidmore and Ferwerda confuse 'use' in the sense of 'exploit' or 'consume as material' with 'use' in the sense of 'benefit from'. As a matter of fact, the Oxford English Dictionary (OED) defines 'resource' as "stock that can be drawn on, available assets, or means of supplying what is needed". Assets and stock can dwindle if they are used faster than their replenishment rate, and if that happens they are used up. We think that the term 'resource' should be limited to this meaning, and thus disagree with Skidmore and Ferwerda the way they apply this key term: 'weather' cannot be used, 'temperature' is a state variable, and 'terrain' cannot increase or decrease. Where light is a non-depletable resource, weather and temperature are environmental conditions. These are variables that describe an organism's habitat, and are therefore sometimes classified as one of the species' niche dimensions, but not its resource. The second problem with the definition of Skidmore and Ferwerda lies in the word 'factor'; according the OED this word has the meaning of "circumstance, fact, or influence; contributing to a result". A consumer cannot use a 'factor'; instead of 'factor', the concept of 'resource' should include on the one hand physical consumable entities, chemical compounds and

H.H.T. Prins and F. van Langevelde (eds.), Resource Ecology: Spatial and Temporal Dynamics of Foraging, 79-80.

elements, for which we use the term 'substance', and on the other hand usable energy. In other words, we propose to modify their definition to "*a resource is usable energy or any biotic or abiotic substance directly exploited by an organism, which includes food, nutrients, water, atmospheric gas compounds, as well as light, and the use of which can lead to the (temporary) exhaustion of that resource (i.e., depletable and non-depletable resources)*". The essence of the concept of 'resource' is that organisms can compete for a resource (such as competing for light, space, nutrients, water, etc.) and that it can be limiting the growth of individual organisms or of populations.

This aside, their chapter provides startling insight into the new techniques that recently have been developed to measure the distribution of the abundance and quality of vegetation. The combination of remote sensing, field sampling and greenhouse experiments in which foliar quality was manipulated has led to the possibility of constructing maps in which an unheard level of detail can be shown regarding the distribution of potential food of herbivores. This will open further avenues of research into testing theories of foraging. Indeed, optimal foraging theory presupposes an omniscient herbivore that has a complete knowledge of all food items, their quality and their distribution. This extent of knowledge can now be tested because the scientist, observing the behaviour and choices of the herbivore, appears to be approaching this omniscience while measuring resources hyperspectrally and using algorithms to transpose this information into maps of individual plant species, individual shrubs or patches of grass, and the concentration of essential elements or nasty deterrents.

This is an important breakthrough, because scientists could test optimal foraging theory relatively easily on consumers. Most predators feed on discrete prey items, and by counting and putting the information in a spatial context, scientists could get quite a good idea about the distribution of the food on offer. From a predator's point of view, different individuals of a similar prey species most of the time offer the same quality. The most important discriminative factor characterising different preys is their individual body mass. Animal ecologists could thus quite easily test predictions from foraging theories by focussing on predators and their prey. The new insights presented here, now offer the potential to further foraging theory considerably because, for the first time, the spatial pattern of food quantity and food quality on a large scale is known.

But … omniscient? The chapter of Skidmore and Ferwerda shows how quickly too much information might be garnered from hyperspectral technology or other types of know-how. The technology offers the possibility of mapping each and every individual shrub in a near-infinite area, and to give each and every individual shrub (or leaf) a description of how much nitrogen, phosphorus, potassium or whatever element it contains and how much tannins, lignin or polyphenols it has. However, the total number of chemical compounds to which a herbivore may react, which it needs or which it finds repulsive may be reckoned in the thousands or even tens of thousands. Too many data do not yield better understanding, and the application of know-how without a clear hypothesis to be tested, may easily devalue into a gimmick. Our task is thus to harness this new way of looking at resources, and to use it for testing ecological theories.

CHAPTER 5A

FORAGING IN A HETEROGENEOUS ENVIRONMENT

Intake and diet choice

EMILIO A. LACA

Department of Plant Sciences, University of California, Davis, One Shields Ave, Davis, CA 95616, USA
E-mail: ealaca@ucdavis.edu

Abstract. Resource heterogeneity and its effects on consumers are crucial in the dynamics of landscapes with large herbivores. Although all elements necessary for a general quantitative theory of resource heterogeneity and foraging behaviour across spatial scales are available, such a theory has not been put forth yet. We need to learn what scales, what resources and what types of heterogeneity are relevant to conserve and manage landscapes with large herbivores. More specifically, what scales, variables and heterogeneity are important in determining intake and diet selection by large herbivores? Large herbivores interact with their resources through a series of nested processes such as ingestion, searching, digestion and resting, which define relevant scales. Empirical relationships between animal performance and average resource abundance are scale-specific. Extrapolations should be based on explicit models to change scale, and will benefit by using concepts and techniques from geostatistics. Heterogeneity and average herbage mass are frequently related, so that measured effects on intake cannot be unequivocally attributed to total herbage mass. Resource heterogeneity can affect intake and behaviour through non-linearity of responses to local conditions, selectivity and changes of local functional response due to global conditions. In general, coarser resolution of heterogeneity allows a greater selectivity. These points are illustrated with examples from the literature and reinterpretation of published and unpublished data.
Keywords. spatial scale; resource distribution; grazing; patchiness; functional response

INTRODUCTION

Integration of what we know about herbivores in conceptual and quantitative models immediately results in the realisation that "spatial patterns of resources may regulate resource uptake and depletion rates within the landscape, with potential implications for demographics, intraspecific competition, and community assembly" (Milne et al. 1992). Heterogeneity is a rich and multidimensional concept (Skidmore and Ferwerda, Chapter 4). Thus, it is useful for making general assertions, but it is too

H.H.T. Prins and F. van Langevelde (eds.), Resource Ecology: Spatial and Temporal Dynamics of Foraging, 81-100.

general to describe well-defined quantitative relationships. Heterogeneity is necessary for life and physical events to happen. The living and the physical worlds are flows and changes that are possible because of, and fuelled by, heterogeneity. The concept is so central to the way we understand and study the world that its study cuts straight across disciplinary and hierarchical organisations of science. In the context of herbivore foraging, a relevant issue is whether there is a general conceptual framework to interpret past research and plan future studies to improve management and conservation of ungulates based on resource heterogeneity (Box 5.1).

Resource heterogeneity is a key factor in ecology; there is room for a synthetic theory to interpret empirical results and to derive hypotheses for conservation

How one operationally defines, measures and thinks of heterogeneity has a major impact on our ability to understand and manage landscapes where large herbivores are an important component (landscapes with large herbivores, LLH). Evidence and quantification of effects of heterogeneity on LLH are abundant in the literature; but is there a synthetic theory available to put empirical results into perspective, and to derive hypotheses of practical relevance for management or conservation? Has vegetation heterogeneity been defined and measured in the most meaningful way to increase our ability to predict animal and landscape responses?

Heterogeneity of an LLH is multidimensional because it simultaneously has values for all scales (heterogeneity is scale-dependent) of time and space, and for all variables or functions of variables considered. Once again, this argues for the need to move beyond statements invoking 'heterogeneity' in general into more specific questions that are organised by a conceptual framework: what scales, what variables and what types of heterogeneity are relevant to understand and manage LLH? More specifically, what scales, variables and heterogeneity are important in determining intake and diet selection by large herbivores? How and how much do they determine diets and intake?

This chapter uses examples from the recent literature to address some of the foregoing questions. Literature sources are, by the most part, restricted to the last 12 years. However, when no new articles explaining important concepts were found, older references are included. This is not a review, but an attempt to present the some relevant concepts and theories in a coherent structure and with some support from the literature, and additional references to guide the reader into the subject. The goal is not to give the reader a prescription, but to pose explicitly the questions and issues that the information brings up.

***Box 5.1.** Components of resource heterogeneity*

Perhaps the simplest way to quantify heterogeneity involves a measure of variance and one of spatial pattern. Consider the heterogeneity of herbage abundance in a 10,000-ha grassland, as measured with a resolution of 0.25 m^2. Imagine that the centre of the quadrat is placed in all points of the grassland and that the mass within it (x = herbage mass within each quadrat) is measured non-destructively. The variance of x would represent the degree of difference among quadrats, regardless of their spatial arrangement. A histogram of the frequency of values of x would be more informative, but less parsimonious. A compromise can be achieved if the distribution of x can be modelled with few parameters. For example, Shiyomi et al. (1991, 1998) modelled the distribution with a gamma function, which requires only two parameters.

Spatial pattern can be summarised into one number: the fractal dimension of the pattern. If herbage mass changed smoothly over space, following a gradient, the fractal dimension would be 2.0. In this case, the amount of herbage in any quadrat could be predicted with certainty using information from surrounding quadrats. On the other extreme, if herbage mass changed completely randomly, taking any value from the distribution regardless of its position relative to other quadrats, the fractal dimension would be 3.0. In this second extreme case, it would be impossible to predict the herbage mass in a quadrat based on the values of adjacent ones. Fractal dimensions between 2.0 and 3.0 would reflect intermediate cases such as a patchy distribution of herbage mass.

Variance and fractal dimension of spatial heterogeneous resources can be used as proxies of variables that are potentially crucial for foragers, particularly for herbivores that depend on sessile resources and have to explore large areas to gather sufficient nourishment every day. Variance of herbage mass and or quality determines the probability of randomly finding a bite of a certain quality and mass. Assuming a unimodal distribution, large variance means that very good and very poor bites are relatively abundant, whereas small variance means that most bites are about the same. A fractal dimension of 2.0 means that all good bites can be found easily and predictably in a gradient towards the best area of the grassland; a fractal dimension of 2.5 means that there are patches of good bites that can be found and exploited by specialised adaptive search mechanisms, whereas a fractal dimension of 3.0 means that bites of all types are finely interspersed, so a systematic search would be the best strategy.

Although variance and fractal dimension can summarise heterogeneity efficiently, this is not always the case, and a more complete and complex set of descriptors may be necessary. The fractal dimension of a given grassland might change as a function of the resolution at which herbivores interact with it. Imagine our 10,000-ha grassland as having a clear W-E gradient over a distance of 10 km, from 10 g 0.25 m^{-2} on a dry hilltop to 100 g 0.25 m^{-2} in a humid valley. Further, consider that due to randomly distributed micro-topography and disturbances, the coefficient of variation of herbage mass at any distance from the hilltop and as measured with the 0.25-m^2 quadrat, is constant and equal to 1.0. A rabbit whose home range is at any point in the gradient would perceive its resource as almost homogeneous with a fractal dimension close to 2.0. The rabbit would search for the randomly distributed good patches, and would not perceive the gradient within its home range because it would be swamped by the random variance. A bison, whose daily movement range could easily span most of the 10 km, would perceive a clear resource gradient with a fractal dimension close to 1.0 over its daily range. The bison can choose where in the gradient to forage, and within that area it can search and select bites.

HETEROGENEITY EVERYWHERE

Spatial heterogeneity has two components (Palmer 1992), i.e., variance (or more generally, probability density function) and spatial pattern or arrangement. For example, a shrub land with a total aboveground mass of 2000 kg ha^{-1} can have 20% of the area covered by shrubs with 10^4 kg ha^{-1} of canopy cover, or it can have 40% cover by shrubs with 5000 kg ha^{-1} of canopy cover. In turn, each one of them can have any spatial pattern imaginable from one block with all shrubs and another empty, to patches to a perfectly regular distribution, as in a plantation. In all cases the average is the same, and within the small or large shrubs, the probability density function is constant across spatial patterns.

Past interpretations of relationships between resource abundance and animal performance may need to be revised with the novel emphasis on heterogeneity

Variance among sample units increases with increasing extent (e.g., area of pasture or region) and decreases with increasing size of the sampling unit (e.g., quadrat area). This appears to be a general property of most regionalised variables (O'Neill et al. 1991) and was described for grazed pastures by Shiyomi (1987). The rate of decrease in variance with increasing quadrat size can be used as a summary characteristic of the spatial distribution of forages. The variance among sampling units is both the degree of heterogeneity and the variety of choices an ungulate has when sampling or perceiving the landscape with a certain resolution. Thus, the evaluation of a given area as habitat for ungulates may strongly depend on the resolution of measurements. On the other hand, the average and extremes of forage quality and mass per unit area will depend on the extent available for animals to choose from. Mobile herbivores can buffer temporal changes in the average availability of forage when grazing large extensions, but are unable to do it in smaller paddocks, even if all average characteristics are the same in both situations (see also Bailey and Provenza, Chapter 2). As a consequence, empirical relationships between animal performance and average resource abundance are scale-specific, and should be used only for the scale at which they were developed, unless some model and theory are used to perform the change of scale. The field of geostatisics provides such models and theories (Wackernagel 1995).

Is it appropriate even to consider heterogeneity as a low-dimensional characteristic of the landscape? In other words, how few independent numbers are necessary and sufficient to characterise completely the heterogeneity that is relevant for large herbivores? This could pose a serious challenge for three main reasons. First, heterogeneity must be considered scale-dependent, and thus it could theoretically be infinitely dimensional. Second, heterogeneity can be defined on any set or function of landscape variables. Third, heterogeneity, like any other landscape variable, can and typically does take different values at different positions in space. This last concept is represented by the idea of 'texture', commonly used in image analysis. However, the continuity of scale dependence can be operationally partitioned into a small number of domains of scale (Bailey et al. 1996; Bailey and Provenza, Chapter 2) without significant loss of precision or generality. Ungulates interact with their forages through a series of nested processes such as ingestion and chewing, searching and walking, digestion and rumination, resting, etc. These processes and the associated behaviours define potential scales of interest. The main variables of interest are likely to be abundance of main plant species, topography, cover and water availability; just a few of the myriad of landscape variables. Finally, there are statistical, modelling and measurement methods to detect and correct for spatial variability in heterogeneity itself. Following one paradigm of spatial statistics, spatial variability can be partitioned into trends and residuals. Responses to large-scale trends or gradients have long been the subject of ecology in general and grazing behaviour in particular, and relatively well-established concepts and facts are available. The novel issues are in the spatial correlations of the residuals, and how herbivores might exploit them.

Heterogeneity is indeed everywhere. The fact that heterogeneity is a constant feature of most herbivore–vegetation systems, even those once thought of as homogeneous, questions the validity of experimental relationships between resource abundance and intake. As shown in Figure 5.1, even if the true response to herbage mass were a ramp function, the observed response to the natural covariance between average and heterogeneity of herbage mass would lead to the observation of a concave-down functional response as a result of the selectivity effect. The exact shape of the functional response in the absence of heterogeneity is not that important for this argument. The main point is that in the presence of heterogeneity, animals can select temporal or spatial patches that are better than the average, thus achieving a higher intake rate than in homogeneous resources of equal average value. Heterogeneity and average herbage mass are frequently related, so that measured effects on intake cannot be unequivocally attributed to total herbage mass, unless heterogeneity is controlled or accounted for through statistical methods. For example, heterogeneity of herbage mass per unit area in pasture composed of *Festuca arundinacea*, *Dactylis glomerata*, *Poa pratensis*, *Agrostis alba*, *Trifolium repens* and other minor species depended on herbage mass and whether the level of biomass had been reached by recent growth or grazing (Shiyomi et al. 1998). Components of resource heterogeneity are discussed in Box 5.1. Shiyomi and colleagues have shown in several studies that the frequency distribution of herbage mass per unit area measured with a grain size of 0.25 m^2 can be described by a gamma distribution (Shiyomi et al. 1983, 1984, 1991). The gamma distribution is a common statistical distribution that describes the intervals between random events that follow a Poisson distribution. Shiyomi and colleagues found that the reciprocal of the square root of the coefficient of variation is strongly dependent on the total or average herbage mass per unit area (Figure 5.2). Within the range of herbage mass they studied, heterogeneity increased with decreasing mass, due to both grazing and growth. Thus, generic functional and production responses, particularly those measured in 'realistic' field conditions should be reconsidered or at least explicitly state that the response measured is likely not caused only by the change in overall resource abundance, but also by changes in heterogeneity.

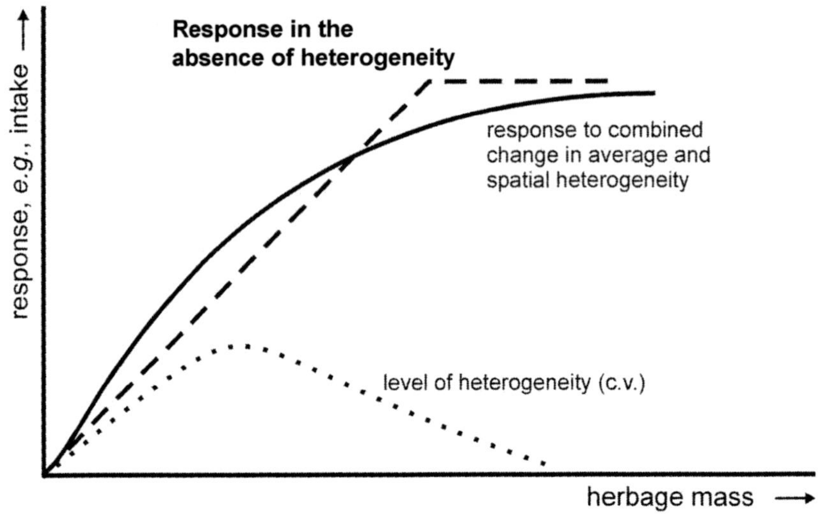

Figure 5.1. *Hypothetical effects of heterogeneity on measured responses to change in herbage mass. The continuous thick line represents the typically observed response. The dashed line represents the hypothetical response to herbage mass in perfectly homogeneous swards. The thin line represents the changes in spatial variance in herbage mass associated with changes in average*

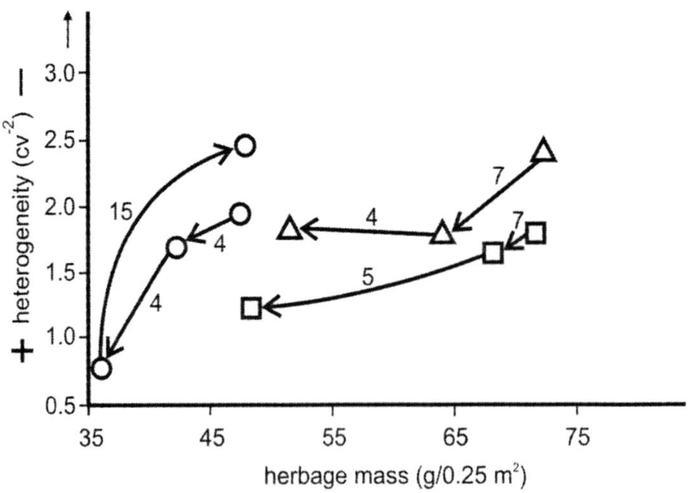

Figure 5.2. *Relationship between sward heterogeneity in mass per unit area measured with 0.25-m^2 quadrats (resolution) and average mass in the whole pasture (extent). Each set of symbols represents a sequence of measurements on sets of parallel transects. Numbers on the lines represent the number of days of grazing (herbage mass declines) or regrowth (herbage mass increases). Based on Shiyomi et al. (1998). Note that heterogeneity increases as the value of the abscissa decreases*

HOW HETEROGENEITY AFFECTS INTAKE AND DIET SELECTION

Heterogeneity is an inherently complex concept, and as indicated above, it cannot be characterised with a single dimension in a quantitatively meaningful and general way. Nevertheless, the concept can be used heuristically as lack of homogeneity, and it can be practically quantified with a few values. Murwira (2003) successfully used two parameters, intensity and dominant scale of vegetation, to determine the relationship between elephant distribution and vegetation heterogeneity. Intensity was defined as the maximum variance in the cover by certain species, and dominant scale was the scale at which the intensity was manifested.

Functional response is affected by forage heterogeneity through selectivity, non-linearity of local effects and integration of information

Regardless of kind and degree, heterogeneity can affect intake and behaviour through three mechanisms: non-linearity of responses to local and instantaneous conditions, selectivity, and change of functional form of local responses due to global conditions. These mechanisms are represented for a hypothetical functional response in Figure 5.3, and are illustrated with examples.

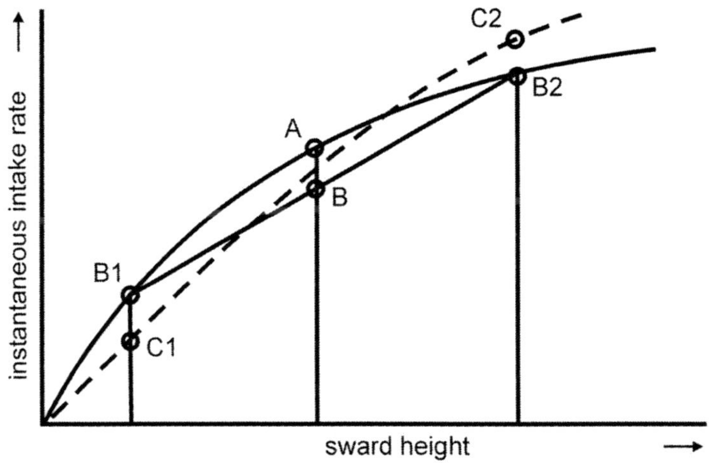

Figure 5.3. *Hypothetical effects of heterogeneity of sward height on intake rate measured at a scale of 10^2 to 10^3 seconds. The continuous thick line represents intake rate in each area of homogeneous sward height. The dashed line represents intake rate within homogeneous patches in a heterogeneous sward. See text for further explanation*

Consider a functional response of instantaneous intake rate to sward height in a leafy pasture. Because of the effects of sward height on bite mass and handling time, this is a concave-down nonlinear response. Point A represents the expected instantaneous intake rate for an intermediate height. Point B represents the expected instantaneous intake rate in a sward that has 50% of the area covered by a short sward (B1) and the rest by a tall sward (B2), such that all average characteristics are the same as for the sward yielding A. The difference between A and B is the effect of the nonlinearity in the absence of selective grazing, i.e., the animal takes 50% of the bites from each sward type. If selectivity is factored in, the response can be any instantaneous intake rate on the continuous curve between B1 and B2, depending on degree and direction of selectivity, and assuming no additional costs for searching due to selective grazing. When searching costs increase due to selectivity, the effect of selectivity is attenuated. Finally, the functional response to the characteristics of each patch in the heterogeneous sward can deviate from what is observed in homogeneous swards. For example, faced with alternating tall and short patches, the animal may take deeper bites in tall patches and shallower bites in short patches than expected on the basis of the 'homogeneous' functional response. The dashed line represents this effect on the functional response. The resulting instantaneous intake rate can vary between C1 and C2, depending on degree and direction of selectivity.

Heterogeneity can affect intake through effects on bite formation and bite dimensions, and effects at larger areas, such as patches, feeding sites and home range (see Fryxell, Chapter 6). Effects at the bite level are probably less responsive to changes in strategy or shape of the functional response, because the constraints on bite dimensions seem to be less subjected to changes in the motivation and behaviour of the forager than to the spatial distribution of forage in the grazed areas. Conversely, heterogeneity has stronger impacts on selection of feeding areas, feeding time and diet selection.

Quality–quantity bivariate heterogeneity

One of the main themes in plant–animal interactions is the trade-off between quantity and quality. In most natural grasslands, the quality of the forage declines as the amount of forage increases due to growth over the season (Prins and Olff 1998). However, patches that are grazed remain in a vegetative stage characterised by low herbage mass of higher quality than patches with more herbage (Fryxell 1991; Wilmshurst et al. 1995). Because ruminants have a limited passage and digestion rate, they choose diets or patches for which cropping rate equals digestion rate (Figure 5.4). The area under the two curves represents feasible combinations of diet quality and intake. This principle explains why grazers prefer short patches in grasslands, why herbivores aggregate to form 'grazing lawns', and patterns of patch selection and body size (Wilmshurst et al. 2000). Heterogeneity over areas that are smaller than the areas visited within a meal interacts with the quality–quantity trade-off by moving up the line of ingestive constraint to the dotted line in Figure 5.4. Herbivores appear to be able to select the preferred patches more easily when the patches are larger (Clarke et al. 1995; Wallis deVries et al. 1999).

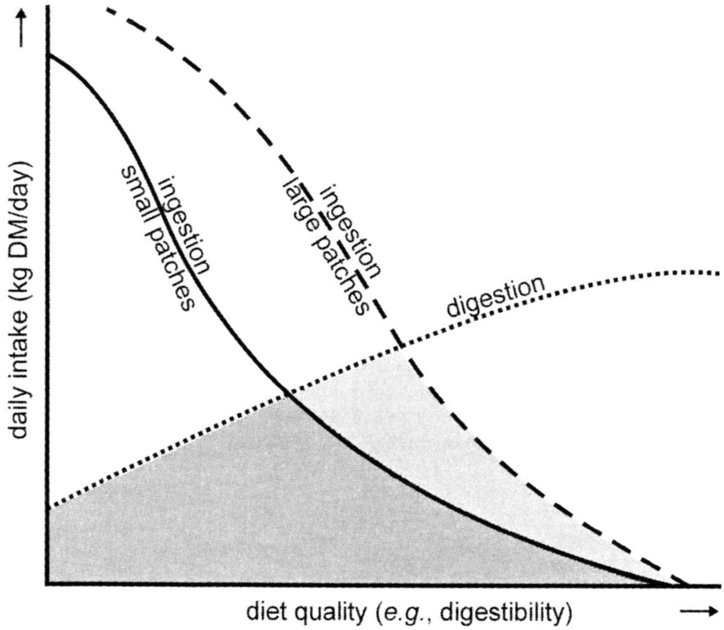

Figure 5.4. *Constraints on daily intake of digestible matter by ruminants. The ingestion curves depend on intake rate and grazing time. Two ingestion curves show the hypothetical effects of spatial distribution of patches of different digestibility. Larger patches make it easier for herbivores to select the preferred choice*

However, the negative relationship between patch quantity and quality observed in natural pastures is not universal (Figure 5.5). Ogura et al. (2002) found a positive correlation ($r = 0.48$) between herbage mass and quality (nitrogen concentration and *in vitro* dry-matter digestibility) and between defoliation rate and herbage mass of patches (r ranged between 0.6 and 0.8) in *Paspalum notatum* pastures early in the season. As the season progressed, the correlations reversed to the more typical pattern where tall, previously ungrazed patches have lower quality and more herbage mass than shorter ones, and animals prefer the shorter patches. Late in the season, the correlation between defoliation and pre-grazing herbage mass declined to about -0.5.

The study of herbivore foraging should encompass situations where quality and quantity have any correlation, not just those where quantity and quality have a negative correlation. Although the positive correlation between herbage mass and quality may have been the exception to the norm in places where herbivores evolved, we need to manage herbivores in disturbed and managed landscapes where conditions are novel. We need to extend our understanding of the spatial and temporal patterns of landscape–ungulate interactions to situations that may escape completely the typical habitats or interactions, because management, local disturbances and global change will likely expose herbivores to novel foraging environments.

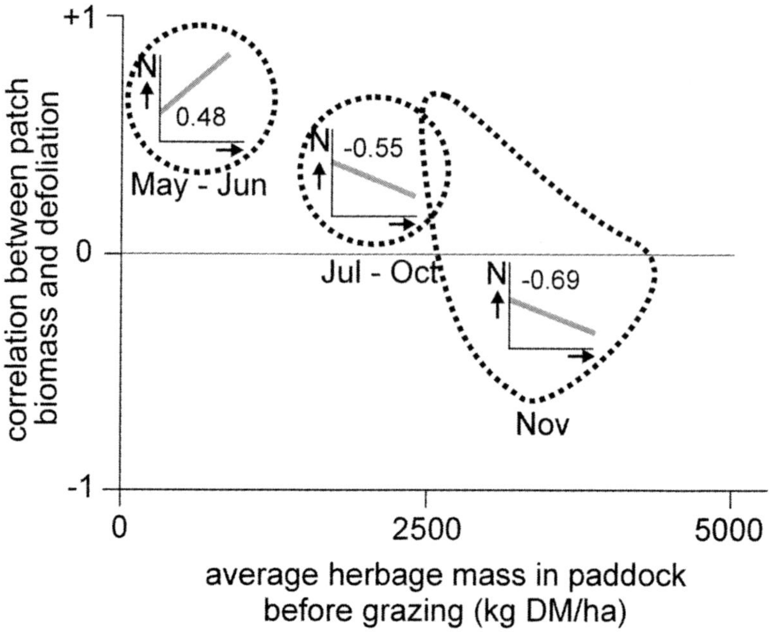

Figure 5.5. *Seasonal patterns of correlation between patch quality, herbage mass and defoliation rate by cattle. Areas encircled by lines represent the scatter of points. Inset graphs are the correlations between nitrogen concentrations and forage mass across patches during each of the three different seasons. Based on Ogura et al. (2002)*

Functional response

Spatial heterogeneity of vegetation determines the functional response of grazers and other ungulates (Drescher 2003). Gross et al. (1993) demonstrated that intake rate by herbivores across a wide range of body sizes is determined by bite mass. Bite mass is determined by the spatial arrangement of the vegetation at a local and very small scale, commensurate with the area of a few bites. Herbaceous swards are described by the proportion of total area covered by canopy, height of canopy, and plant mass per unit canopy volume, called 'bulk density'. For any given average herbage mass available over an area, tall swards yield larger bites and greater intake rate (Laca et al. 1994a). Therefore, the response of intake to herbage mass should be steeper for swards that increase in height than for those that increase in cover or bulk density (Figure 5.6). If spatial heterogeneity changed with herbage mass, the observed functional response could have unexpected shapes. For example, consider a grassland composed of 20% short patches and 80% tall patches of equal quality, where herbivores select only the tall patches and are limited by intake rate. An increase of herbage mass due to an increase of height of the short patches would not result in any change in intake. In more general terms, the functional response depends on the kind and degree of heterogeneity present in the vegetation.

Considering that simple rules and restrictions can result in extremely complex patterns of response, it should be no news that the functional response can have a variety of shapes, and that at any given level of resource abundance, intake rate is strictly dependent on the spatial distribution of the resource (Drescher 2003), particularly on the relationship between scale of measurement and spatial variance in resource abundance.

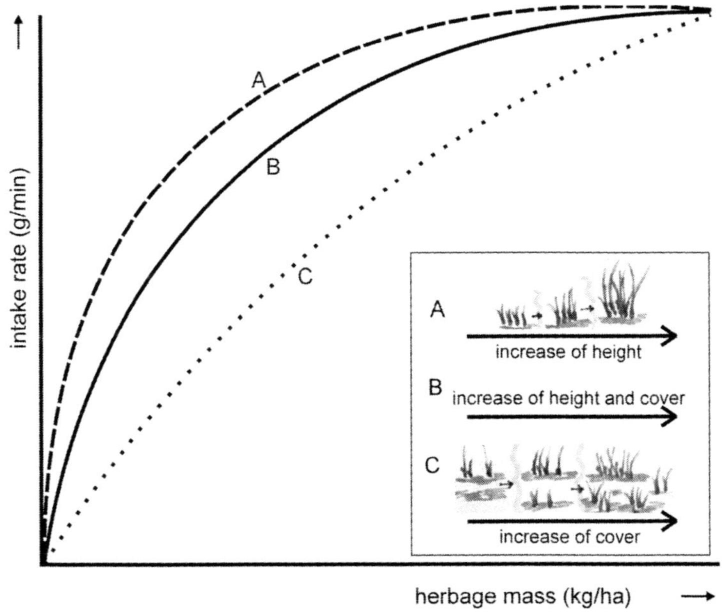

Figure 5.6. *Hypothesised effects of heterogeneity of herbage spatial distribution on the functional response that relates intake rate to herbage mass available*

INTAKE RATE AND SPATIAL HETEROGENEITY OF FORAGES

Intake rate of grazers over periods of 10^2 to 10^3 s is strongly influenced by bite mass over a wide range of values. In turn, spatial arrangement of plant parts in the volume occupied by the sward determines bite mass. The response of bite mass to sward height is typically linear with a slope directly proportional to sward bulk density, whereas intake rate responds to bite mass with diminishing slope. These relationships have been developed mostly on homogeneous micro-swards, where an artificial canopy is created by manually attaching plant parts to a wooden frame. How well do these relationships describe what happens in heterogeneous swards? Is the response of animals grazing a collection of patches of different characteristics equal to the weighted sum of the

Intake rate and bite dimensions of large ruminants respond to heterogeneity at resolutions as fine as 20 cm

responses exhibited when they graze homogeneous areas with characteristics equal to each of the patches in the heterogeneous collection?

Figure 5.7 (Laca unpublished data) shows that the relationship between micro-sward (Ungar 1996) structure and bite mass changes depending on heterogeneity; a result that weakens the use of functions derived in homogeneous swards (Laca et al. 1992). The largest difference was observed between the bite depth in swards with heterogeneity at a fine scale and the one predicted based on responses to homogeneous swards. In this experiment, cattle exhibited selectivity for tall patches but not for density. When patches were at the level of one bite (10 cm), bite depth reflected grazing to the constant residual height (small cross in Figure 5.7) that

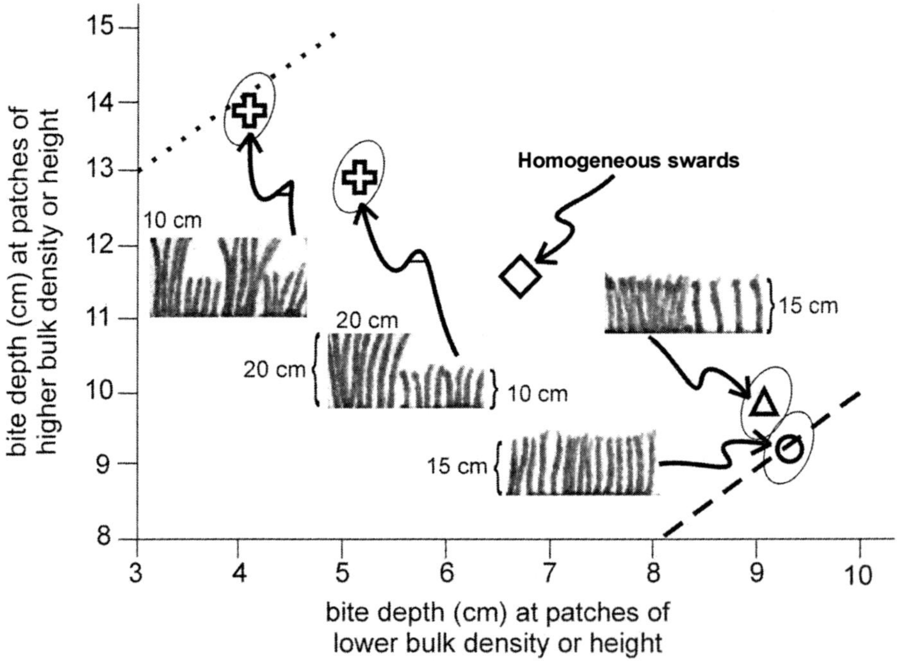

__Figure 5.7.__ Effects of small-scale heterogeneity of sward height and density on bite dimensions of cattle. The two crosses represent swards with heterogeneity in height in patches of 10 or 20 cm. The triangle represents swards with heterogeneity in density in patches of both 10 and 20 cm. The circle is the homogeneous 15-cm-tall control. The diamond is the calculated bite depth for homogeneous swards of 10 and 20 cm of height. All swards had exactly the same average mass, height and density of grass. Ellipses represent approximately one standard error. Dotted line: expected relationship between bite depth in tall and short patches if animals grazed both patches to a constant residual height. Dashed line is the line where bite depth at patches of lower bulk density (x-axis) equals bite depth at patches of higher bulk density (y-axis).

would be expected in a homogeneous sward of equal average height (residual height was ca. 6 cm, as in the sward represented by the circle). As the area of patches increased from 10 to 20 cm, bite-depth combination approached the prediction based on homogeneous swards with height equal to each of the patches present. Overall, the results make intuitive sense: small-scale heterogeneity was integrated or smoothed over by the animal prior to the response resulting in a 'response to the spatial average'. As scale of heterogeneity increased, the response became closer to what would happen in separate homogeneous swards.

Ginnett et al. (1999) determined that bite depth responds to vertical heterogeneity in bulk density by comparing grazing behaviour of steers in swards with structures as depicted in Figure 5.8. The vertically heterogeneous sward used can be considered the smallest resolution of heterogeneity in sward height possible.

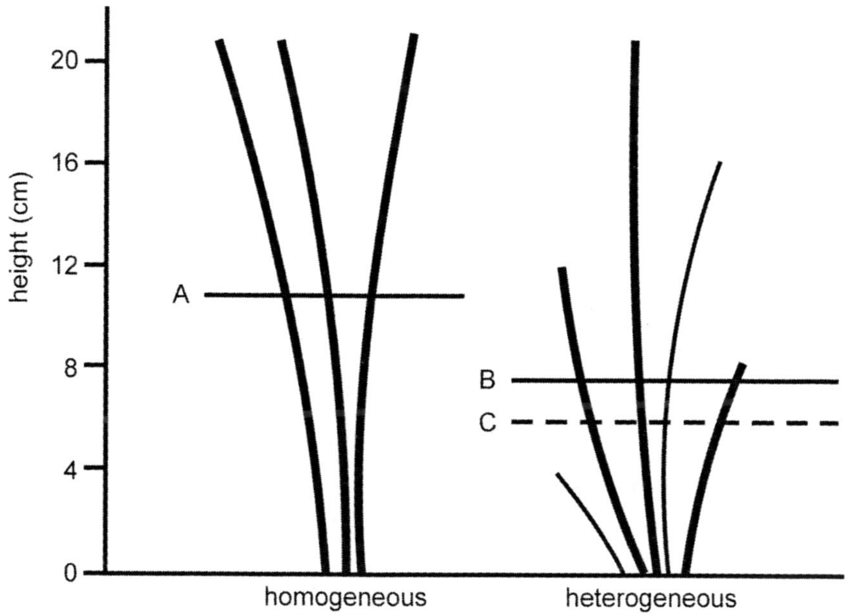

Figure 5.8. *Effect of sward heterogeneity on bite depth by cattle. Both swards have the same mass per unit area and the same maximum height. The continuous horizontal lines A and B show the observed bite depth. Line C shows the expected bite depth in the heterogeneous sward if animals had responded to its average height of 12 cm*

It is interesting to note that the description of heterogeneity becomes hard as the scale of heterogeneity approaches the grain of the forage (one leaf). Ginnett et al. (1999) refer to the treatment consisting of leaves of several lengths in each tiller both as 'variable height' and 'variable bulk density'. In their experiment, bite depth in the heterogeneous swards was 63.7% of the total sward height, leaving a residual height of 7.3 cm. Bite depth in the heterogeneous swards was significantly greater than in the homogeneous sward of equal total height, but not as deep as expected on the basis of average sward height. The consequence of these responses was that the heterogeneous swards resulted in a sigmoid instead of a concave-down depletion curve. This difference in heterogeneity generated significant differences in the expected residence time and depletion at the patch level. In the heterogeneous sward, animals exhibited local, instantaneous behaviour that was intermediate between responses in the homogeneous sward and what was expected if they responded to the average of the heterogeneous one.

DIET SELECTION AND SPATIAL DISTRIBUTION OF PLANTS

Studies of diet selection by herbivores have focused on the changes in diets as a function of changes in the relative proportion of forage components (for example, leaf and stem, or grass and clover), largely ignoring the spatial arrangement of the components and its potential correlation with their relative abundances. Obviously, these studies assumed that at least at some relevant scale, components of forage are separate in space. Otherwise, diets would always be identical to herbage composition. Some studies, however, provided a basis to assess the role of heterogeneity on diet selection.

Diet composition changes as a function of spatial distribution of dietary options: coarser resolution of heterogeneity allows a greater degree of selectivity

Clarke et al. (1995) studied the response of sheep and red deer to the spatial distribution of grass in a heather matrix, keeping the proportion of area covered by each vegetation type the same across treatments. The three treatments were 1 large, 4 medium or 12 small patches of grass covering 1/6 of the total area. Sheep spent more time grazing on grass than on heather, but the degree of selection decreased as patch size decreased. Deer also spent more time grazing on grass than on heather, but they always spent more time on heather than sheep, and their selectivity did not change as a function of grass patch size. These results agree with those obtained by Wallis de Vries et al. (1999) with cattle grazing patches of grass that differed in quality and quantity of forage. It appears that in general, coarser resolution of heterogeneity, or separation of forage options over larger units, allows a greater degree of selectivity.

Hester et al. (1999) conducted a similar experiment with sheep and red deer grazing mosaics of grass and heather, but size of patches was not controlled. Sheep, deer or both grazed 1-ha plots of a semi-natural mosaic containing about 15% of the area covered by grass patches. Grass patches were classified as small (1-6 m^2), medium (6-30 m^2) or large (> 30 m^2). Sheep selected small grass patches, whereas deer showed no consistent selectivity. The greater amount of edge in the small patches resulted in less impact by sheep in the small patches than in the large ones. Given that sheep have greater difficulty moving through heather than through grass, and that they were able to select more grass when patch size was larger (Clarke et al. 1995), it is not clear why they selected the smaller patches in this study.

Selection for a particular dietary component that is preferred (e.g., clover vs. grass) increases with increasing spatial separation between alternatives. Spatial separation of grass and clover resulted in much greater selectivity than mixed swards across a range of relative abundances of grass and clover (Clark and Harris 1985; Ridout and Robson 1991). Even in mixed swards, sheep diets had more clover than in the pastures (Figure 5.9).

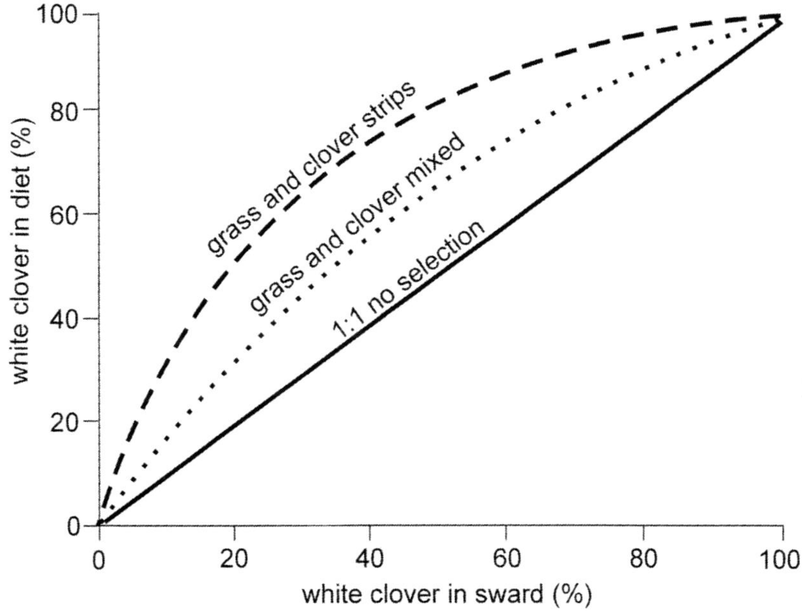

Figure 5.9. *Effect of relative abundance of white clover on diet selection by sheep grazing in paddocks with strips of pure grass and clover or mixed swards. Based on Clark and Harris (1985) and Ridout and Robson (1991).*

Marotti et al. (2002) tested the effects of spatial segregation between forages on diet selection and the mechanisms involved (Figure 5.10). Sheep grazed paddocks with a mixed ryegrass–white clover sward ('mixture'), ryegrass alone, clover alone, or clover and grass side-by-side ('choice'). This design allows comparisons to determine the effect of separation and the mechanisms by which these effects take place. As indicated above, the mechanisms can be selectivity for one of the options, changes of response to local conditions based on global conditions, and non-linearity of responses to local conditions. The design allows comparisons to detect selectivity and changes of response, but in order to detect non-linearity effects it would have to be extended to include more than one level of clover content in the treatments.

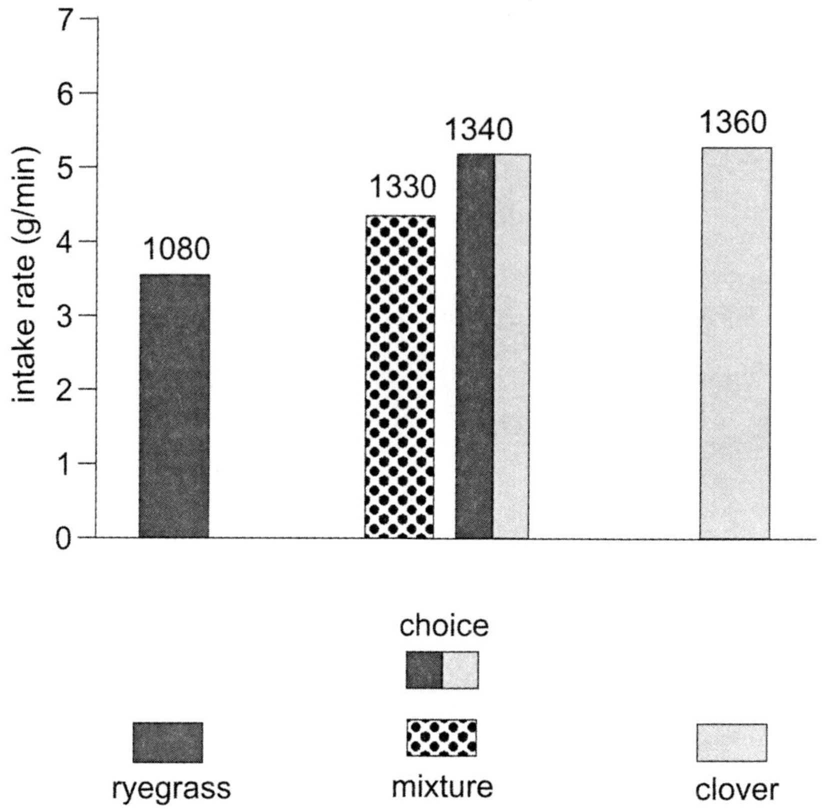

Figure 5.10. *Effect of spatial separation of clover and grass on selectivity and intake rate by sheep (based on Marotti et al. 2002). Sheep grazed paddocks with pure grass (dark bar), pure clover (light grey bar), finely mixed grass and clover (spotted bar), or half of the paddock covered with grass and half with clover (dark and light grey bar). Sheep individual daily intake (g d^{-1}) is given above the bars.*

Intake rate and bite mass were significantly smaller in pure grass than in pure clover swards. Intake rate in mixed swards was intermediate between pure swards, but in the choice treatment values were very similar to those observed in the pure clover sward. Sheep grazing a paddock with grass and clover in different areas achieved greater daily intake and intake rate than when grazing swards where similar amounts of grass and clover were interspersed (Marotti et al. 2002).

The study by Parsons et al. (1994) included several levels of clover content in the treatments, but only the 'choice' treatment was explored. This was partly compensated for by combining the results from Parsons et al. (1994) with those from Milne et al. (1982), which consist of responses of sheep grazing behaviour to changes in clover–grass proportions in mixed swards (Figure 5.11).

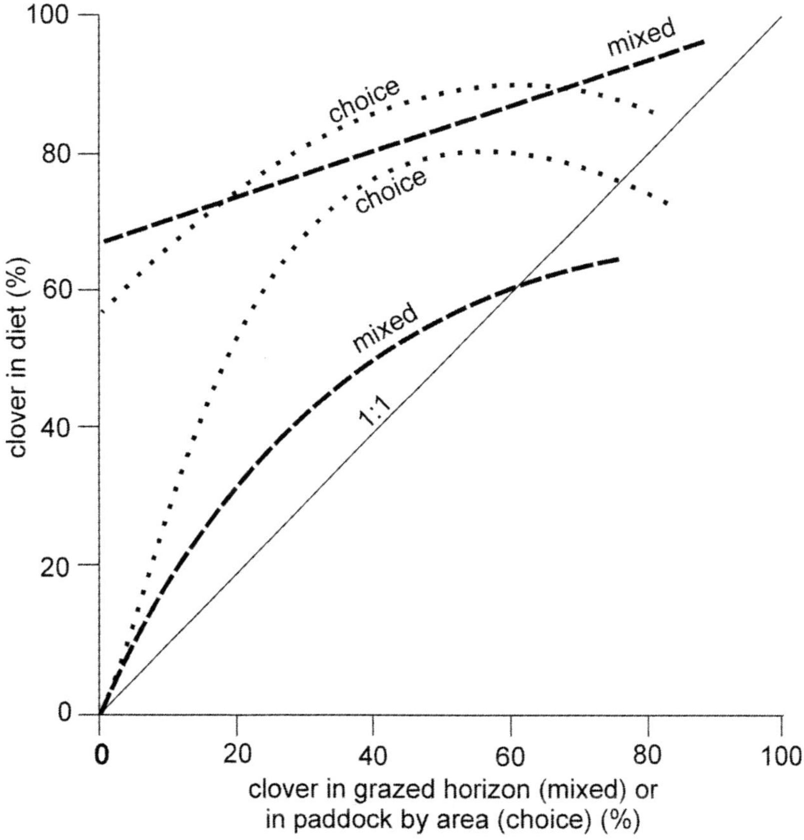

Figure 5.11. *Effects of clover content of the grazed horizon of mixed swards (Milne et al. 1982) or in paddocks containing adjacent areas (choice) of pure clover and pure grass (Parsons et al. 1994) on grazing behaviour of sheep. Full lines are proportion of clover in the diet. Dashed line: bite mass of sheep grazing mixed swards as a percentage of bite mass projected for swards with 100% clover. Dotted line: bite mass of sheep grazing-choice swards as a percentage of bite mass in the pure clover sward*

Several points are remarkable about the responses depicted in Figure 5.11. First, sheep apparently selected a mixed diet (see Prins and Van Langevelde, Chapter 7); and proportion of clover in the diet increased with increasing proportion in the pasture. Arguably, sheep in the choice treatment could have selected any desired diet composition without additional costs for a wide range of proportion of area covered by each species. Why did they not select their favourite diet composition in all treatments that offered at least a minimum area of each species? Parsons et al. (1994) discussed several possible explanations for the changing proportion of clover in the diet, including maximisation of intake rate, novelty, balancing of nutritional needs, preference for rarity and sampling. No single mechanism explains the observed pattern. Second, sheep selected more clover in the choice than in the mixed swards. Considering that mixed and choice treatments are extremes of a continuum, it is interesting to consider how the curves would change from the 'mixed' to the 'choice' as spatial separation and pattern change from completely uniform at all scales (perfectly mixed) to heterogeneous at all scales to completely to adjacent areas of pure grass or clover. I know of no experimental studies that have addressed these distributions and animal responses in manipulative fashion. Note that degree of heterogeneity at the smallest scales perceived by sheep is constrained by the size of plants or plant parts. In other words, under a magnifying glass the pasture always looks like a mosaic of pure clover or pure grass patches. Third, it is likely that the mixed treatment was not completely mixed, but showed some degree of natural patchiness. As indicated in the section on pervasiveness of heterogeneity, although we know that in the 'mixed' treatment grass and clover were more mixed than in the choice treatment, we do not know exactly how patchy the mixed pastures were. More importantly, we do not know if patchiness changed in a structured way as clover content increased. Covariance between clover content and spatial distribution would make it impossible to determine whether the response seen in the mixed curve is due to the change in clover content or to the change in pattern. Finally, the dotted and dashed lines (Figure 5.11), representing bite mass as a percentage of the value on pure clover, are more similar between treatments than diet compositions. The dotted line for the choice treatment was derived assuming that bite mass on the clover and grass parts of the pasture remain constant, regardless of the proportion of clover. The continuous line for the mixed treatment, derived by regression of the measured bite mass on sward composition, is consistent with the assumption that regardless of sward composition, bite mass on clover and grass was constant.

But, can animals gauge the mixing of their diets? Apparently, yes. Not only do ruminants respond to the mixing of dietary components, they also respond to it in a spatial fashion. Animals are able to select diets and spatial location for grazing not only based on positive diet characteristics, but also on amelioration of negative post-ingestive consequences (see also Bailey and Provenza, Chapter 2). Villalba and Provenza (2002) found that lambs preferred foraging locations where tannin-containing food was near food boxes with polyethylene glycol (PEG). Tannins are known to produce negative post-ingestive consequences and constitute an anti-quality factor, but PEG binds to tannins and reduces their negative effects. The implications of this finding are far-reaching in relation to impacts of heterogeneity because it demonstrates that sheep were able to integrate the consequences of mixing spatially separate foods.

Animals can gauge the mixing of their diets

Presumably, ruminants also integrate positive consequences of mixing complementary forages, such as grasses and clover, at temporal scales that allow better rumen functioning than if the same daily diet were split into meals of pure grass or pure clover. Rumen fermentation and microbial growth depend on an almost simultaneous supply of labile, slowly fermentable and N-rich compounds. For optimal rumen function, the mixing has to be at the scale of minutes, definitely shorter than meals. If energy- and protein-rich forages are segregated at large scales, mixing them at the appropriate scale will be more difficult than if they are interspersed at some intermediate level. Thus, the preference to have mixed diets within meals imposes a scale on foraging behaviour that is commensurate with the area grazed in 15-30 minutes. If complementary forages are in patches segregated at large scales, animals should prefer to forage on edges or they would forego the benefits of mixed diets in the short term. As scale of patchiness increases, separation of dietary options eventually has to result in negative effect on the diet, by making it impossible for animals to get a diet mixed within meals (see Prins and Van Langevelde, Chapter 7).

Box 5.2. *Testable hypotheses for future research*

Hypothesis 1. Functional and production responses are likely not only related to the change in overall resource abundance, but also to resource heterogeneity.
Hypothesis 2. Faced with alternating tall and short patches, herbivores take deeper bites in tall patches and shallower bites in short patches than expected on the basis of the functional response measured in 'homogeneous' resources.
Hypothesis 3. In heterogeneous resources, consumers exhibited local, instantaneous behaviour that is intermediate between responses in homogeneous resources and what is expected if they responded to the average of the heterogeneous one.

SYNTHESIS

Heterogeneity of forage resources is the norm, and it should be expected to have implications for intake and diet selection by herbivores. Herbivores respond to heterogeneity by selecting a subset of the options available and by potentially responding in ways that cannot be predicted on the basis to responses derived in homogeneous vegetation, even if developed for each and all of the options. This happens because animals are able to integrate the characteristics of the forage and respond in non-linear fashion to the integrated values. The integration can happen from the perceptual level to cognitive and to physiological levels. Some hypotheses for future research are formulated in Box 5.2.

Modelling and geostatistics help to develop theory of foraging in heterogeneous resources, but it will be impossible to extrapolate to scales not studied

Although in practical terms heterogeneity impacts will not produce responses that are significantly different from predictions based on the study of homogeneous vegetation, the theoretical development of the field will require models that dwell on the variety of scales at which animals integrate information and forage inputs. The fact that heterogeneity is a complex and multidimensional characteristic of the food environment effectively burdens our ability to determine experimentally the responses that are necessary to make management plans or impact assessments. We should borrow and re-cast concepts used in geostatistics and related fields to guide the development of practical principles for the management of herbivores in fragmented landscapes (see also Stein and Georgiadis, Chapter 3, and Skidmore and Ferwerda, Chapter 4).

CHAPTER 5B

COMMENTS ON "FORAGING IN A HETEROGENEOUS ENVIRONMENT: INTAKE AND DIET CHOICE"

MICHAEL DRESCHER[#], HERBERT H.T. PRINS [##] AND AREND M.H. BRUNSTING [##]

[#] *Ontario Forest Research Institute, 1235 Queen St. East, Sault Ste. Marie, Ontario, P6A 2E5, Canada*
[##] *Resource Ecology Group, Wageningen University, P.O. Box 47, 6700 AA Wageningen, The Netherlands*
E-mail: michael.drescher@mnr.gov.on.ca

In Chapter 5, Laca poses that currently there is no general quantitative theory of large-herbivore foraging behaviour in landscapes with heterogeneous resources. Though his stated goal here is not to put forward such a theory, he aims to present a number of relevant concepts and theories, and to place them in a coherent framework. Based on these, he poses some questions and hypotheses in an attempt to fill apparent knowledge gaps.

Laca starts his chapter by giving a general description of some concepts of spatial heterogeneity of vegetation. He explains how herbivore size and mobility can determine an animal's perception of heterogeneity, and the minimal and maximal scales on which an animal may respond to it. Then, he argues that the multidimensional character of heterogeneity can be reduced to a very limited number of dimensions that need to be understood in order to sufficiently describe herbivore foraging behaviour. He continues by illustrating mechanisms through which forage resource heterogeneity affects forage intake and diet selection. The spatial scales of the experiments he uses to support his points vary from a single feeding station to pastures of several hectares. Most of these studies investigate the effects of grass height on intake in mono-specific swards, and of proportion and spatial arrangement of two qualitatively different forage species on diet choice and intake.

H.H.T. Prins and F. van Langevelde (eds.), Resource Ecology: Spatial and Temporal Dynamics of Foraging, 101-103.

Perception and mobility implicitly are quite central to Laca's coverage of the effects of heterogeneity on forage intake and diet choice. Perception essentially sets a lower limit, while mobility sets an upper limit to the heterogeneity that a herbivore can respond to. Laca is touching upon these points in Box 5.1 with his example of the rabbit and the bison. However, he largely misses the opportunity to stress clearly that perception and mobility are likely affected by animal size and thus species-specific.

In Laca's example of the mixed swards of grass and clover (Figure 5.11), smaller herbivores like sheep can still select a higher-quality diet. But a large herbivore whose ingestive apparatus is several times the size of the distance between the clusters of clover, would not be able to select the clover from in-between the grass. In such an arrangement of things, a large herbivore's perception of the spatial variation in forage quality between clover and grass would be of no advantage. Therefore, we propose that it is unlikely that a large forager will perceive heterogeneity at a very small scale. An example of the varying selection ability between differently sized herbivores is browsing kudu, which are able to select small clusters of high-quality leaves while black rhino browse whole branches (Wilson and Kerley 2003).

Laca's example of the rabbit and the bison in Box 5.1 is also an illustration of the effects of body size on mobility and its subsequent effect on the perception of large-scale spatial variation in resource heterogeneity. Generally, smaller herbivores are likely to have smaller territories and home ranges than larger herbivores (Mysterud et al. 2001). This may not only be the result of reduced travelling speed in smaller herbivores, but also due to their increased need for cover, which is modified in some species by their tendency to congregate in larger herds. The perception of resource heterogeneity above the spatial scale of the territory or home ranges would be of no advantage to a herbivore. As there is no advantage from the perception of this heterogeneity, we propose that it is unlikely that a herbivore has developed a sensory system to perceive heterogeneity at that scale. Real-world examples for the described differences in home ranges and response to resource heterogeneity can be found for many antelope species of different size (Garland et al. 1993), where smaller antelopes tend to stay put during the dry season while larger antelopes may follow the rain (Fryxell, Chapter 6; Fryxell et al. 2005).

Treating a forage resource as a surface, as done by Emilio Laca, might be appropriate in the situation of herbivores grazing on pastures or short-grass savannas, where sward bulk density is high or distances between plants are small. However, forage resources actually exist in 3-dimenional space, and viewing them as a plain surface might often be an oversimplification. Ruyle et al. (1987) have shown that in some situations cattle graze on grass tufts by approaching them from the side instead from above, apparently to avoid grass stems. Also Drescher (2003) found that free-ranging cattle in tall-grass savannas grazed tall grass tufts from sideways. Laca's concept of the relations between bite size and sward height alone appears too simple to describe forage intake in such grasslands. Instead, as shown by Drescher et al. (2006a) forage accessibility, i.e., the ease with which preferred forage parts can be harvested as affected by the spatial arrangement of plant parts increases in importance for understanding grazers' intake. It might be possible to grasp the structural complexity of forage resources in 3 dimensions with a measure of fractal geometry as proposed in Box 5.1 of Chapter 5. However, the success of such an approach is uncertain, as demonstrated by a failed attempt to relate forage intake to fractal geometry of grasslands by Drescher et al. (2006b). It might be that a measure of fractal geometry will not be informative, because structural complexity in natural grasslands co-varies with forage mass, sward height, and many other variables. At least, it will need very careful experimental set-ups to isolate the effect of structural complexity from the effects of other variables.

Another aspect that Laca only refers to in passing but that deserves more attention is resource heterogeneity in the temporal dimension (see also Owen-Smith, Chapter 8). Temporal resource heterogeneity can be brought about by ageing and depletion of the forage resource. However, the majority of studies cited by Laca to illustrate his points do not take these aspects into account. This approach implies that Laca assumes that herbivores can always avoid the negative effects of forage resource depletion by moving between available resources, or that herbivores do not respond to temporal changes in resource abundance and quality. We believe that neither one would be a correct assumption. Though many large herbivores seem to track the spatially and temporally shifting abundance and quality of forage resources, other herbivores do not seem to do this, or to a much lesser degree. In this latter case, one would expect that herbivores show behavioural and physiological changes in response to the temporal heterogeneity in the forage resource. Examples of such behavioural and physiological changes are suggested by studies on seasonal changes in selectivity in goats (Duncan et al. 2005) and rumen fermentation in sheep (Sankhyan et al. 2001).

We do not agree with Laca's statement that "in practical terms heterogeneity impacts will not produce responses that are significantly different from predictions based on the study of homogeneous vegetation". We believe that Laca himself and also we in this comment illustrate numerous concepts of the effects of resource heterogeneity on large-herbivore foraging behaviour. But it is especially the intermediate scales where knowledge about foraging behaviour is currently lacking. Advances in our understanding can be expected from the field of animal cognition. For example, work about the spatial memory of herbivores has not only shown their great capacities (e.g., Edwards et al. 1996), but also some limits of their abilities (Dumont et al. 2000). We want to close by agreeing with Laca that a general theory of large-herbivore foraging in landscapes of heterogeneous resources is needed to put past and more recent study results into perspective and to generate relevant hypotheses. We hope that Laca's chapter in this volume and our comments will bring us closer to reaching this aim.

CHAPTER 6A

PREDICTIVE MODELLING OF PATCH USE BY TERRESTRIAL HERBIVORES

JOHN M. FRYXELL

Department of Zoology, University of Guelph, Guelph, Ontario, N1G 2W1, Canada
E-mail: jfryxell@uoguelph.ca

Abstract. All animals are faced with substantial variation in resource abundance over time and space. Patch-use theory, often based on optimality principles, can be useful in gaining insight into possible evolutionary solutions to this puzzle. A key consideration in applying patch-use theory to large terrestrial herbivores is that local variation in the nutritional quality of food is often inversely related to local resource abundance. Trade-offs between resource quality and abundance can change traditional models of patch use in important ways, some of which are explored in this chapter. I consider two aspects of patch-use decisions: which patches to visit and how long to stay in a patch, once visited? Empirical data for large herbivores often suggest that optimality principles are useful in explaining which patches are used in a landscape, but are less successful at explaining how long herbivores choose to stay in a particular patch. I end the chapter by exploring emerging challenges in applying patch-use principles to landscape ecology of large herbivores.
Keywords. patch selection; giving up; patch departure; short-term versus long-term intake; constraints on foraging rates; functional response

INTRODUCTION

Trophic interactions by definition involve fluxes over time in the abundance of both resources and consumers. While general ecological theory was largely founded on the notion of well-mixed, homogeneous resources, this abstraction is at odds with ecological reality. For example, the foods required by all large herbivores are patchily distributed, regardless of whether those herbivores are grazers that feed predominantly on graminoids and sedges, or browsers that feed on forbs, shrubs or the lower branches of trees. An indiscriminate forager would almost always prove to be less efficient in acquiring vital nutrients or energy than one that is more selective, simply because the indiscriminate forager would be just as likely to feed in patches with low rates of gain as in more rapidly-yielding patches.

H.H.T. Prins and F. van Langevelde (eds.), Resource Ecology: Spatial and Temporal Dynamics of Foraging, 105-123.

While it is debatable whether the fitness consequences of foraging indiscretion are inevitably serious, it nonetheless remains incontestable that appropriate decision-making should have selective advantages, at least sometimes. In this chapter, I consider the underlying nature of spatial variability in plant resources available to large terrestrial herbivores, review the potential constraints that may guide appropriate decision-making, apply optimal and sub-optimal models of decision-making in the face of such spatial variability, and evaluate the empirical evidence for such decision-making. My objective in this chapter is to explore the use of patch-use models as a cornerstone for a new approach to trophic interactions, one that considers fluxes in resource and consumer abundance over space as well as time.

Patch-use models can be used as a cornerstone for a new approach to trophic interactions

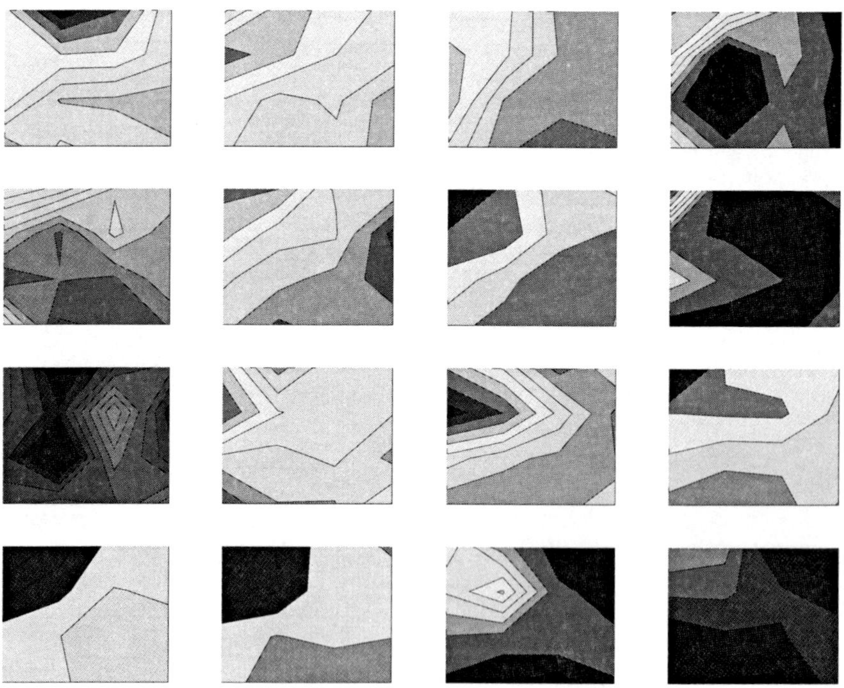

***Figure 6.1**. Spatial variation in grass biomass on the Serengeti Plains over the course of 2 growing seasons (1995 and 1996), based on 220 measurements per census of grass and cover taken every km during transects driven within a 40 × 40-km grid (Fryxell et al. 2004)*

RESOURCE VARIABILITY IN SPACE

At almost any spatial scale imaginable, there is pronounced variation in plant abundance and quality, even in rather featureless systems. Consider, for example, the Serengeti Plains of Tanzania. Growing conditions on the plains are tilted in favour of grasses in a variety of ways. There is a calcareous pan several feet below the surface that is too shallow to permit deep root penetration by tall trees, but deep enough to permit unrestricted rooting by graminoids. Rainfall is relatively low, sporadic, and temporally restricted to a short growing season of 4 or 5 months (McNaughton 1985). The net result is a sward of low-growing grasses and sedges mixed in with low-lying forbs and shrubs, and only a thin scattering of trees, particularly near watercourses. If there is a uniform resource base anywhere in the world, this should be it. Nonetheless, sampling of graminoid abundance over a 1600-km^2 study area clearly demonstrates considerable spatial variability at both fine and coarse spatial scales (Figure 6.1). Nor does the situation get better if one looks at ultra-fine scale, because there is obvious variation in biomass density and quality among grass tufts and even as one progresses from the soil surface to the growing point. If one repeats the exercise throughout the growing season, the patterns of graminoid abundance shimmer and shift in space from one census to the next (Figure 6.1). From a grazing herbivore's point of view, plant resources are highly variable in both time and space (Skidmore and Ferwerda, Chapter 4; Owen-Smith, Chapter 8).

Resource availability is perhaps slightly more consistent from the point of view of a browser, particularly those that feed on low branches of emergent woody plants, because growth rates are less pronounced and the spatial distribution of stems changes over a decadal, rather than annual, time frame. Even browsers, however, face considerable variation in the ratio of rich versus nutritionally-poor plant tissue within and among plant ramets. In seasonal environments, of course, usable tissue availability of woody plants varies considerably over time (Owen-Smith, Chapter 8).

There is no spatial scale at which resources are uniform for large herbivores

In sum, there is no spatial scale at which resources are uniform for large herbivores. Meaningful patch choices are possible at the scale of the feeding station, the foraging bout, daily home range, and seasonal home ranges.

ALTERNATE MODELS OF MASS, ENERGY, AND NUTRIENT INTAKE

Short-term intake of food or energy

Before one can consider patch-use decisions, further consideration of constraints on foraging rates is needed. The functional response is the cornerstone principle of all foraging models (Holling 1959; Spalinger and Hobbs 1992; Laca, Chapter 5). It specifies the pattern of food intake with respect to food abundance. The precise manner in which this occurs depends, not surprisingly, on the mode of feeding and the distribution of food items in the environment.

The simplest (and oldest) way to represent this process is to imagine that food occurs in discrete chunks (bites) that are distributed randomly across the environment. We furthermore imagine that each forager wanders aimlessly across a featureless landscape, feeding continuously, with no other distractions or needs for shelter, social interaction, mating, or predator avoidance. These assumptions may seem ludicrous at first glance, but they may not be far off the mark for large herbivores that are commonly faced with sparse resource abundance. Indeed, it is not uncommon to find conditions under which large herbivores forage for 10-12 hours per day.

Spalinger and Hobbs (1992) started from the basic recognition that terrestrial herbivores differ from most other heterotrophs in being able to move from one prey 'encounter' to the next while they are processing the results of the last successful 'attack'. In other words, herbivores can walk while they chew. This subtle fact can have a surprisingly large impact on foraging because of its consequences for the rate at which a foraging herbivore encounters food. In conventional predators, once a prey item has been found, the predator must invest a further period of time in 'handling' the item before search can be renewed. In contrast, larger herbivores can move onward as soon as they have made a bite, processing the bite as they move on to the next feeding station. This shortens the intervals between bites considerably, particularly when the forager can see the next bite as it departs from the last one. Nonetheless, foraging reduces the velocity with which individuals move across the landscape.

If an animal is foraging in a desert-like landscape, then there can be an appreciable distance between bites. As before, the rate of encounter with bites (λ) equals the velocity (v) multiplied by the foraging radius (w) and the density of bites per unit area (D). For the reasons mentioned above, velocity is compromised to a certain degree by each bite taken, so that effective velocity equals the maximum possible velocity (v_{max}) minus the bite frequency (vwD) multiplied by the velocity reduction per bite (δ). Experimental work by Shipley et al. (1996) nicely illustrates that there are profound changes in movement velocity between feeding stations for terrestrial mammals faced with experimental swards. After rearranging the terms to solve for v, Spalinger and Hobbs (1992) would predict that a forager would have an average velocity of $v = v_{\max} / (1+ \delta wD)$. It therefore follows that $\lambda = vwD = v_{max}wD / (1+ \delta wD)$. Hence, the food intake rate would equal bite size (S) multiplied by the encounter rate with bites (λ)

In contrast to other consumers, herbivores can walk while they chew, which has a large impact on the rate of encountering food

$$X(S) = \frac{v_{\max} wDS}{1 + \delta wD} \tag{1}$$

Note that there is no explicit handling time *per se* as the herbivore makes each bite. The successful forager loses velocity with each bite, but other than that there is no direct time investment in processing bites. The Spalinger-Hobbs equation predicts that intake will be linearly related to bite (i.e. stem) size, but the relationship to stem density will be curvilinear, with decelerating shape.

One troublesome element in the Spalinger and Hobbs (1992) functional response is that one cannot actually measure bite size without explicit reference to the animal. If one is prepared to assume that bites are synonymous with ramets, then one can make *a priori* predictions about bite rates and intake rates in relation to measurable ecological variables, ramet density and ramet size, which when multiplied together simply yield plant biomass ($V = SD$). It readily follows that cropping rates are a linear function of plant biomass under these conditions (by substituting V for SD in the equation for $X[S]$)

$$X(V) = \frac{v_{\max} w V}{1 + \delta w D} \tag{2}$$

The scenario in which bites are widely spaced was termed process-1 foraging by Spalinger and Hobbs (1992). It is directly comparable to process-2 foraging, in which the herbivore can actually detect each bite at some distance away, making a beeline between bites rather than searching blindly. This has rather an obvious, but minor, effect on feeding rates: encounter rates are increased relative to that of comparable process-1 foragers. A more important distinction can be drawn with situations in which bites are densely distributed across the landscape. This kind of ecological circumstance, termed process-3 foraging by Spalinger and Hobbs, implies that herbivores have insufficient travel time between bites to process the bite obtained at the preceding station. Under extreme bite densities, the rate of intake would therefore be completely dictated by the rate of clearance of bites from the mouth before a new bite could be taken, because the animal takes no time to move between bites. Hence, in process-3 foraging, intake is completely constrained by handling time rather than bite encounter.

What ecological factors might influence the all-important handling-time constraint? Bite size clearly must play a predominant role (Black and Kenney 1984; Illius and Gordon 1987; Spalinger et al. 1988; Ungar et al. 1991; Shipley and Spalinger 1992; Gross et al. 1993; Bradbury et al. 1996; Wilmshurst et al. 1995; 1999b). Other factors that also influence the handling time are plant toughness or the amount of protection afforded by spines and thorns (Cooper and Owen-Smith 1986). Spalinger and Hobbs suggest that handling time can be usefully decomposed into cropping of bites versus chewing those bites. These are mutually exclusive activities, so the time-budgeting logic that underlies other functional response behaviours can be applied to process-3 foragers.

Food intake can be constrained by searching and ingestion

One might predict that cropping rates would tend to decline with bite size because chewing of larger bites tends to lengthen the interval between bites, for the simple reason that the rate of input to the mouth cannot exceed the rate of output. Assume that there is a maximum rate of chewing (R_{max}) that decreases by each bite taken. It then follows that intake rate $X = R_{max} - \beta B$, where β is the effect of each bite on the rate of chewing and B is the bites per unit time. If we presume that intake rate can be estimated by multiplying together bite rate (B) and bite size (S), then substitution of $B = X/S$ for B in the expression $X = R_{max} - \beta B$ and rearranging terms to solve for intake as a function of size $X(S)$ yields the following functional response equation for process 3 herbivores (Spalinger and Hobbs 1992)

$$X(S) = \frac{R_{\max} S}{\beta + S} \tag{3}$$

It may be useful once again to derive an equivalent expression for intake in relation to a measurable ecological variable such as biomass V, by converting $S = V/D$

$$X(V) = \frac{R_{\max} V}{\beta D + V} \tag{4}$$

One interesting feature of this line of reasoning is that the same pasture could change from process-1 to process-3 conditions through simple growth processes. When ramets are small, chewing time is so short that there is no conflict with cropping. Hence, the functional response is linear, increasing proportionately with each unit increase in plant abundance. At ramet sizes above this threshold, the situation reverts to a process 3, and intake is curvilinearly related to plant abundance. The net effect of changing mechanistic constraints on intake is a discontinuous functional response, with the discontinuity at the point of transition between process-1 and process-3 foraging (Figure 6.2a). Similar conclusions have emerged from alternate mathematical formulations in which herbivore search can overlap with processing (Parsons et al. 1994; Farnsworth and Illius 1996). Mathematically, this switch in mechanistic constraints can be represented by the following piecewise function for food intake

$$X(V) = \min\left[\frac{v_{\max} w V}{1 + \delta w D}, \frac{R_{\max} V}{\beta D + V}\right] \tag{5}$$

A substantial body of experimental work corroborates the key predictions of the Spalinger-Hobbs model. Grazers presented with sward conditions likely to produce process-3 conditions usually show a smoothly decelerating functional response, as predicted (Wickstrom et al. 1984; Short 1985; Hudson and Frank 1987; Wilmshurst

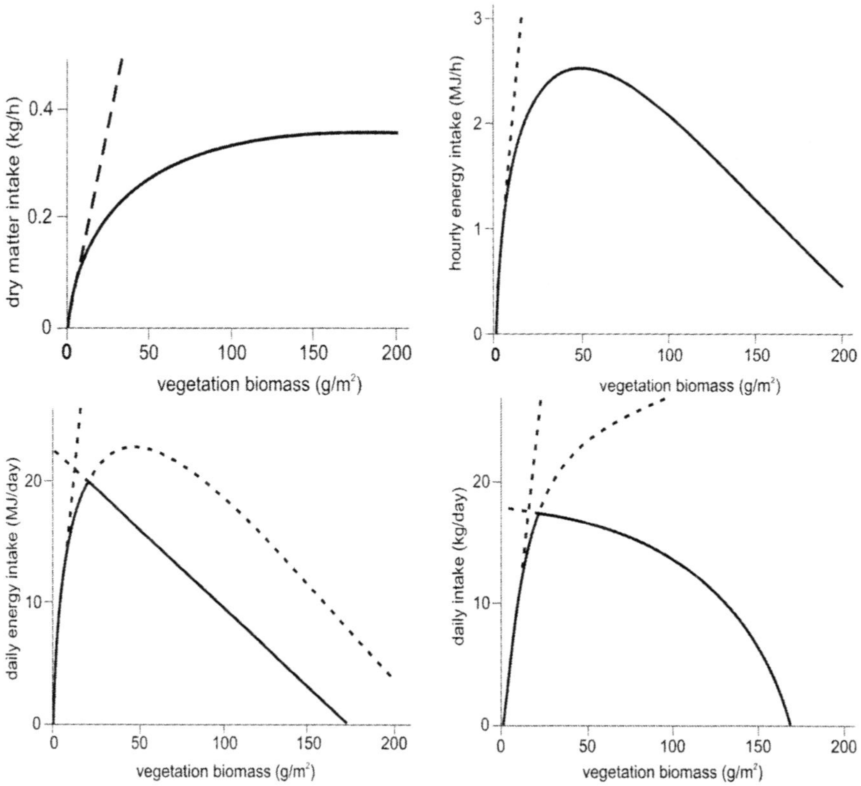

Figure 6.2. *Estimated rates of intake of dry matter per hour (a), digestible energy per hour (b), digestible energy per day (c), and dry matter per day (d) for Thomson's gazelles, based on the experimental results of Wilmshurst et al. (1999)*

et al. 1995; 1999a; Bergman et al. 2000). On the other hand, browsers often show very poor or no relationship to biomass per se (Trudell and White 1981; Spalinger et al. 1988), but do show positive relationships to bite size (Wickstrom et al. 1984; Spalinger et al. 1988). More tightly controlled comparisons have been enabled in recent years by the use of experimental swards mounted on plywood boards at different ramet spacing, ramet height, and biomass levels. Such experiments provide strong support for the prediction that intake should be positively related to plant size (Shipley and Spalinger 1992; Gross et al. 1993a, 1993b; Drescher 2003; Hobbs et al. 2003). Gross et al.'s (1993a) paper is particularly instructive, in that they experimentally controlled for the effects of plant size, plant spacing and plant biomass, which co-vary in most natural systems.

Based on a large set of these observational and experimental data, Shipley et al. (1994) calculated allometric coefficients for the key parameters in the Spalinger-Hobbs (1992) model, most of which were predicted with a remarkably high degree of precision. Key parameters relate of course to maximum rates of chewing and bite dimensions. These characteristics are strongly affected by the shape and dimension of the dental arcade, which themselves scale allometrically with body size (Illius and Gordon 1987; Gordon et al. 1996). Controlled studies in experimental swards clearly demonstrate that bite dimensions depend strongly on the spatial distribution of plant tissues and sward height (Laca et al. 1992, 1994b), with profound impact on rates of depletion of swards (Laca et al. 1994a).

More recently, Hobbs and co-workers set up experimental trials to test directly the predictive ability of the Spalinger-Hobbs equations relative to other mathematical formulas for the functional response (Hobbs et al. 2003). Results from these experiments indicate that the mechanistic models of Spalinger and Hobbs (1992) were the best predictors of short-term food intake.

It is a relatively simple matter to expand the functional response to model short-term intake of specific nutrients or energy $Y(V)$, by simply multiplying the functional response by quality of the ingested forage $Q(V)$

$$Y(V) = \min\left[\frac{v_{\max} wVQ(V)}{1+\delta wD}, \frac{R_{\max} VQ(V)}{\beta D + V}\right] \quad (6)$$

Because digestible energy or nutrient concentration usually declines with plant biomass, primarily due to maturational changes in tissue as plants increase in size, $Y(V)$ will often be a dome-shaped function of plant biomass (Figure 6.2b).

Digestive constraints and long-term intake of energy or mass

Over a longer time frame, such as a day, the rate of intake can be limited by digestion rather than ingestion (Belovsky 1978; Demment and Van Soest 1985; Fryxell 1991; Illius and Gordon 1992; Newman et al. 1995; Laca and Demment 1996; Hodgson et al. 1997). This is an important topic of continuing physiological research, and detailed discussion of the digestive kinetics is well beyond our review. Suffice it to say that clearance of digesta from the tract can be a rate-limiting step: more food cannot be ingested than gets cleared from the digestive tract. The time it takes to process material in the digestive tract tends to be inversely related to the nutritional quality. Forage that is high in lignin and cellulose but low in cell contents tends to digest more slowly than material of higher nutritional quality. As a consequence, the potential for digestive constraints to limit daily intake is more pronounced in poorer forages than in better forages. That is not to say that there is no potential for physiological adjustment to poor nutritional quality. Indeed, there is some evidence that both small (Gross et al. 1985) and large (Owen-Smith 1994) herbivores can adjust digestive capacity or passage rate to some degree in response to declines in food quality, but perfect compensation seems to be rare or impossible. Otherwise, farmers would raise their livestock on sawdust.

Over a longer time frame, the rate of intake can be limited by digestion rather than ingestion

In an ecological setting, this has interesting consequences. A herbivore feeding in patches of poor forage might have daily intake that is constrained by quality of food, whereas the same animal feeding in patches of high quality might have daily intake that is constrained by food abundance. Hence, it is plausible to postulate that daily intake could be regulated by either ingestive or digestive constraints (Belovsky 1978; Fryxell 1991; Newman et al. 1995), depending on the ecological circumstances.

Such trade-offs often come into play in comparing the energetic gain obtainable from grass patches of different maturational stage. As graminoids mature, the proportion of poorly digestible tissue increases in order to meet the structural needs of an erect versus prostrate growth form. Hence it is common, although by no means ubiquitous, for nutritional quality to decline with ramet height or biomass of the sward (assuming similar plant spacing). These maturational changes in quality suggest that daily energy gain in herbivores might be limited by ingestive constraints when plants are small, but by digestive constraints when plants are taller.

Wilmshurst et al. (1999a) tested this prediction for Thomson's gazelles in the Serengeti ecosystem of Tanzania. Digestibility of leaf and sheath tissue sampled from areas occupied by territorial gazelles declined by half with a fourfold increase in grass abundance. Feeding trials on captive animals presented with forage of varying maturational stage suggested that *ad libitum* daily intake of energy increased sharply with digestible energy content. Functional-response trials using experimentally controlled grass swards on plywood boards clearly demonstrated a positive relationship between instantaneous intake and plant abundance. As predicted by the Spalinger-Hobbs model, instantaneous intake was significantly related to stem density per unit area for small ramets, but constant for large ramets. Daily energy gain potentially obtainable from the instantaneous functional response was then compared to the energy gain dictated by the *ab libitum* feeding trials. This comparison showed that ingestive processes regulate intake for only the shortest of swards commonly encountered on the Serengeti Plains. Similar patterns have been corroborated for two other wild herbivores: elk (Wilmshurst et al. 1995) and woodland bison (Bergman et al. 2000, 2001). The logical conclusion is that daily rates of energy gain in mammalian herbivores can be controlled by constraints on digestion, rather than ingestion.

Daily energy gain in herbivores might be limited by ingestive constraints when plants are small, but by digestive constraints when plants are taller

Our conceptual understanding of the processes involved in digestion by wild ungulates lags far behind our conceptual understanding of ingestion, no doubt because it demands physiological experimentation that is intrinsically costly and logistically challenging for large, wild organisms. Fortunately, there is a substantial amount of information available on domesticated ungulates that can be taken advantage of to formulate predictive models of digestive limitation (Illius and Gordon 1992; Meissner and Paulsmeier 1995). This work suggests that daily intake of energy is proportionate to the digestible energy content of forage, which scales with plant biomass. We can accordingly combine the ingestive and digestive constraints into the following piecewise formula, after scaling up the short-term functional response to a daily time scale, by multiplying by the maximum daily feeding time (t_{max} = 9h in the case of Thompson's gazelles):

$$Z(V) = \min\left[\frac{v_{\max} w V t_{\max} Q(V)}{1+\delta w D}, \frac{R_{\max} V t_{\max} Q(V)}{\beta D + V}, \alpha - \psi V\right], \tag{7}$$

where α is the maximum daily rate of energy consumption, and ψ is the rate of decline in the daily rate of energy consumption with each unit change in digestible energy content of the ingested forage. Note that we have presumed a linear function for the digestive constraint in relation to plant biomass V, based on the experimental data on Thomson's gazelles gathered by Wilmshurst et al. (1999a). This pattern is shown in Figure 6.2c. Note that we are now postulating a multiple piecewise formula, with the leftmost piece constrained by stem density, the middle piece constrained by bite processing, and the right-hand piece by digestion.

We can extend these results to consider a fourth alternative gain function: daily intake of dry matter (Figure 6.2d). This is simply obtained by dividing the formula for daily energy gain by energy content of the ingested forage

$$W(V) = \min\left[\frac{v_{\max} w V t_{\max}}{1+\delta w D}, \frac{R_{\max} V t_{\max}}{\beta D + V}, \frac{\alpha - \psi V}{Q(V)}\right] \tag{8}$$

We now have 4 different foraging objectives that might conceivably influence patterns of patch use by large herbivores. They might prefer resource patches producing high rates of short-term intake of dry matter ($X[V]$), short-term intake of energy or nutrients ($Y[V]$), daily intake of dry matter ($W[V]$), or daily intake of energy or nutrients ($Z[V]$). To be honest, we have no *a priori* reason to expect one objective to dominate above all others, for all species, under all circumstances. For example, short-term intake of dry matter or energy might dominate behaviour of an animal faced with minor energetic stress or faced with other important needs, such as mating, predator avoidance or social needs. On the other hand, we might expect long-term energy or nutrient intake to dominate behaviour of animals facing significant energy shortfall. One could argue these circumstances, no doubt, long into the night. A more useful approach is to go back to nature, to see which objective (model) is most consistent with observed patterns of herbivore behaviour.

The actual pattern of patch use by herbivores depends, of course, on the degree to which herbivores select particular patches and time that herbivores tend to spend in each patch, at probably any spatial and temporal scale. In principle, it should be possible to consider each of these decision variables in isolation, but this is rarely done, particularly for free-ranging animals. Rather, most researchers score overall patch use, and interpret patterns with respect to one or the other decision variable.

PATCH PREFERENCES

Now that we are armed with some understanding about the constraints affecting the rates of dry-matter, nutrient or energy intake, we can consider patterns of patch preference. One common approach to testing such problems is to generate an experimental arena of patches of different resource levels. We have performed this kind of experiment twice, on wapiti (*Cervus elaphus* L.) and wood bison (*Bison bison athabascae*), making predictions about expected patterns of patch use on the basis of direct estimates of key foraging parameters relating to both the ingestive and digestive constraints.

Different foraging objectives might influence patterns of patch use by large herbivores

Parameter estimates for wapiti had indicated that net energy gain should be maximized when feeding on grass swards of 100-110 g m^{-2} (Wilmshurst et al. 1995). We constructed large experimental mosaics of grasses ranging in biomass from 80 to 300 g m^{-2}. The patches that fell closest to the long-term rate-maximizing value were selectively used most heavily, with degree of use proportionate with net energy gain (Wilmshurst et al. 1995). By coincidence, an identical experiment was conducted simultaneously by another research team on red deer in Norway (Langvatn and Hanley 1993), with results also indicative of long-term nutrient or energy maximisation. These independent studies suggest that the behaviour of wapiti conformed to a matching rule. In other words, patches yielding twice as high a rate of energy gain were used twice as often as patches of lesser gain.

In a second study, we constructed similar mosaics of sedges ranging in biomass from 107 to 419 g m^{-2}. Prior experimental work (Bergman et al. 2000, 2001) had indicated that daily energy gains should be maximised at a sward biomass of 10 g m^{-2}, hence the shortest swards in our mosaic should have yielded the highest daily energy gain. Our results were totally inconsistent with the energy-maximising model: bison preferentially grazed in patches with a biomass of 217 g m^{-2}, well above the predicted value (Bergman et al. 2001). Why might this happen? Our interpretation was that instead of maximising daily energy gain, bison were maximising the short-term rate of energy gain, i.e. acting as though ingestive constraints were the sole determinant of fitness (Figure 6.2). This suggests that animals were basing their decision on a different time frame than we were, valuing instantaneous rates of energy gain more than daily rates of gain.

We have no idea why bison might differ from wapiti in their evaluation of short-term versus long-term gains. Maximising short-term gains allows foragers to minimise the time required to meet an arbitrary energetic target, while reserving time for other activities that might enhance fitness, such as social behaviour, grooming or avoidance of potential competitors or predators. Bison may be more sensitive than wapiti to foregoing such activities, perhaps because social interactions are so important to future fitness or because of feeding competition that can accompany life in large herds (Manseau 1996). In any case, the key point is that the predicted outcome of this particular optimal-foraging model depends on the time frame under consideration – i.e. it is scale-dependent.

There is similar ambiguity in patch preference studies reported in the literature. In tightly controlled experimental trials, Laca et al. (1993) and Distel et al. (1991, 1995) showed that patch use by cattle was strongly linked to instantaneous rate of energy gain, but the experimental design in this case probably generated similar levels of plant quality despite contrasting levels of plant abundance. Wallis de Vries (1996) performed detailed calculations of daily and instantaneous energy gain for wild cattle. Opportunistic field data showed that long-term energy gain was a better predictor of patch use by cattle than short-term gain (Wallis de Vries and Daleboudt 1994). In later trials with a mosaic of patches with high (>600 g m^{-2}) and moderate (>300 g m^{-2}) biomass, cattle showed a strong preference for the patches of lower biomass, with correspondingly lower rates of hourly energy intake but high rates of daily energy intake. In small-scale trials on manipulated vegetation patches, wild Svalbard reindeer preferred the patches with highest plant abundance, but lowest plant quality (Van der Wal et al. 2000), early in the growing season. Later in the summer, reindeer showed no preference among patches.

Some herbivore species prefer patches with highest nutritional quality but low abundance, whereas others have reversed preferences

The bottom line is that some species prefer patches with highest nutritional quality but lowest abundance, whereas others have reversed preferences. There is an inherent difficulty, however, in interpreting most of the experiments in the published literature. If trials are conducted only on swards below the hump in the daily-intake function, then daily energetic intake is maximised by selecting patches with high abundance – that is, animals should apparently prefer quantity to quality. The opposite would be true for trials conducted for sward abundance to the right of the hump in the intake curve. Clearly, such trials call for non-linear model evaluation (Hobbs et al. 2003). Without careful parameter estimation, it is difficult to know which situation might apply, because the shape of daily energy intake in relation to plant biomass varies with herbivore body mass (Wilmshurst et al. 2000). To make modelling even more problematic, maturational changes in plant quality can vary enormously across sites (Albon and Langvatn 1992). The sensitivity to local parameter values and the non-linear form of the alternate gain functions present sizeable challenges to predictive patch-use modelling in large grazing herbivores.

Rigorous quantitative tests of patch preference by ungulates have been largely confined to tightly controlled experimental trials at small spatial scales. It is less clear whether these experiments can be used to predict patterns of space use of free-ranging herbivores at large spatial scales. This is important, because resource heterogeneity occurs at all spatial scales in the environment and we cannot say *a priori* at which spatial scale resource selection by grazers might occur (Senft et al. 1987a). Until we know the answer to this question, optimal foraging theory has little to add to current approaches to population management and conservation of large herbivores.

Very little work has addressed this key question in large grazing mammals. Schaefer and Messier (1995) performed one of the most detailed analyses to date on habitat (i.e., resource) preference of musk oxen at a multitude of spatial scales ranging from the population level to that within feeding stations of individuals. They found that patterns of food selection were generally consistent across spatial scales, although there were some reversals at different scales for marginal species. Nellemann (1997) found that musk oxen in Greenland preferred areas of high graminoid abundance (over 100 g m^{-2}), suggesting that they valued short-term over long-term intake rates.

Resource heterogeneity occurs at all spatial scales and we cannot say a priori *at which spatial scale resource selection by grazers might occur*

In an elegant set of experiments, Wallis de Vries and co-workers (1999) evaluated the effect of scale on decision-making by cattle feeding on a mosaic of patches of high and moderate grass biomass. In one set of trials, each patch in the mosaic measured 2 × 2 m, whereas in another set of trials the patches measured 5 × 5 m. Their results showed that selectivity was demonstrably higher in the coarse-grained than in the fine-grained environment, and as a consequence animals maintained higher levels of energy gain. Surprisingly, however, there was little evidence that animals altered the tortuosity of foraging trajectories to keep themselves longer in favoured patches or that they fed longer in favoured patches. Animals sought out preferred patches, but once there did little to stay in preferred patches.

Ward and Salz (1994) measured use of patchy madonna-lily plants by dorcas gazelles in the Negev Desert. During the dry part of the year, all live plant material was below ground, necessitating digging by the gazelles, whereas emergent plant tissue was fed upon during the growing season. Gazelles concentrated feeding activity in areas with high lily-bulb density, as evidenced by short move lengths between feeding stations and depth of digging. At the level of single bulbs, which could also be considered a 'patch', requiring an extended period of digging versus cropping, gazelles selected plants with large leaves during the growing season, but preferred small bulbs during the dry season. The latter was interpreted as an adaptive response to increased energetic costs relative to minor rewards associated with digging up large, deep bulbs.

Wilmshurst et al. (2000) evaluated patterns of habitat selection by radio-collared wildebeest in Serengeti, to test whether wildebeest preferred short swards (as expected if animals are maximising daily energy gain) or tall swards (as expected if animals are maximising short-term gain). They found that the spatial distribution at a large spatial scale was concentrated in areas of short grass, as predicted by the daily maximisation model, but sward selectivity seemed to be more strongly related to grass greenness than grass height at smaller spatial scales.

Seasonal patterns of migration by red deer in Norway show that use of specific ranges by deer was linked to seasonal and spatial variation in forage quality (Albon and Langvatn 1992). Animals wintered close to the coast, where nutritional quality of graminoids and herbs was higher than that of dormant plants in the summer range. Migration to higher elevations was apparently timed to coincide with the emergence of nutritious immature plants. By migrating seasonally between the coast and the mountains, animals maintained a considerably higher nutritional plane than would be possible by sedentary behaviour in either coastal or upland areas.

In a more recent study (Fryxell et al. 2004), we used direct experimental data (Wilmshurst et al. 1999a) to parameterise the four foraging functions shown in Figure 6.3 (short-term food intake, short-term energy intake, long-term food intake and long-term energy intake) for Thomson's gazelles in Serengeti National Park. We then evaluated the ability of each of these models to predict the spatial distribution of gazelles across a 40 × 40-km expanse of the Serengeti Plains. Predictions of gazelle spatial distribution were generated in relation to samples of grass abundance collected at several hundred sample points spread around the study area. At the same sites where grass abundance was sampled, we also counted all large herbivores in a semi-circle with a radius of 1 km. This exercise was repeated at roughly bi-weekly intervals during the growing seasons of 1995 and 1996, yielding 16 separate replicates of gazelle density relative to plant biomass. We then regressed observed gazelle density against relative fitness (Figure 6.2) predicted by each of the foraging models. Results clearly demonstrated that Thomson's-gazelle distribution was best predicted by long-term rate of energy intake, with animals preferring swards of 20-30 g m^{-2}. These data offer perhaps the strongest evidence to date that foraging gain can be used to predict patch preference at large spatial scales.

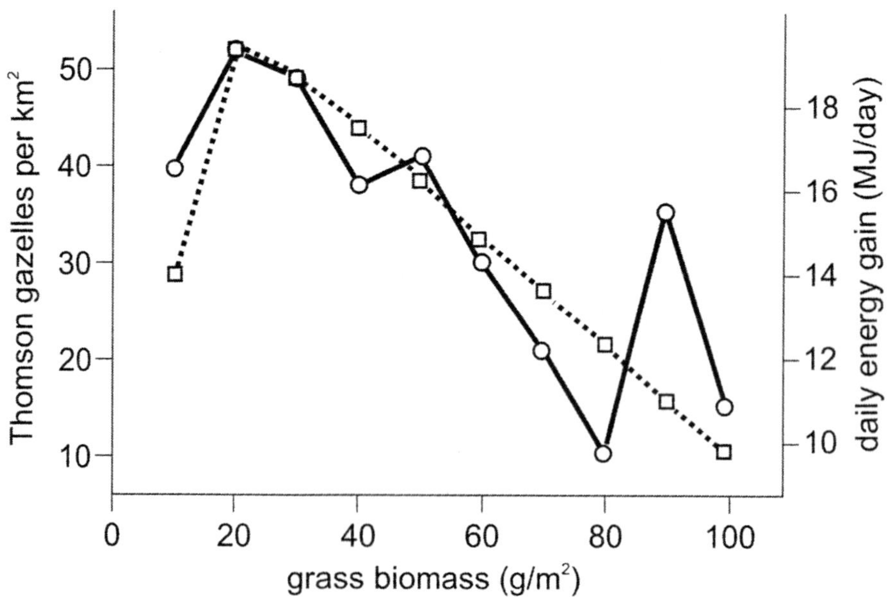

Figure 6.3. *Observed (solid line and filled symbols) versus predicted (dotted line and open symbols) densities of Thomson's gazelles, based on a daily energy-matching strategy (Fryxell et al. 2004)*

PATCH DEPARTURE RULES

A second bio-mathematical approach to understanding patch use is to consider whether foragers depart patches in a predictable manner. There is a well-developed body of theory that relates to patch departure, based on the so-called marginal-value theorem (Charnov 1976). The basis of the model is straightforward. Imagine that an animal forages indiscriminately within a patch for a given period of time. The longer a forager stays in a particular patch, the more the abundance of food items (bites, for a herbivore) within that patch declines. Changes in food abundance translate into slower rates of short-term food intake, according to the functional response, so there have to be diminishing energetic or dry-matter returns the longer a forager stays in each patch.

In principle, one could keep score of the cumulative gain obtained by the forager over time since it left the previous patch. For simplicity, we first concentrate on gain with respect to dry matter, rather than energy or nutrients. There is an initial period of time spent on travelling from the previous patch, during which no gain occurs. After the forager settles in the patch, gain increases at fast rate. Over time, however, the rate of gain decelerates, ultimately levelling off at an asymptote (maximum value) set by the total resource abundance initially present in the patch.

Now, let's calculate the long-term rate of yield, by dividing the cumulative gain at any particular residence time ($G[t]$) by the total elapsed time since the forager left the last patch (search time [τ] plus time in the current patch [t]). Charnov (1976) elegantly showed that the optimal decision for the forager would be to stay in the patch until the instantaneous rate of food gain ($dG[t]/dt$) equals the long-term rate of yield ($G[t]/[\tau+t]$). This is graphically shown by drawing onto the cumulative gain curve, the tangent going through the origin (Figure 6.4a). The optimal patch residence time is obtained by finding the point of intersection of this tangent and the cumulative-gain curve, and projecting downwards to the horizontal axis (Figure 6.4b).

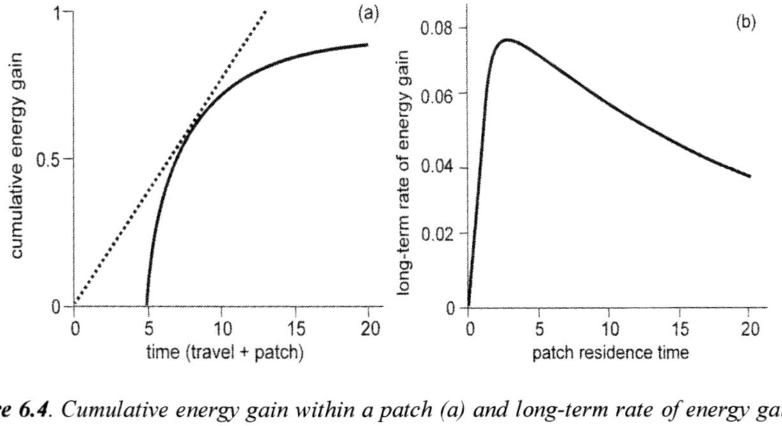

Figure 6.4. *Cumulative energy gain within a patch (a) and long-term rate of energy gain (b) as a consequence of different patch residence times. The optimal decision is to stay in the patch until the marginal rate of cumulative energy gain (decelerating curve in a) equals the tangent to the curve rooted at the origin*

The marginal-value theorem predicts that the more widely-spaced patches are, the longer should foragers stay in each patch. That is, the greater the time investment to relocate, the less picky should a 'clever' animal be. A relevant example of this kind of scenario is moose feeding on saplings or small trees. Astrom et al. (1990) tested whether animals spent longer at each sapling when saplings were widely spaced than when they were close together, having first shown that cumulative gain from a given sapling tended to decline over time. Results demonstrated that animals were in some sense sensitive to changing gain rates over time, as predicted.

Rigorous tests of the marginal value theorem with grazing ungulates are rare. Some experimental trials with cattle clearly suggest that patch departure can be well predicted by the marginal-value theorem, at least sometimes. Laca et al. (1993) and Distel et al. (1995) conducted an elegant series of trials with livestock grazing on small sward patches of given height (varying between 5-15 cm height) and bulk density (sparse versus dense), created by mowing an initially uniform pasture to desired levels. Because the sward was uniform before cutting, forage quality should vary little among patches, whereas intake rates would positively scale with height and bulk density. Results of these trials clearly suggest preference by livestock for the patches with highest instantaneous intake. As predicted by the marginal value theorem, residence time within patches increased with distance between adjacent patches and with the degree of variation between good and poor patches.

Rigorous tests of the marginal-value theorem with grazing ungulates are rare

Using the same data on Thomson's gazelles described earlier, we also tested alternate behavioural responses to the shifting mosaic of grass abundance across the Serengeti Plains (Fryxell et al. 2004). We found that gazelle distribution was best predicted by a mechanistic model that animals leave a given 10 × 10-km patch when the long-term rate of energy gain in that patch fell below the mean value recorded across the rest of the landscape. Moreover, animals settled in adjoining patches at a rate proportionate to the rate of daily energy gain. These empirical results offer support for the marginal-value theorem. Departure was scaled to local rates of gain relative to the expectation elsewhere in the environment, as predicted. On the other hand, emigrating animals did not settle evenly in adjoining patches, but rather the tendency to settle was matched to relative energy gains. In other words, both patch departure and patch selection rules apparently influence patterns of spatial redistribution of Thomson's gazelles from week to week.

As indicated earlier, plant abundance varies spatially at every meaningful scale. This implies that food resources do not occur necessarily in discrete patches that are readily definable (Skidmore and Ferwerda, Chapter 4). The marginal-value theorem can be extended to situations in which resource abundance varies continuously over space (Arditi and Dacorogna 1988; Focardi and Marcellini 1995). The outcome of this modification is that animals should exhibit a threshold response, grazing at every site encountered with a rate of food intake exceeding the expectation derived over the home range. Animals should locally deplete resources until intake rates reach the marginal value, at which point the grazer should move on.

There is limited evidence consistent with this continuous version of the marginal-value theorem. Wallis de Vries et al. (1999) found no evidence that cattle foraged longer in preferred patches, despite showing a strong preference in seeking those patches out. On the other hand, Jiang and Hudson (1993) showed data on captive wapiti suggestive of a threshold response at the feeding-station level. Focardi et al. (1996) made detailed observations of foraging trajectories of fallow deer and patterns of foraging over the course of those trajectories in relation to local resource abundance. In captive deer, feeding in a fenced-in area, there did seem to be a threshold response to local food abundance, in accordance with the model. In a wild population of deer in another area, however, there was no apparent evidence of such a threshold response. The reason for this discrepancy is unclear, but may stem from differences in the motivational state of animals, differences in the degree of familiarity of foragers with resource distribution, and/or greater demands on informational processing by free-living animals living in the wild site. It is interesting that here again the evidence for strategic behavioural response is most compelling in captive, rather than wild, herbivores.

When resource abundance varies continuously over space, the marginal-value theorem predicts that animals should exhibit a threshold

SYNTHESIS

A new generation of spatially-explicit models of herbivore movement behaviour has emerged in the past decade (Turner et al. 1993; Moen et al. 1997; Grünbaum 1998; Farnsworth and Beecham 1999; Fryxell et al. 2005), capitalizing on rapid growth in micro-computing power needed to consider detailed spatial processes. To varying degrees, these models incorporate patch-use criteria, either in the form of patch preference rules that influence patterns of movement with respect to neighbouring patches (Turner et al. 1993; Farnsworth and Beecham 1999; Illius and O'Connor 2000), patch departure rules (Moen et al. 1997), or both (Fryxell et al. 2005). A general result that has emerged from such modelling is that decision criteria (optimal versus sub-optimal), spatial scale on which decision-making is based (local versus regional), and motivational objectives (daily versus hourly rate maximisation) have profound impact on modelled rates of intake, animal performance and sustainability in heterogeneous environments.

For example, we have recently shown that Thomson's gazelles in Serengeti National Park may require much larger grazing areas (on the order of 2500 km^2) than one might have expected in order to sustain themselves during the inevitable periods of drought and superabundance of rainfall that they experience (Fryxell et al. 2005). This shows that spatially-explicit modelling of large herbivores, based in large part on patch-use decisions, can have important implications for understanding ecological interactions and in shaping wise management decisions (Fryxell et al. 2005).

A key impediment to such progress, however, is a pronounced gulf between the sophistication of models versus empirical data. Part of the problem is in finding appropriate ways to compare the explanatory power of alternate models, although recent advances in information-theoretic approaches have great promise (Burnham and Anderson 2002). It is also enormously difficult to gather data at relevant scales to parameterise spatially-explicit models. Movement rates and transition probabilities among patch categories (e.g., high, medium and low vegetation biomass) are notoriously difficult to measure in the field. As mentioned earlier, appropriate experimental measurement of animal performance criteria (intake rates of energy versus dry matter at hourly versus daily time frames) is similarly rare. Without these parameters, models are limited to showing the potential importance of biological features, not using models as tool to enhance understanding of even more complex ecological processes at the community or ecosystem scale. Until this occurs, utility of spatially-explicit models for management purposes will be necessarily limited. We see particular need for strong research linkages among modellers, spatial statisticians, GIS specialists, experimental behavioural ecologists and field ecologists in tackling spatially-explicit foraging processes in large herbivores. Without such a team approach, solutions of these intractable problems will be slow in coming.

Modelling shows that decision criteria, spatial scale on which decision-making is based, and motivational objectives have profound impact

***Box 6.1**. Testable hypotheses for future research*

Hypothesis 1. Forage processing in the digestive tract may be more commonly limiting to energy intake than forage availability.
Hypothesis 2. Herbivores that value social needs or security ahead of energy gain should choose patches to maximise short-term intake rather than long-term intake.
Hypothesis 3. Energetic gain influences patch use more than other constraints, such as predation or risk of parasitism.

Several key hypotheses would seem to be of particular relevance. Some hypotheses for future research are formulated in Box 6.1. The first hypothesis is so central to other questions with respect to herbivore patch use that it must be of high priority.

Enormous strides have been made in the past two decades in understanding and predicting patterns of patch use by large herbivores, although there is insufficient replication and insufficient consistency in experimental methods to allow definitive conclusions. Most trials suggest that grazing herbivores trade forage quality against abundance, in a manner that favours long-term rates of energy gain. Due to co-variation between digestible energy and nutrient composition, energy maximisation probably succeeds in maximising intake of important nutrients as well.

This can be achieved either through seeking out patches of high energy gain or by staying in such patches whenever encountered. On balance, the experimental evidence for optimal patch departure is much less compelling than is the evidence that animals seek out adjacent patches with the high rates of energy gain. The pattern of selection is rather all-or-nothing, but much more commonly proportionate scaling of patch selectivity with energy gain. As a consequence, energy matching is a consistently more realistic description of observed patch-use patterns than is energy optimisation.

Predictive modelling of herbivore patch preferences has often proven highly successful at a small spatial scale (1-100 m^2), using manipulated patch conditions in experimental arenas. On the other hand, predictive modelling of patch use at the larger spatial scale (100-10,000 m^2) is needed for management decisions. Such large-scale patch-use modelling is in its infancy, although recent work points to enormous strides in this direction (Turner et al. 1993; Moen et al. 1997; Illius and O'Connor 2000; Farnsworth and Beecham 1999; Fryxell et al. 2005). Given the success of predictive modelling at smaller spatial scales, we suspect that there are enormous opportunities for application of behavioural models of patch use at larger spatial scales. The current limitation is not so much ingenuity in formulation of models, but rather finding innovative and informative ways to link models meaningfully with empirical data. Such a step is needed to winnow out non-useful models and to apply herbivore movements to more complex ecological interactions (i.e., predation, competition and host disease) on spatially-realistic landscapes.

ACKNOWLEDGEMENTS

We thank the Natural Sciences and Engineering Research Council for its generous support of our research. Part of this chapter was originally presented as a contribution to the XIX International Grasslands Congress.

CHAPTER 6B

COMMENTS ON "PREDICTIVE MODELLING OF PATCH USE BY TERRESTRIAL HERBIVORES"

IGNAS M.A. HEITKÖNIG[#], MICHAEL DRESCHER[##] AND WILLEM F. DE BOER[#]

[#] *Resource Ecology Group, Wageningen University, P.O. Box 47, 6700 AA Wageningen, The Netherlands*
[##] *Ontario Forest Research Institute, 1235 Queen St. East, Sault Ste. Marie, Ontario, P6A 2E5, Canada*
E-mail: ignas.heitkonig@wur.nl

Fryxell's aim (Chapter 6) is to evaluate the current understanding of forage intake and patch selection by herbivores across temporal and spatial scales, where resource (food) heterogeneity is large. His approach starts at a description of the functional response, i.e., the food intake response of consumers to quantitative changes in the resource supply. Re-developing several models of food and energy intake applicable at the detailed level of feeding station, he places bite size or bite processing central to short-term food procurement. This is then developed into longer-term (daily) food or energy intake functions, where digestive rather than bite-size or bite-processing constraints may operate.

In developing both instantaneous- and daily-intake functions, he makes use of so-called multiple piecewise formulas. Essentially, they read as minimising functions, where the objective function, i.e., food or energy intake in the short or the long term, is the minimum of either summed bite size, summed bite procurement, or digestion (long-term only) achieved. For instance, daily energy intake can be calculated as the intake of lab-determined energy from ramets over one day of feeding, or as the energy intake from ingestive processes, or as the energy intake of lab-determined total daily food, the latter subject to digestive constraints. If the energy intake calculated from, say, daily food digested is lower than any of the other calculated intake values, then digestion is considered to be the rate-determining

H.H.T. Prins and F. van Langevelde (eds.), Resource Ecology: Spatial and Temporal Dynamics of Foraging, 125-127.

parameter. The merit of this approach is that it clearly shows which of the most rate-limiting functions determines resource intake at various temporal scales. Indeed, the 'forage maturation hypothesis', which explains foraging behaviour with instantaneous and daily time-scale minimum functions, has also been corroborated by recent work on cattle foraging in complex tropical swards (Drescher et al. 2006b). However, the drawback of this likelihood approach is that the demonstration rests on one major assumption, namely that bite size is known and is considered to be synonymous with ramets.

Bite size is one of the most evasive variables in foraging ecology. It can be estimated and deduced, but not appropriately measured. Not only is this problematic in swards with one tissue (leaf) of one height in production grasslands, but even more so in complex swards, the other 50% of the world's grazing areas, where leaves mingle with bite-deterrent stems, in various densities and heights, and at various stages of maturation. One of the consequences of Fryxell's assumption is that intake increases to an asymptote with increasing resource biomass. In this case, an increase in ramet mass evokes one maximum functional response. In reality, herbivores select – or at least prefer – leaves from among leaf-and-stem swards, and mature swards show increased stem biomass that may in fact reduce resource intake at high values (type-4 functional response). That the basic assumption on ramets being equal to bite size appears to be supported by most of the results from the experimental studies cited in this chapter says more about the experimental conditions that were created than about the harvesting processes in nature's grasslands. Drescher (2003), and Drescher et al. (2006a) showed that stems in swards consistently depressed the functional response across the tested resource biomass (up to 220 g m^{-2}).

Another assumption in the reasoning developed in this chapter is that 'scaling up' from instantaneous to daily intake is a matter of multiplication by time. However, the digestive constraint does not emerge from scaling up, but is separately introduced as an add-on to the reasoning. This is, for obvious reasons, understandable, but it emphasises the scaling problem when it comes to food intake predictions. Linked to this, and still on the subject of temporal scales, various experiments with large herbivores show conflicting results in terms of the rate-limiting resource ingestion. While this may be a herbivore species effect, it is not satisfactory from a theoretical point of view to see the ecology of scale being subject to effects of natural history or taxonomy. One issue here may be that the tests on rate-limiting factors have not been rigorously executed, and that too many assumptions (like ramets as bite size, limited or no variability in bite size or bite rate, etc.) were included. This does not necessarily call for more details in the testing, but rather for more tractable testing *per se*. This underlines the problems that ecologists face in their quest for a tractable, low-scale – high-scale explanation of resource consumption.

Moving from the temporal to the spatial scale, Fryxell then addresses an arbitrary 'next' level, one above feeding station, and terms it 'patch', in line with many others. Patches are not spatially defined, but are presumed to consist of feeding stations. In experimental situations, patches are created and defined in structure and space. In natural situations, one patch is left and another one entered when some distance needs to be covered before entering a next feeding station. This makes contrasts among patches arbitrary, even though we may pre-define how patches should differ. Even though Fryxell refers to a recent study (Fryxell et al. 2004) to lend support for theoretically derived patch departure rules at the landscape level (10×10 km), that study basically correlated field data with several foraging and animal-movement models to determine the best fit, and thus the most likely strategy employed by the animals in question, Thomson's gazelles. Variation among sub-populations of gazelle was large, but the best fit appeared to be with the model describing movements of Thomson's gazelles from patch to patch ruled by their daily rate of energy intake. We do not downplay the value of such studies, but we do note that the likelihood method should be seen as delivering an interim product, a rough approximation of the possible underlying processes determining patch use and patch departure rules. Correlating field data with model predictions does not provide the rigorous testing of hypotheses that we still require in this field.

Like with other chapters described in this book, Fryxell's chapter on predictive modelling of patch use by terrestrial herbivores provides a thought-provoking, up-to-date review on theory development and underlines the need for rigorous testing of these theories.

CHAPTER 7A

ASSEMBLING A DIET FROM DIFFERENT PLACES

HERBERT H.T. PRINS AND FRANK VAN LANGEVELDE

Resource Ecology Group, Wageningen University, P.O. Box 47, 6700 AA Wageningen, The Netherlands
E-mail: herbert.prins@wur.nl

Abstract. Resources are unequally distributed over the landscapes and it is only seldom that food of a herbivore at a given spot exactly matches its requirements. However, because non-sessile animals can move, they can assemble a diet from different patches that, in its total, does meet the intake requirements. Because herbivores of different sizes have different requirements for energy and nutrients, a linear-programming model that takes into account the different satisficing requirements of herbivores of a range of body masses (or of reproductive status) yields new insights into the causality of the differential way that these animals use the same landscape. Depending on landscape configuration and extent, and especially grain size of the distribution of resources, our model predicts that lactating females are much more constrained than other animals of the same species vis-à-vis the array of patches in the landscape. We also predict that small ruminants should be much rarer than large ruminants, and conclude that small ruminants can only survive under most circumstances if they are specialised feeders or if they live in a fine-grained landscape. We further conclude that natural selection favours ruminants with a large body mass to those with a small body mass if nutrient acquirement is the dominant selection force.
Keywords. landscape grain; resource distribution; diet selection; body mass; linear programming, pleistocene extinctions

INTRODUCTION

Resources are rarely homogeneously distributed so animals have to move in their search for food. This movement takes place at several scale levels: from steps between foraging stations, to daily movement in home ranges, to even seasonal migratory movements. For the better understanding of foraging in a spatial context, it is useful to introduce the concept of 'patch'. Patches are defined as localities (areas) that are more or less homogeneous with respect to a measured variable. It is, indeed, only rarely that in a given patch with food, a foraging animal can satisfy all its nutrient and energy requirements. For example, a patch of vegetation often does not satisfy the nutrient and energy requirements of herbivores. Even if one ignores

H.H.T. Prins and F. van Langevelde (eds.), Resource Ecology: Spatial and Temporal Dynamics of Foraging, 129-155.

the issues of patch depletion or patch size, and if one assumes that the instantaneous intake rate can be maintained for a prolonged period, it is clear that the vegetative parts of plants are barely sufficient to meet the requirements of an animal simply because nutrients are encapsulated in a matrix of carbon-based fibre. This matrix can be mechanically broken up, so as to gain access to the protein-rich cell content; the cell wall then is no source of energy but only an impediment to the process of nutrient acquirement. This is the case for animals with a simple digestive system, such as geese. The matrix can also be degraded with the aid of microbes. This process of fermentation yields energy but this energy has to be shared by the host with the microbes, which extract a price in the form not only of energy lost but also in the form of amino acids that are degraded (see Van Soest 1982). Ruminants are typically animals that rely on fermentation of their food.

The ultimate aim of the foraging herbivore is to cover its energy requirements (for maintenance, locomotion, possibly for pregnancy and lactation too, and perhaps as buffer for lean periods), and to maintain its store of minerals and amino acids in the form of its body tissues and skeleton. Ideally, any quantity of vegetation thus has a perfectly balanced mineral composition; that is, the ratio between nitrogen, phosphorus, calcium and potassium or even iron, zinc, magnesium and cobalt in the plant material (after digestion) perfectly reflects that in the animal's body and its depletion rates. Simultaneously, that ideal quantity of vegetation also yields a positive energy balance for the herbivore, it has the right balance between energy and proteins (Prins and Beekman 1989), and it even has all the essential amino acids in the proper relative amounts. In other words, the ideal patch of vegetation contains plant material that has all characteristics of animal tissue, and a herbivore would be better off as carnivore! For example, the calorific value and nitrogen content of the bodies of herbivores and carnivores are roughly equal. The calorific value of the herbivore's body, however, is approximately 1.5 times higher than that of the plants it eats, while the nitrogen concentration is 2.5 times higher (Table 7.1) (Crawley 1983 p. 184). Plant material is, however, much more available and easier to find, especially leaves and twigs that form the bulk of a plant's mass (see the so-called fibre curves of Demment and Van Soest 1985) than meat.

In reality, the vegetative parts of the vegetation hardly ever reflect the herbivore's ideal diet, and most of the primary production (about 90% on average) is not eaten by herbivores, whether invertebrate or vertebrate (Crawley 1983, p. 14). Plant production is selected for a maximum plant growth rate under the given conditions of availability of light, water and soil nutrients. Maximum growth is necessary so as to outcompete the neighbouring individuals of the same or other plant species. For example, plants invest in the carbon-based matrix referred to above, namely in cellulose and lignin, so as to grow tall to intercept the light before a neighbouring plant can make use of it. In nutrient-rich systems plants can and should invest more in stems than in leaves or roots, but in nutrient-poor systems the optimum is to invest in roots and leaves (Tilman 1988, p. 107 *et seq.*; Gleeson and Tilman 1994; Grace 1995). Because plants cannot move, they have to make best use of the available soil nutrients, even if these do not perfectly reflect what the plant

Table 7.1. *Relative proportion of elements (g g^{-1}) in the crust of the Earth, the average in plants (Van Soest 1982), the marginal levels minimally needed for animals (Robbins 1993) and the average in animals (Robbins 1993). n.r = not reported.*

%w/w	Crust: Granite	Crust: Basalt	Average crust	Average plants	Marginal level for animals in food	Average animals
Si	51.90	41.68	46.79	0.03 - 20	n.r.	n.r.
Al	15.22	15.17	15.20	n.r.	n.r.	n.r.
Fe	9.38	15.89	12.63	n.r.	0.005	0.034
Mg	3.91	6.78	5.35	0.20	0.2	0.11
Ca	6.80	11.68	9.24	0.03 - 3.0	0.4	2.91
Na	5.34	4.21	4.78	0.01 - high	0.05 - 0.18	0.46
K	4.87	2.30	3.58	1.5 - 3.0	0.6	0.90
Mn	0.19	0.45	0.32	0.01	0.001	0.002
Zn	trace	trace	trace	0.003 - 0.01	0.002	0.010
Cu	trace	trace	trace	0.001	0.001	0.002
S	0.09	0.00	0.05	0.12 - 0.3	0.2 - 0.6	n.r.
P	0.24	0.36	0.30	n.r.	0.2	1.88
C	0.17	0.00	0.08	n.r.	n.r.	n.r.
N	0.00	0.00	0.00	4.00	n.r.	10.00
Rest	1.89	1.49	1.69	n.r.	n.r.	n.r.

needs. In other words, at any given locality, the production of plant material in a patch of vegetation is nearly always limited by some nutrient. In most terrestrial systems, nitrogen is the limiting factor for plant growth but quite often plant growth is limited by both N and P (DeAngelis 1992, p. 41, Table 3.2; Ludwig 2001). The basic cause is that neither micro-nutrients nor macro-nutrients are distributed homogeneously in space, nor do they occur in the 'right' balance.

If, thus, a herbivore aims at ingesting a perfectly balanced quantity of plant material from a given patch, the chances are high that it cannot do so: at this locality the plant material lacks in, for instance, phosphorus, and that locality lacks in, for example, calcium. The ratio between Ca and P (g g^{-1}) in a mammal is 2:1. In forage this can range from 6:1 (e.g., in red clover) to 1:8 (e.g., in peanuts) (Robbins 1993, p. 40, Table 5.2). The solution for the herbivore is threefold. The first is differential rates of absorption from plant tissue through the digestive tract into the animal's body. The second is different rates of secretion. The third is by blending the intake from these 'imperfect' patches into a diet that satisfies the needs of the herbivore. The herbivore can blend its intake by moving from patch to patch whereby, ideally, ingested plant material low in, for instance, phosphorus from one patch is compensated by food high in phosphorus from another patch.

Plants and especially the vegetative parts of plants are rarely of sufficiently high quality to meet all requirements of animals

A herbivore thus has to assemble a diet from different plant species, from different patches, and sometimes from different seasonal ranges (e.g., dry and wet season ranges), and only the assembled diet can meet all of its requirements (Prins and Beekman 1989; Simpson et al. 2004). The aim of this chapter is to explore this problem of how the herbivore has to move between different patches that are imperfect from the herbivore's point of view, but that in combination can satisfy the herbivore's requirements.

Box 7.1. *Water consumption of large herbivores*

Ungulates, like all other organisms, need water for their survival. They have two sources of water, namely surface water and plant leaf water. The daily water needs of ungulates are about 4% of their body mass (du Toit 1996). These needs are not a function of body mass (see Table in this box; consumption data after Delany and Happold 1979, Table 11.3).

Table 11.3. *Water needs of ungulates*

	Body mass	Ambient T 22^{0}C	Ambient T 22-40^{0}C	Average
	kg	litre/100 kg	litre/100 kg	
Dikdik	5	5.59	7.72	
Thompson's gazelle	25	2.20	2.74	
Impala	52	2.49	2.93	
Grant's gazelle	55	2.08	3.86	
Oryx	169	1.88	3.00	
Waterbuck	211	5.98	no data	
Blue wildebeest	276	2.99	4.81	
Hartebeest	465	2.98	4.04	
Eland antelope	476	3.74	5.49	
African buffalo	636	3.43	4.58	
Average		3.34	4.35	3.84

Measuring stable isotopes of oxygen in body tissues or bone enables establishing the proportions of plant leaf water and free-standing water in the total water intake of an animal (e.g., Johnson et al. 1998). Some ungulates can cover all their requirements from water in plants, Oryx for instance, but others need free-standing water if the dry-matter content falls below a certain threshold. Impala need surface water if the dry-matter concentration of their food plants is higher than 67% (du Toit 1996), and buffalo always need surface water (Prins 1996). Oryx are thus not restricted in their search for food by surface water, impala are restricted during the dry season, while buffalo are always found within a couple of hours walking distance from a river or a waterhole.

In contrast to many other investigations into the foraging ecology of herbivores, we not only look at the two usual properties of a resource, namely vegetation quantity and vegetation quality, but we take a third important property into account, namely the grain size of the landscape. Grain size is a relatively new concept that describes the spatial array of patches of vegetation in a landscape (e.g., Murwira 2003; Skidmore and Ferwerda, Chapter 4).

We first review existing knowledge on requirements of herbivores and how these requirements are allometrically scaled. We then develop a model to investigate diet and patch selection of herbivores in environments that differ in the grain size of the food, that is, the distances between patches of grassy vegetation differ from close together to large inter-patch distances. We look for conditions where herbivore species can meet their energy and nutrient requirements and therefore can exist (we assume that other resources such as water are not limited, and that predator or parasite avoidance does not play a role). These conditions may differ for herbivores with different body sizes. As the possibility to meet the energy and nutrient requirements determines the presence and absence of species of different body sizes, the heterogeneity in food may determine the structure of herbivore communities. We therefore explore the effect of inter-patch distance and diet assembly on community structure. Finally, we discuss several hypotheses derived from our modelling study that may explain dramatic changes in herbivore assemblages such as the Pleistocene extinctions.

Small herbivores need, proportionally to body mass, more energy and nutrients than large herbivores, but not more water

In our analysis of interpatch movements we ignore the water requirements of the animals. We do this on purpose because, first, water requirements scale to body mass with a factor 1 (Box 7.1). In other words, on average ungulates need water as a fixed proportion of their body mass (for further reading, see Wallis et al. 1997; Williams et al. 1997; Evans et al. 2003). This makes water less interesting a parameter to include in our model, which investigates the effects of body mass. The second reason is that interspecific variation is very large indeed (Box 7.1). This second reason defeats the purpose of our investigation because it does not facilitate finding general rules to predict movements of animals. We thus acknowledge the fact that searching for water may be more important, on the short term, than searching for food, but in the present analysis we ignore this.

REQUIREMENTS OF HERBIVORES AND SIZE CONSTRAINTS

Body tissues of herbivores are the same as those of other animals, so the ultimate needs of herbivores are similar to those of carnivores when minerals, amino acids and vitamins are considered. Herbivores that make extensive use of fermenting microbes, however, have much less stringent needs vis-à-vis vitamins or specific amino acids (Robbins 1993, p. 17; Van Soest 1982, p. 246). In this chapter, we thus concentrate on ruminants, because the dietary needs of large ruminants can adequately be described in terms of digestible fibre for energy and of macro-nutrients. Requirements are very well known for domestic ruminants and to a lesser extent for some wild herbivores.

Energy requirements

Both theoretical considerations and measurements point out that smaller animals need, proportionally to body mass, more energy. There has been a controversy about the scaling factor with which energy expenditure scales with body mass, but it is now generally assumed that energy scales with a factor 0.75 (Moen 1973, p. 116; Hudson 1985; Robbins 1993, p. 123 *et seq.*; Nagy et al. 1999; West et al. 2002, 2003).

Daily basal metabolic rate E_{BMR} [kJ d^{-1}] is taken as [1]:

$$E_{BMR} = 293 \times W^{0.75} \quad (1)$$

where W is body mass [kg]. Body mass is the single most important factor explaining variation between species within a taxonomic class (it explains between 93 and 95%) (Nagy et al. 1999). Body mass is also important because it is related to speed of locomotion, foraging radius and home range size. The energy expenditure for each kilometre of walking per unit mass, E_w, for ruminants [kJ kg^{-1} km^{-1}] is related to body mass (Robbins (1993, p. 133; see also Moen 1973, p. 349) as

$$E_w = 10.75 \times W^{-0.316} \quad (2)$$

For climbing, the energy expenditure for each kilometre per unit mass, E_c [kJ kg^{-1} km^{-1}], appears to be independent of body size. Moen (1973, p. 349) reports 27.36 kJ kg^{-1} km^{-1}, whereas Robbins (1993, p. 137) gives 25.10 kJ kg^{-1} km^{-1}. We take the mean value of these two as

$$E_c = 26.23 \quad (3)$$

The energy expenditure per day including moving over a certain distance, E_{BW} [kJ d^{-1}], is then

$$E_{BW} = E_{BMR} + \left(E_w + E_c \times H\right) \times D \times W \quad (4)$$

where H is the vertical height ascended expressed as percentage of km on level, D the distance travelled [km d^{-1}].

Energy expenditure for standing, E_s, may be taken as 20% above E_{BMR} (Fancy and White 1985; Robbins 1993, p. 129), while running, E_r, may be 8 times more expensive than E_{BMR}, foraging E_f can be taken as 54% above E_{BMR} and, finally, ruminating E_h as 24% above E_{BMR} [all kJ d^{-1}], thus

$$E_s = 0.2 \times E_{BMR} \quad (5)$$

$$E_h = 0.24 \times E_{BMR} \quad (6)$$

$$E_f = 0.54 \times E_{BMR} \tag{7}$$

$$E_r = 8.0 \times E_{BMR} \tag{8}$$

Energy expenditure is strongly influenced by pregnancy and by lactation. We do not follow Moen's (1973, p. 353) approach but take a simplified one. Hudson (1985) gives the gestation length, L_d [d], for artiodactyls and for ungulates in general as

$$L_d = 120.31 \times W^{0.16} \text{ (for artiodactyls)} \tag{9a}$$

$$L_d = 109.97 \times W^{0.19} \text{ (for ungulates)} \tag{9b}$$

Oftedal (1985) calculated the costs for peak lactation for ungulates with a single young, E_{PL} [kJ d^{-1}], and then for the whole period of lactation, E_L [MJ], as

$$E_{PL} = 669 \times W^{0.70} \tag{10a}$$

$$E_L = 38.6 \times W^{0.81} \tag{10b}$$

Also for ungulates with a single young, he determined the total costs for pregnancy, E_P, and lactation, E_L, so the costs for reproduction, E_R [all MJ]. He found these relationships to be different for animals lighter than about 450 kg, and animals heavier than that as

$$E_R = E_P + E_L = 7.64 \times W^{0.90} + 38.6 \times W^{0.81} \text{ (for 4 – 450 kg)} \tag{11a}$$

$$E_R = E_P + E_L = 23.5 \times W^{0.68} + 38.6 \times W^{0.81} \text{ (for > 450 kg)} \tag{11b}$$

Since we know the length of gestation (eqn 9), we can calculate the average energy expenditure per day for this period. We can do the same for the average cost of lactation, because the age at weaning, L_w [d] (given by Peters 1983, p. 282), is

$$L_w = 34 \times W^{0.15} \tag{12}$$

Much work has been done on measuring energy expenditure (field metabolic rate, FMR) of animals under field conditions. A good overview can be found in Nagy et al. (1999). Their review shows that desert mammals have lower energy expenditures than mammals under mesic conditions (see also Tieleman and Williams 2000).

Nutrient requirements

Just as energy expenditure scales with body mass, so do nutrient requirements. It should be realised that, although ruminants have a large store of macro-nutrients in body tissues and the skeleton (Table 7.2), loss rates do not adequately reflect the problem animals face when they forage for macro-nutrients. With a loss rate for Ca of 0.25 g kg^{-1} d^{-1}, it would take 3.2 years before the calcium store would be finished. Of course, the animal would have severe difficulties before that time, and that is why we concentrate on the amounts of nutrients the animal needs to maintain balance.

Table 7.2. *Proportion of minerals in whole animals (mg g^{-1} d.w.) (from Robbins 1993)*

Element	White-tailed deer	Short-tail shrew	Cotton mouse	Golden mouse	Old-field mouse	Fox squirrel	Blue tit	Coal tit	Gold crest	Meadow pipit	Rook	Average
Ca	3.09	3.44	4.05	3.74	1.6	2.56	3.28	3.31	2.84	2.04	2.04	2.91
P	2.26	1.72	1.67	1.92	1.86	1.80	2.04	2.08	1.88	1.65	1.75	1.88
K	0.95	n.r.	n.r.	n.r.	1.2	1.07	0.58	0.63	0.58	1.27	0.94	0.90
Na	0.39	0.42	0.24	0.36	0.43	0.84	0.37	0.39	0.40	0.83	0.45	0.46
Mg	0.09	0.14	0.12	0.14	0.06	0.13	0.10	0.11	0.11	0.13	0.09	0.11
Fe	0.016	0.050	0.020	0.024	0.038	n.r.	n.r.	n.r	n.r.	0.040	0.048	0.034
Zi	0.007	0.012	0.010	0.011	0.013	n.r.	n.r.	n.r	n.r.	0.011	0.010	0.010
Mn	0.003	n.r.	n.r.	n.r.	0.001	n.r.	n.r.	n.r	n.r.	0.001	0.003	0.002
Cu	0.003	n.r.	n.r.	n.r.	n.r.	n.r.	n.r.	n.r	n.r.	n.r	0.001	0.002

It can be assumed that there is a constant tissue turnover, and for nitrogen this can be expressed as the average nitrogen intake to achieve nitrogen balance, N_b [g d^{-1}] (see Moen 1973, p. 334; Robbins 1993, p. 180 *et seq.)*: $N_b = 0.650 \times W^{0.75}$, but this equation does not take into account that when an animal expends much energy, it also loses more nitrogen due to enzymatic turnover. However, because the ratio of endogenous urinary N in mg to kcal in the equation for basal metabolism is 2 (Moen 1973, p. 334), which thus equals $N_b = 0.140 \times W^{0.75}$ [g d^{-1}], we can deduct a relationship between energy expenditure and nitrogen needs. The energy expenditure for basal metabolism (eqn 1) is 293 $W^{0.75}$ kJ d^{-1}, i.e., 70 $W^{0.75}$ kcal d^{-1}, meaning that the constant for the N intake for an animal functioning at basal metabolic rate only is $0.140 = (70 \times 2)/1000$ g d^{-1}. As a matter of fact, for ruminants, the constant is 0.093 and for non-ruminants it is 0.160 (Robbins 1993, p. 180). Because the amount of energy an average animal spends is $2 \times E_{BMR}$, we have taken the nitrogen intake to achieve nitrogen balance,

Deduction of a ruminant's nitrogen needs from its known energy expenditure

N_b, to be proportional to this. We then assume that if the animal spends 4 × E_{BMR}, it needs twice the amount of nitrogen to stay in balance, hence N_b is per multiple of E_{BMR}

$$N_b = \gamma \times 0.325 \times W^{0.75} \tag{13a}$$

where γ is the coefficient to scale the required amount of nitrogen to achieve nitrogen balance, N_b, with the energy required for maintenance, E_{BMR}. The coefficient γ [-] is defined as

$$\gamma = \frac{E_{tot}}{2 \times E_{BMR}} \tag{13b}$$

where E_{tot} is the total daily energy expenditure, including energy needed for maintenance, walking, foraging, etc. (eqn 4). Equation (13a) includes nitrogen lost through abrasion of tissues in the gastrointestinal tract and appears to be a constant of 5 g N per kg dry-matter intake. For growth, the nitrogen retention has been estimated to vary between 2.4% and 3.5% of the body mass gain (see for further details on pregnancy and lactation Moen 1973, p. 343 *et seq.*; Robbins 1993, p. 177 *et seq.*).

Eqn 13a tallies well with the estimate for the digestible protein requirement at maintenance, DP [g d^{-1}], for ruminants (Lloyds et al. 1978, p. 425 *et seq.*) as $DP = 3.150 \times W^{0.75}$, taking into account the conversion factor of 6.25 with which N has to be multiplied to calculate protein. On the basis of this we calculate the amount of protein to achieve protein balance DP_b per multiple of E_{BMR} as

$$DP_b = \gamma \times 2.03 \times W^{0.75}; \tag{14}$$

for γ see equation (13b).

For calcium, not much is known about the daily quantities needed. Robbins (1993, p. 37) reports a loss of 22 to 28 mg Ca kg^{-1} d^{-1}, and maintenance of a Ca balance at an intake of 73 mg Ca kg^{-1} d^{-1} (with a net retention of 30-39%). The ARC (1980, p. 186), however, is sceptical about the use of net retention. A regression on advisory dietary requirements ARC (1980, Tables 5.3 and 5.6, columns with no growth) for cattle and sheep yields the amount of calcium needed to achieve calcium balance, Ca_b [g d^{-1}], as

$$Ca_b = 0.024 \times W^{1.00} \tag{15a}$$

$$Ca_b = 0.048 \times W^{1.00} \text{ (while pregnant)} \tag{15b}$$

$$Ca_b = 0.096 \times W^{1.00} \text{ (when lactating)} \tag{15c}$$

Dietary calcium to phosphorus ratios between 1:1 and 2:1 are best for proper absorption (Robbins 1993, p. 38; ARC 1980, p. 201). Daily requirements for phosphorus to achieve phosphorous balance, P_b [g d^{-1}], can be deduced from ARC (1980, Tables 5.14 and 5.17, columns with no growth). A regression on the sheep and cattle data yields

$$P_b = 0.020 \times W^{1.00} \tag{16a}$$

$$P_b = 0.040 \times W^{1.00} \text{ (while pregnant)} \tag{16b}$$

$$P_b = 0.080 \times W^{1.00} \text{ (when lactating)} \tag{16c}$$

The minimum intake necessary to balance sodium, Na_b, is 9 mg Na kg^{-1} d^{-1} (Robbins 1993, p. 44). This is approximately the same prediction as based on ARC (1980, Table 5.36)

$$Na_b = 0.009 \times W^{1.00} \tag{17a}$$

$$Na_b = 0.018 \times W^{1.00} \text{ (while pregnant)} \tag{17b}$$

$$Na_b = 0.036 \times W^{1.00} \text{ (when lactating)} \tag{17c}$$

For pregnant animals eqns (15-17) the requirements were doubled, and for lactating females they were quadrupled (see ARC 1980).

Foraging time and intake requirements

Finally, because nutrient turn over and energy expenditure is proportionally higher in smaller ruminants than in larger ones, the throughput rate of the food through the gut is lower while the stomach is larger in large ruminants as compared to small ones. This implies that large ruminants can acquire a blended diet more easily than smaller ones. Different relations are important to model food intake and food acquisition for differently sized herbivores. Foraging time, T_f, as proportion of a day and time for foraging and ruminating, T_{f+h} [both have no units] are body-mass-dependent (Hudson 1985)

$$T_f = 0.24 \times W^{0.08} \text{ (for ungulates)} \tag{18a}$$

$$T_{f+h} = 0.52 \times W^{0.09} \text{ (for ruminants)} \tag{18b}$$

Regarding daily food intake, I [kg d^{-1}], there has been some controversy in the literature, and sometimes it is taken to scale with body mass, e.g., for African herbivores intake has been reported to be $0.058 \times W^{0.80}$ (Hudson 1985) but it can better be taken as a constant proportion of body mass, so scaling with $W^{1.00}$ (Arnold 1985; Van Soest 1982; Prins 1996, p. 264) as

$$I = 0.025 \times W^{1.00} \text{ (for ruminants)} \quad (19)$$

Because in ruminants the rumenoreticular volume [litre] relates to metabolic body mass, and not to body mass, and has been found to be equal to $-3.49 + 0.77 \times W^{0.75}$ (Bunnell and Gillingham 1985; Demment and Van Soest 1985), fermentation in the stomach of small ruminants has to be higher than in large ones. Indeed, the rumenoreticular volume of, for example, a 45-kg-small impala (*Aepyceros melampus*) in relation to its metabolic body mass is 0.57 litres per kg of metabolic mass (9.9 litre with a metabolic mass of 17.37 kg), while in a 620 kg large buffalo (*Syncerus caffer*) this is 0.74 litre per kg of metabolic mass (92 litre with a metabolic mass of 124.2 kg). Therefore, small ruminants thus need a higher-quality food to enable a higher rate of fermentation.

It is fundamentally easy to imagine why the spatial array of patches in a landscape is of importance for the survival of herbivores

For ruminants, the daily intake rates are often constrained by the rate of digestion and passage through the rumen (Voeten and Prins 1999). The digestibility rate of food is correlated with the cell wall content, measured in the vegetation type i as the percentage neutral detergent fibre NDF_i [g kg^{-1}]. Reid et al. (1988) calculated from a feeding trial of cattle on a C_4-grass diet that the maximum daily $NDFM$ intake [kg d^{-1}] can be calculated as

$$NDFM = 66.7 \times 10^{-3} \times W^{0.75} \quad (20)$$

All these considerations enable us to predict that small ruminants are more constrained by patch differences and differences in grain size of the vegetation than larger ones. This can easily be imagined when one patch has, for instance, a deficit in phosphorus and an excess of nitrogen, while another patch has the opposite: assume that both patches would exactly compensate each other. If the distance between these two patches is 10 km, then it would take a buffalo (620 kg) approximately two hours to reach the second patch after visiting the first. Because the average throughput rate of buffalo food is approximately 36 hours, the forage from the two patches can be well blended. However, for a dikdik (*Madoqua spec.*) (5 kg) these two patches represent two ‘different worlds’ because the average home range of a dikdik is measured in hectares: the two patches are simply too widely separated for the dikdik to achieve a

The scaling of stomach size and the necessary associated rate of fermentation of food in the rumen explains why small ruminants need better-quality food

blended diet, and it ends with food that does not meet all its requirements. Inter-patch distance is thus of considerable interest to understand the possibilities the landscape offers for a herbivore to survive. In another context, Crawley (1983, p. 150) draws attention to the interaction between an animal's dispersive abilities and the plant pattern: if plants occur in a density with inter-plant distances shorter than the ranging distance of caterpillars, there will be a high herbivore survival. However, if this distance is too long, then plants will generally survive while the herbivore goes extinct.

MODELLING DIET AND PATCH SELECTION AS FUNCTION OF AMOUNT, QUALITY AND SPATIAL DISTRIBUTION OF FOOD

To simply model diet and patch selection in landscapes with different grain size, we consider a model with two patches that can contain food that differs in amount and quality. Also, the distance between the two patches is variable. To do this, we use an approach analogous to the linear-programming model introduced by Belovsky (1978). Central in this approach is that the selection of its diet (Belovsky 1978), the selection of patches (Ludwig et al. 2001) and the migratory movements (Voeten and Prins 1999) are all based on the constraints for herbivores to meet their requirements. Often such linear-programming models are used to determine the best diet possible where the animal meets all its requirements. Instead of looking at the maximum, we will use this approach to study whether herbivores can meet all their nutritional and energy requirements by selecting food from one of the two patches or a combination of both. Extension of the two-patch model is possible, but will increase the computational efforts. Our model is largely based on Voeten (1999) and Ludwig (2001).

We formulated minimum requirements for nutrient, protein and energy intake and a maximum value for fibre intake based on the requirements of herbivores and size constraints as formulated before. The energy intake necessary for maintenance, E_{BMR} (eqn 1), walking, E_w (eqn 2), foraging, E_f (eqn 7), and ruminating, E_h (eqn 6) [all kJ d^{-1}] is formulated as the constraint

$$E_{BW} + \left(E_f + E_h\right) \times T_{f+h} \leq \sum_{i=1}^{k} G_E \times \%DOM_i \times c_{GD} \times X_i \tag{21}$$

where E_{BW} is given by eqn (4) (assuming a flat area where $H = 0$), T_{f+h} is the fraction of the time needed for foraging and ruminating (eqn 18b), G_E is the energy content of tropical grasses per mass unit dry weight [kJ kg^{-1}], $\%DOM$ the digestibility of organic matter of the vegetation type i, c_{GD} is the fraction of the digestible energy of grasses that is converted into metabolic energy [-], k is the number of vegetation types (in our model k = 2). The left-hand side of eqn (21) is the total daily energy expenditure of the animal, E_{tot}. The parameter X_i is the decision variable representing the intake of the animal in patch i. The sum of the intake per patch should be less than or equal to the daily intake (eqn 19):

$$\sum_{i=1}^{k} X_i \leq I \tag{22}$$

In Box 7.2, the model for wildebeest (*Connochaetes taurinus*) is given as an example.

The third constraint is the digestible-protein requirement at maintenance DP_b (eqn 14). This can be formulated as

$$DP_b \leq \sum_{i=1}^{k} aDP_i \times X_i \tag{23}$$

The available digestible protein aDP [mg g^{-1}] for tropical grasses of type i can be calculated from crude protein CP_i [mg g^{-1}] (Prins 1987a) as

$$aDP_i = 0.91 \times CP_i - 32.2 \tag{24}$$

The constraints for nutrient requirements of herbivores concern the amount required to achieve a balance for nitrogen N_b, phosphorus P_b, sodium Na_b and calcium Ca_b, see eqns 15-17. The constraint equations for maintenance level are (Murray 1995)

$$N_b \leq \sum_{i=1}^{k} aN_i \times X_i \tag{25a}$$

$$P_b \leq \sum_{i=1}^{k} aP_i \times X_i \tag{25b}$$

$$Na_b \leq \sum_{i=1}^{k} aNa_i \times X_i \tag{25c}$$

$$Ca_b \leq \sum_{i=1}^{k} aCa_i \times X_i \tag{25d}$$

where aN_i, aP_i, aNa_i and aCa_i [all mg kg^{-1}] are the concentrations of these nutrients measured in vegetation type i. Since often the Na concentration in the vegetation is not sufficient to meet the requirements of the animal (for example, see Voeten and Prins 1999; Ludwig 2001), we assume that the animals find alternative sources for Na and ignore the Na requirement in our model.

Based on the rate of digestion and passage through the rumen (eqn 20), the maximum daily intake rate for herbivores is

$$NDFM \geq \sum_{i=1}^{k} NDF_i \times X_i \tag{26}$$

For the values of the vegetation parameters in the two patches, we used data obtained by Ludwig et al. (2001). They measured the vegetation parameters for several grass species under tree canopy, just outside tree canopy and in open grassland. The parameter values are enumerated in Table 7.3.

Finally, to grasp the issue of getting the right time horizon into our model, we assume that a ruminant needs to keep the food mass in its rumen well within the constraints set by the requirements of the microbes. We thus assume that the herbivore allows a deviation of the allowable solution of the linear modelling of maximally 10%. We thus reasoned that its rumen contents can be emptied for maximally 10%, to be filled up again by 10% with food from a different patch, to maintain a well-blended food mass in its rumen from which it derives its needed energy and macro-nutrients. Because the retention time of particles can well be described with a Michaelis-Menten function, and because the retention time, T_R [hours], of particles is known (that is, the time after which 95% of the particles have left the rumen) through Demment and Van Soest (1985) as

$$T_{R,0.95} = 7.67 \times Dig \times W^{0.346} \text{ (95\% disappeared)} \tag{27a}$$

thus

$$T_{R,0.1} = 0.045 \times Dig \times W^{0.346} \text{ (10\% disappeared)} \tag{27b}$$

The latter provides the maximum time an animal can walk between patches of food. In this equation, *Dig* stands for digestibility of the food [ranging between 0 and 1, no units]. In the present analysis we have taken a disappearance of 10% only, but we can adopt other thresholds too, if necessary. We now calculate the maximum allowable inter-patch distance taking the time it maximally walks to be equal to $T_{R,0.1}$.

Table 7.3. *Digestibility of organic matter (%DOM), neutral detergent fibre (NDF) and nutrient concentrations of grass species growing under and just outside canopies of large Acacia tortilis trees and in open grassland patches (Ludwig et al. 2001). Variance of data was analysed with a GLM with grass species and location (open grassland and under and outside canopies) as fixed factors. Species with the same letter are not significantly different (Tukey HSD, P > 0.05) (* P < 0.05, *** P > 0.001)*

DATA Location	Species	*%DOM* (%)	*NDF* (%)	*CP* $mg\ g^{-1}$	*P* $Mg\ g^{-1}$	*K* $mg\ g^{-1}$	*Ca* $mg\ g^{-1}$	*Mg* $mg\ g^{-1}$	*Na* $mg\ g^{-1}$
Under	*Panicum maximum*	65.65^{ab}	70.60^{ab}	141.1^{b}	1.77^{ab}	40.62^{a}	4.12^{c}	3.25^{b}	0.10^{a}
tree	*Cynodon spp.*	70.33^{a}	61.45^{a}	165.1^{a}	1.80^{ab}	38.41^{a}	6.91^{a}	2.44^{bcd}	0.12^{a}
canopy	*Cenchrus ciliaris*	59.78^{ab}	70.12^{ab}	136.8^{b}	1.74^{ab}	44.83^{a}	4.34^{c}	1.95^{def}	0.21^{a}
Outside	*Cynodon spp.*	65.07^{ab}	66.49^{ab}	126.8^{bc}	2.38^{bc}	36.07^{a}	6.70^{ab}	2.58^{bcd}	0.17^{a}
tree	*Cenchrus ciliaris*	57.96^{ab}	71.45^{ab}	105.5^{cd}	2.43^{bc}	41.89^{a}	3.30	1.87^{def}	0.14^{a}
canopy	*Digitaria macroblephera*	62.64^{ab}	72.91^{ab}	74.5^{e}	2.85^{cd}	38.35^{a}	3.75	2.26^{cde}	0.20^{a}
	Chloris virgata	66.80^{ab}	70.40^{ab}	104.6^{cd}	2.59^{cd}	37.56^{a}	4.91	3.03^{bc}	0.87^{b}
	Urochloa mosambicencis	69.13^{a}	64.77^{ab}	104.0^{cd}	3.59^{e}	44.28^{a}	5.28^{ab}	4.00^{a}	0.13^{a}
	Heteropogon contortis	56.97^{ab}	70.15^{ab}	81.5^{e}	1.46^{a}	17.17^{b}	3.30^{c}	1.64^{def}	0.14^{a}
Open	*Heteropogon contortis*	51.68^{b}	74.00^{b}	67.6^{e}	1.65^{ab}	15.08^{b}	3.31^{c}	1.22^{f}	0.13^{a}
grassland	*Sehima nervosum*	54.66^{ab}	74.65^{b}	60.8^{e}	1.46^{a}	12.02^{b}	4.67^{c}	1.47^{ef}	0.14^{a}
STATISTICS									
Species	F	2.25^{*}	2.42^{*}	14.09^{***}	5.45^{**}	8.79^{***}	7.75^{***}	13.51^{***}	18.43^{***}
Location	F	1.19	1.53	6.91^{***}	9.07^{***}	0.57	0.72	0.91	0.02
Species × Location	F	0.26	0.60	0.29	0.09	0.01	0.66	0.27	1.15

***Box 7.2.** Illustration of the diet and patch selection model for the wildebeest (Connochaetes taurinus) (Voeten 1999; Ludwig 2001; see also Murray and Brown 1993, Murray 1995)*

The energy intake necessary for maintenance, E_{BMR}, and walking, E_w, and foraging and ruminating, E_f and E_h, is formulated as the constraint (eqn 21). We use the equations 1, 2, 4, 6, 7, and 18b to obtain the following constraint ($H = 0$):

$$E_{BMR} + E_w \times W \times D + (E_f + E_h) \times T_{f+h} \leq \sum_{i=1}^{k} G_E \times \%DOM_i \times c_{GD} \times X_i$$

that becomes

$$293 \times W^{0.75} + 10.75 \times W^{-0.316} \times W \times D + (0.54 + 0.24) \times 293 \times W^{0.75} \times 0.52 \times W^{0.09} \leq \sum_{i=1}^{k} 19 \times 10^3 \times \%DOM_i \times 0.82 \times X_i$$

when G_E is 19 × 10^3 kJ kg^{-1} (Crampton and Harris 1969) and c_{GD} is 0.82 (Van Soest 1994). This equation can be re-arranged as

$$293 \times W^{0.75} + 10.75 \times W^{0.684} \times D + 118.8 \times W^{0.84} \leq \sum_{i=1}^{k} 15.6 \times 10^3 \times \%DOM_i \times X_i$$

For wildebeest with the average weight of 143 kg, this means

$$19.8 \times 10^3 + 320.3 \times D \leq \sum_{i=1}^{k} 15.6 \times 10^3 \times \%DOM_i \times X_i \, .$$

The second constraint is that the maximum distance D_{max} (eqns 23, 27b, 29) is limited by the time period needed for foraging as

$$D_{\max} \leq V \times (1 - T_m) \times 24$$

that becomes

$$D_{\max} \leq 1.188 \times W^{0.21} \times \min(1 - 0.52 \times W^{0.09}, 0.045 \times Dig \times W^{0.346}) \times 24 \, .$$

For wildebeest, the maximum distance [km] that can be covered with walking during 1 day is

$$D_{\max} \leq 16 \text{ (for } Dig = 0.8\text{)}.$$

Box 7.2 (cont.)

Box 7.2 (cont.)

The third constraint for the digestible-protein *DP* requirement for maintenance is (eqns 14, 23 and 24)

$$\frac{E_{tot}}{2\times E_{BMR}}\times 2.03\times W^{0.75}\leq\sum_{i=1}^{k}(0.91\times CP_i-32.2)\times X_i$$

where E_{tot} is the left-hand side of the first equation. For wildebeest, this constraint becomes

$$60.4+1.11\times D\leq\sum_{i=1}^{k}(0.91\times CP_i-32.2)\times X_i$$

The constraint equations for the requirements of nitrogen, phosphorus, sodium and calcium (eqns 15, 16, 17 and 25) are:

$$\gamma\times 0.325\times W^{0.75}\leq\sum_{i=1}^{k}aN_i\times X_i \quad \text{and} \quad 0.02\times W\leq\sum_{i=1}^{k}aP_i\times X_i$$

$$0.009\times W\leq\sum_{i=1}^{k}aNa_i\times X_i \quad \text{and} \quad 0.024\times W\leq\sum_{i=1}^{k}aCa_i\times X_i$$

For wildebeest, these equations become:

$$9.7+0.18\times D\leq\sum_{i=1}^{k}aN_i\times X_i \quad \text{and} \quad 2.86\leq\sum_{i=1}^{k}aP_i\times X_i$$

$$1.29\leq\sum_{i=1}^{k}aNa_i\times X_i \quad \text{and} \quad 3.43\leq\sum_{i=1}^{k}aCa_i\times X_i$$

The digestion constraint is (eqns 20, 26):

$$66.7\times 10^{-3}\times W^{0.75}\geq\sum_{i=1}^{k}NDF_i\times 0.025\times W$$

Box 7.2 (cont.)

Box 7.2 (cont.)

For wildebeest, this equation becomes:

$$2.76 \geq \sum_{i=1}^{k} NDF_i \times 3.6$$

Combining these constraints leads to the feasible region as depicted in Figure 7A.

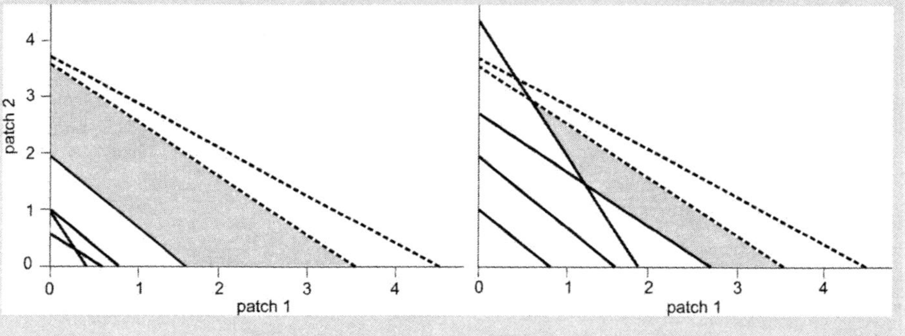

Figure 7A. *Graphical representation of the energy and nutrient constraints, predicting whether wildebeest can meet their daily requirements for energy, nutrients and protein by selecting forage from under tree canopies or in open grassland patches. Each line indicates the minimum food intake required to meet nutrient, energy or protein requirements (solid lines). The maximum daily intake and the intake to meet the maximum digestibility due to the fibre content are given as maximum constraints (dashed lines). The shaded part (the 'feasible region'; see Box 7.3) indicates all possible combinations of food sources that meet all nutrient, energy and protein requirements (left-hand figure with distance D = 0 km and figure a the right with distance D = 180 km)*

The maximum distance travelled per day, D_{max}, is the product of the speed of locomotion, V [km hr^{-1}], and the maximum number of hours an animal can move when not foraging and ruminating is then

$$D_{\max} = V \times T_m \times 24 \tag{28}$$

where T_m is the fraction of the day that the animal is moving. We set $T_m = 1 - T_{f+h}$. If T_m is larger than $T_{R,0.1}$, the maximum fraction of the day that the animal is moving becomes $T_{R,0.1}$. Thus, T_m = min(1-T_{f+h}, $T_{R,0.1}$). The speed V is important for understanding how animals can amalgamate their intake from different patches of vegetation. Hudson (1985) provides the formula for speed for mammals in general as $V = 5.5 \times W^{0.24}$, but this yields too high speeds for understanding the problem of normally walking animals. Peters (1983, p. 89) gives the optimal rate of moving, which formula we adopt, as:

$$V = 1.188 \times W^{0.21} \tag{29}$$

Model analysis and results

We analysed the diet and patch selection model in several ways. First, we calculated the extent of the feasible region. The feasible region is the set of combinations of food from different patches limited by the constraints as set by the animal's requirements (see Box 7.3). The extent of the feasible region is an indication of the tolerance of the animal to collect its food under the conditions of the vegetation amount and quality (such as nutrient concentration and fibre percentage) and the requirements of the animal. A larger extent implies that the combination of patches provides a larger tolerance. Second, we determined the contribution of each patch in the absence of the other patch. Therefore, we calculated for example the possible intake of vegetation in patch 1 [in kg] that can meet the requirements of the animal in the absence of patch 2, that is, the distance between the lowest maximum requirement and highest minimum requirement on the axis of patch 1. This indicates the relevance of each patch in the diet composition of the animal. In Box 7.3, the extent of the feasible region and the contribution of each patch in the absence of the other patch are illustrated.

First, we determined the effect of body size on the extent of the feasible region (see Figure 7.1). Here, we assumed that the 2 patches are close together. It appears that the extent of the feasible region increases with body size. For lactating animals, the extent of the feasible region is smaller, implying that lactating animals are more constrained by the possible combination of patches in the landscape. For lactating females it is thus much less easy to find an array of patches that serves their needs than for non-lactating individuals, and this differential increases with body mass. However, this differential handicap of demands for lactation is not fully set off by the benefits of increased body mass.

In Figure 7.2, the effect of body size on the contribution of one patch in the absence of the other patch is given. For small animals, patch y (in Box 7.3 named patch 2) does not meet all the requirements (in Box 7.3 this is C2), so they have to concentrate on patch x. This means that with decreasing body mass, it becomes rarer and rarer that a ruminant can make use of one patch only. However, simultaneously, it becomes rarer and rarer that an increasingly smaller ruminant can compensate for this by making use of another patch *and* satisfice the requirements. Thus the chances that an array of patches satisfices the requirements of a ruminant decrease with decreasing body mass. Hence it follows that ruminants become rarer with decreasing body mass. Figure 7.3 gives the extent of the feasible region as function of the distance between

Body mass is a critical parameter to understand the way ruminants can make use of a landscape

Box 7.3. *The feasible region in a linear-programming problem*

Two of the most basic concepts associated with a linear-programming problem are feasible region and optimal solution. For defining these concepts, we use the term point to mean a specification of the value for each decision variable (Winston 1994). The feasible region for a linear-programming problem is the set of all points satisfying all the constraints. For a maximisation problem, an optimal solution to a linear-programming problem is a point in the feasible region with the largest objective function value. Similarly, for a minimisation problem, an optimal solution is a point in the feasible region with the smallest objective function value. The extent of the feasible region is illustrated in figure 7B.

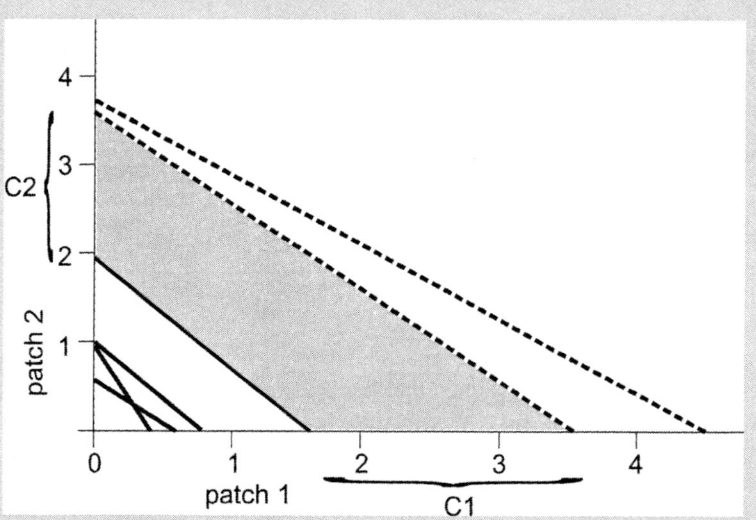

Figure 7B. *The extent of the feasible region in a linear-programming problem is the shaded region bounded by the minimum (dashed lines) and the maximum (solid lines) requirements of the animal. The contributions of each patch in the absence of the other patch are C1 for patch 1 (in the absence of patch 2) and C2 for patch 2 (in the absence of patch 1)*

the patches for wildebeest solely based on the energy and nutrient requirements. Here, the time after which 10% of the particles has left the rumen, $T_{R,0.1}$ (eqn 27b), is not included. The point where the line meets the x-axis is the maximum distance that an animal can move between patches based on diet assembling over a range of body mass. When the maximum distance is also determined by $T_{R,0.1}$, then the maximum distance over which wildebeest (143 kg with Dig = 0.8) can move between patches is 16 km.

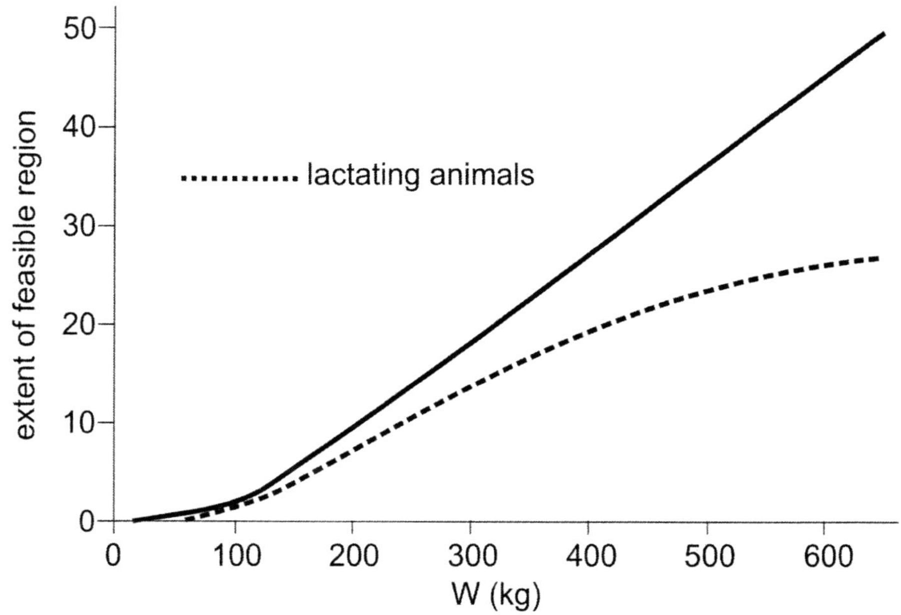

Figure 7.1. *Extent of the feasible region as function of body mass (dashed line is for lactating animals). Parameter values based on the vegetation measurements of patch 1 under trees and patch 2 in open grassland (see Table 7.3), the distance between the patches D =0 km*

Figure 7.4 shows the effect of the inter-patch distance on the extent of the feasible region with and without considering the time after which 10% of the particles has left the rumen, $T_{R,0.1}$ (eqn 27b). It appears that energy is not the limiting factor for animals because the maximum allowable inter-patch distance in Figure 7.4 is unrealistically high (top line in the figure); this implies that plants can provide more than sufficient energy. When including the throughput rate ($T_{R,0.1}$) in the constraints, the maximum allowable inter-patch distance becomes much smaller (bottom lines in Figure 7.4). Recall that rumen micro-organisms are not primarily limited by energy but by macro-nutrients (Van Soest 1982) for which we have included the parameter $T_{R,0.1}$. This parameter represents the necessity for the rumen micro-organism to have a well balanced food mass in the rumen. When including the constraints set by the lactating female's requirements for Ca and P, no feasible region could be found. This indicates that plants cannot satisfice these nutrient requirements. Lactating females thus have to find other sources for these nutrients during this time of their life cycle, or they have to mobilise them from their skeleton. This confirms that plant biomass, generally speaking, is only marginal from the animals' perspective.

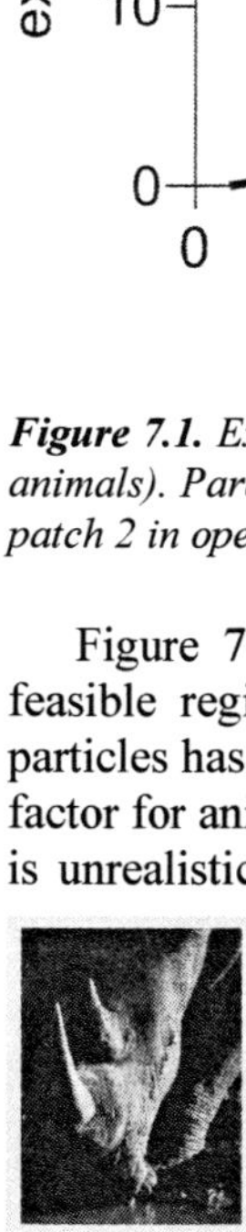

Often in grass the amounts of calcium and phosphorus are too low to satisfy the needs of females when producing milk

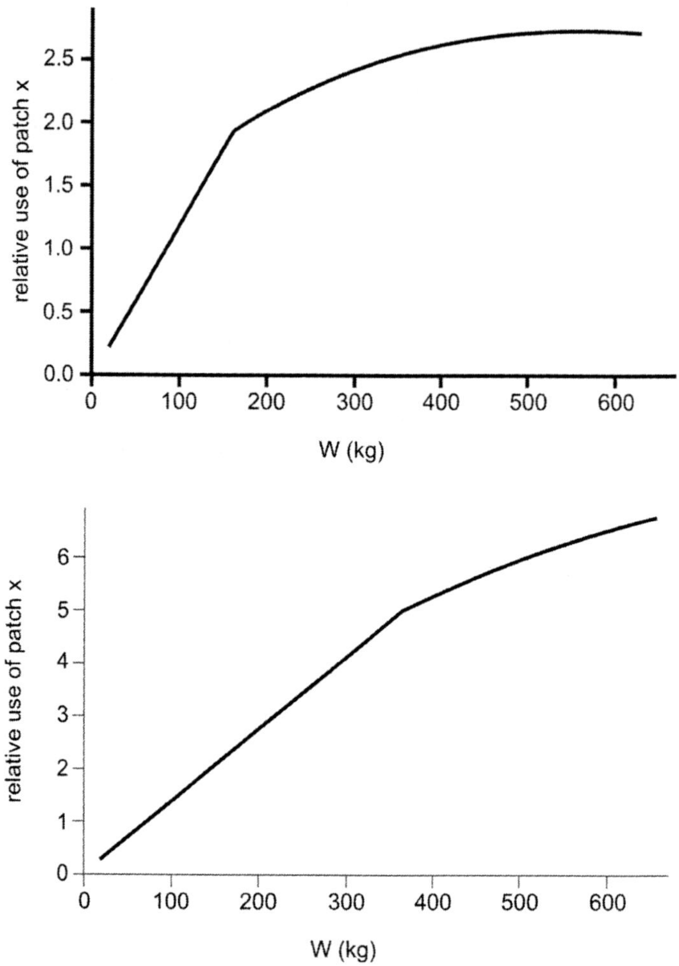

Figure 7.2. *The contribution of one patch in the absence of the other patch as function of body mass. Parameter values based on the vegetation measurements of patch x under trees and patch y in open grassland (see Table3), the distance between the patches D = 0 km*

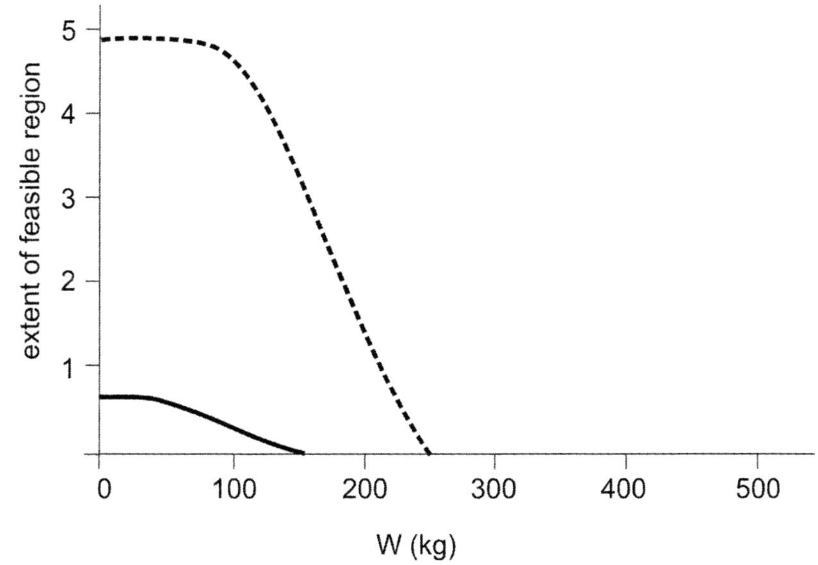

***Figure** 7.3. Extent of the feasible region as function of the inter-patch distance. Parameter values based on the vegetation measurements of patch 1 under trees and patch 2 in open grassland (see Table3), the body mass is 50 kg (solid line) and 143 kg (dashed line). The maximum distance that a wildebeest of 143 kg can walk is 16 km when the time after which 10% of the particles have left the rumen,* $T_{R,0.1}$ *(eqn 27b), is included (Dig = 0.8)*

DISCUSSION

In this chapter, we present a modelling approach to explore the conditions for the presence or absence of foraging animals based on the energy and nutrient requirements that should be met. We did not only look at vegetation quantity and vegetation quality, but we took a third property into account, namely the grain size of the landscape. Grain size describes the spatial array of patches of vegetation in a landscape (e.g., Murwira 2003, Skidmore and Ferwerda, Chapter 4). In our approach, the spatial distribution of patches with acceptable food and areas without, or with vegetation that is non-food from the herbivores' perspective, is of critical importance for understanding how herbivores can use the vegetation. This spatial distribution determines usage just as properties of the animals, such as body mass and digestion system. We explored this for herbivores since a lot of information on requirements is available, but it can also be applied to other forager types that make use of spatially dispersed food.

Spatial distribution of food and 'non-food' (just as, for instance, an animal's physiology) co-determines usage

We stress that the requirements of an animal often force it to forage from different patches with resources when the availability and quality of the resource is spatially heterogeneously distributed. Since we believe that food in the real terrestrial world is always heterogeneously distributed, movement and searching, and thus the selection of patches to obtain a well balanced diet, will always be necessary for a herbivore. A herbivore thus has to assemble a diet from different patches, and from different plant species, and only the assembled diet can meet all of its requirements. The herbivore has to move between different patches that are imperfect from the herbivore's point of view, but that in combination can satisfy the herbivore's requirement. We show that the requirements of the animal set limits to the amount and spatial distribution of the diet components. The model makes predictions when herbivore species can meet their requirements or not, and thus when it can be present or not.

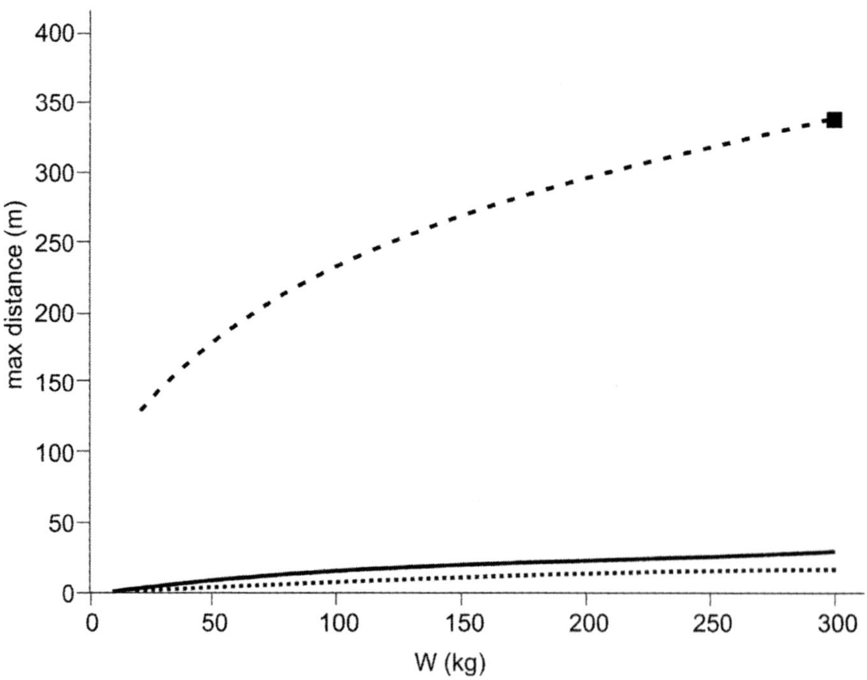

Figure 7.4. *The maximum distance based on diet assembling over a range of body mass: when the time after which 10% of the particles have left the rumen,* $T_{R,0.1}$ *(eqn 27b), is not included (dashed line); when* $T_{R,0.1}$ *is included for Dig = 0.8 (solid line); and Dig = 0.6 (dotted line)*

Since nutritional requirements and energy balance are allometrically scaled, we can express the conditions for presence or absence as a function of body size (Figures 7.1 and 7.4). Figure 7.1 shows that large ruminants have a much wider tolerance for assembling a diet from different patches than smaller ruminants. Alternatively stated, smaller ruminants have much less freedom in assembling a diet from patches with different quality than larger ones. Based on this finding, we predict that, firstly, smaller ruminants appear to be much more specialised than larger ones. We secondly predict that it is easier for large ruminants to assemble a satisficing diet from different patches than it is for smaller ones (see also Figure 7.2), and hence, thirdly, we predict that large herbivores are more common than small ones. The critical assumption behind this last prediction is that at a meta-scale ('landscape', 'ecosystem') the juxtaposition of patches of vegetation with different quality is more common than the concurrence of patches with the same quality that simultaneously are of sufficiently high quality to meet the requirements of the small ruminants. Based on the availability, quality and spatial distribution of different resource types, we can thus predict the number of species that can be present in a given situation and relate the number of species to minimum body size.

Our model predicts that large herbivores should be more common than small ones

If it is true that the steppe-tundra during the Pleistocene was more fine-grained than the succeeding steppes, tundras or boreal forests, as suggested by Guthrie (1990), due to increased temperature, and perhaps increased precipitation too, which led to a segregation of plant species that occurred together before, then we would predict the opposite from what has been observed. This steppe-tundra would namely have been a better place for smaller ruminants than the succeeding coarse-grained vegetation types. Indeed, we would then expect that large herbivores, and especially the megafauna, have survived better than the smaller ruminants. This is in contrast to what generally has been assumed.

***Box 7.4.** Testable hypotheses for future research*

Hypothesis 1. Given equal conditions of disturbance and predation and equal body mass, individuals of ungulate species that are adapted to dry conditions and which are not dependent on surface water, can assemble an optimal diet on a shorter time horizon than individuals of species that need surface water.

Hypothesis 2. Given equal conditions of disturbance and predation and equal body mass, individuals of ungulate species that are adapted to dry conditions and which are not dependent on surface water, need smaller body reserves for nutrients and/or energy than individuals of species that need surface water.

Hypothesis 3. Given equal conditions of disturbance and predation, individuals of ungulate species that are not adapted to dry conditions and which are dependent on surface water during the dry season but not during the wet season, can assemble an optimal diet on a shorter time horizon during the wet season than during the dry period.

Hypothesis 4. Given equal conditions of disturbance and predation and need for surface water, individuals of large ungulate species can assemble an optimal diet much more easily than individuals of small species.

From our analysis, we can draw some other conclusions too. First, increasing inter-patch distance, measured in absolute terms, increases the difficulty for a herbivore to assemble a diet sufficiently able to cover its requirements (Figure 7.3). Larger herbivores, however, are less constrained than smaller ones (Figure 7.3). Increasing inter-patch distance puts a heavier constraint on acquiring sufficient energy and nitrogen from the food than on obtaining adequate amounts of other macro-nutrients. From this we can deduce that in coarse-grained landscape especially the N:P ratio is of importance: in such a coarse-grained landscape nitrogen is more limiting to the herbivore (and the N:P ratio should thus be high) than in a fine-grained landscape. Because smaller herbivores are more nitrogen-limited than larger ones and not *vice versa*, our analysis puts even more emphasis on the differential survival for large ruminants as compared to small ones. Our models would have predicted extinction of the *smaller* fauna at the end of the Pleistocene. Much attention has been devoted to the extinction of the megafauna. One should not forget, though, that numerous ‘ordinary’ (body mass mid-range) herbivores went extinct too. Cases in points are camelids, horses and pronghorn species in North America (e.g., Anderson 1984). Before we can draw a final conclusion, though, we have to explore how our conclusions hold for non-ruminant herbivores, and it may be that our model results do not hold for mammoths or woolly rhinoceroses. This we have not done yet. There is no convincing evidence for overkill of large herbivore populations by early Man (see Anderson 1984; Graham and Lundelius 1984; Murray 2002). Late Pleistocene extinction of large mammals is certainly real (Graham and Lundelius 1984; Table 3.6 in Prins 1998). Perhaps the unusually high mortality of the end of the Pleistocene is best regarded as a natural consequence of high faunal turn-over caused by major oscillations during the Pleistocene and the Holocene in climate and environmental heterogeneity (Gingerich 1984), together with our explanation about the increasing difficulty smaller ruminants would have faced to assemble a diet from an increasingly coarser-grained environment. This would have turned out to be even more difficult for lactating females (Figure 7.1). Added to this then was the increased leaching of the soil in the wetter parts of the temperate zone during the Holocene, which further exacerbated the difficulties of meeting a balanced and sufficient diet in a coarse-grained landscape.

In coarse-grained landscapes nitrogen is more limiting than in fine-grained landscapes; this affects smaller ruminants more than larger ones

We can thus deduce that is more difficult for a small herbivore to live in a coarse-grained environment than in a fine-grained one. This is enforced by the finding that for a lactating female this is even more difficult (Figure 7.1). In our equations, we have shown that the nutritional requirements of pregnant and, especially, lactating females are much higher than those of animals that are not reproducing. From our model it is clear that this has important implications for the fit of animals of different body masses in the landscape. Our fourth conclusion is then that, given the same body mass, non-reproducing individuals can cope with longer inter-patch distances than lactating females can cope with (Figure 7.1). We found that for lactating females it is much less easy to find an array of patches that serves their needs than for non-lactating individuals. We also found that this differential increases with body mass, but that this differential handicap of demands for lactation is not fully set off by the benefits of increased body mass. We can thus conclude that this must have led to a strong selection pressure favouring larger body masses in ruminants in the course of evolution. A larger size simply means that it is easier to satisfy the nutritional requirements in a landscape. Again, we have not investigated yet whether this holds also for non-ruminant herbivores, but we suspect this to be the case because generally speaking evolution shows an increase in body mass of, for example, equids and other hindgut fermenters.

Lactating females have difficulties finding landscape configurations that match their needs; natural selection favours high body mass

With our set of formulas and given the assumption underlying equation 27b we can construct a graph (Figure 7.4) that shows that the maximum interpatch distance beyond which a herbivore cannot assemble a balanced diet anymore is dependent on body mass. This has important implications if the dominant scale of the landscape (*sensu* Murwira and Skidmore 2004) changes. Indeed, Murwira (2003) have shown that due to man this dominant scale has changed in, for instance, the communal lands of Zimbabwe. The same has happened in many other landscapes where the impact of humans changed the scale of the landscape from fine-grained to coarse-grained; cases in point are the countries of the European Union, where agricultural policy caused these changes. We thus predict that small herbivores face increasing chances of extinction when the dominant scale of the landscape increases. In fragmented landscapes, animals may thus be constrained by the possibilities to assemble their diet from different places. This provides an alternative explanation for the effects of the fragmentation of habitat on survival, which is different from the explanation that dispersal is constrained as is often suggested, especially in the context of metapopulation theory (Hanski 1999, 2001). Concurrently, large herbivores are more extinction-prone in such a landscape because large herbivores are more easily exterminated by man than smaller ones. We suggest further research based on the hypotheses derived in this chapter; see Box 7.4.

NOTES

[1] In published sources many formulas are reported using the calorie as unit. Throughout this chapter we have recalculated these by converting to Joule, by using one cal as 4.184 J.

CHAPTER 7B

COMMENTS ON "ASSEMBLING A DIET FROM DIFFERENT PLACES"

IAIN J. GORDON

CSIRO, Davies Laboratory, PMB PO Aitkenvale, Qld 4814, Australia
E-mail: iain.gordon@csiro.au

Prins and Van Langevelde (Chapter 7) use a linear-programming model to assess the extent to which ruminants of different size are able to satisfy their nutrient, energy and protein requirements from a landscape composed of two 'food' patches that differ in their relative densities of these important nutritional variables. Amongst their findings they conclude that, overall, small species are less able to balance their nutritional requirements when patches are widely dispersed than are large species, and are, therefore, more likely to be found in fine-grained (i.e., more closely dispersed food patches) than in course-grained ecosystems. Whilst this is an interesting and testable hypothesis, I will argue that it is the 'foodscape', not the landscape, that foraging animals respond to. My view is that the conclusions are an artefact of the model description rather than an actuality of the real world in which ruminants forage for a living. I posit that the dispersion of food in the landscape is a species-specific construct with the result that the foodscape of two species foraging in the same landscape will differ because of their differing views of food and their differing ability to select that food from the array of non-food on offer (see also Underwood 1983).

Belovsky (1978 and others) was the first person to use the linear-programming methodology to describe the diet selection of herbivores; in his studies he used the constraints of foraging time and digestive capacity to determine the optimal mix of two food types in the diet. He then went on to test his predictions against actual diet composition with, often, remarkable results. Whilst there have been many arguments and debates about the application of linear programming in Belovsky's studies (e.g., Hobbs 1990; Owen-Smith 1996), it still provides a valuable framework for conceptionalising simple foraging decisions and outcomes. I applaud Prins and Van Langevelde (Chapter 7) for using linear programming to develop hypotheses concerning the effects of food patch distribution on the ability of animals of

H.H.T. Prins and F. van Langevelde (eds.), Resource Ecology: Spatial and Temporal Dynamics of Foraging, 157-158.

different size to balance their diet. In their model, Prins and Van Langevelde add a distance-travelled constraint to those originally developed by Belovsky (e.g., foraging time and digestive capacity). They then model the effects of varying the distance between patches on the 'feasible region'[1] in the linear-programming model for animals of different size. As already mentioned, they conclude that small species are only likely to be found in fine-grained landscapes.

I would argue, however, that body size will have an effect, not only in the movement costs between patches, but also on the ability of animals to select the nutritionally relevant food from the non-food matrix in the landscape (see also Jarman 1974). I (Gordon and Illius 1988) and others (Janis and Erhardt 1988) have shown that the foraging-apparatus size scales allometrically with body size in herbivores; this shows that small animals have mouths that will allow them to be more selective than will large animals (a gazelle vs. a buffalo; a roe deer vs. a moose). Field studies have also shown that small animals consume a higher-quality diet than large animals, even when feeding in the same landscape (Jarman and Sinclair 1979; Gordon and Illius 1996). This suggests that smaller species are able to make more fine-scale selection of food than are large species, which require relatively large patches of acceptable food to forage from. In effect, small species forage in a more fine-grained foodscape than do large animals. Counter to this, large animals are able to digest poorer-quality food items more efficiently than are small ones (Illius and Gordon 1991), and so will have more of the food on offer that is acceptable to them than are smaller animals. As such the model of Prins and Van Langevelde, by varying the distance between acceptable food patches, without taking body size differences in food dispersion into account, is not reflecting the reality of food distribution for small and large species. I, therefore, disagree with the conclusions drawn by Prins and Van Langevelde and argue that the grain size of a foodscape is defined by the animal itself and not by human-defined arbitrary food patches that can be seen and measured by ground survey on remote sensing. Until we can see the foodscape through the eyes of the animal we will not be able to clearly define hypotheses about the way in which the structure of the landscape affects animals of different size.

I may be wrong, Prins and Van Langevelde may be right, or we both may be wrong, but the wonder of science is to stand on the shoulders of others and scan the horizon.

NOTES

[1] The area in a linear-programming model that satisfies an animal's nutritional requirements, bounded by the constraints.

CHAPTER 8A

EFFECTS OF TEMPORAL VARIABILITY IN RESOURCES ON FORAGING BEHAVIOUR

NORMAN OWEN-SMITH

Centre for African Ecology, School of Animal, Plant and Environmental Sciences, University of the Witwatersrand, Wits 2050, South Africa
E-mail: norman@gecko.biol.wits.ac.za

Abstract. Trait plasticity in physiology, behaviour, morphology and life history enables organisms to survive and populations to persist despite temporal variability in environmental conditions and resource availability. Through non-linear responses, the effect of adverse periods outweighs that of benign conditions, following Jensen's inequality. This chapter considers how large mammalian herbivores adjust broader aspects of their foraging behaviour to cope with variability over different temporal frames: within a day, day versus night, between days, over seasonal cycles and between years. It outlines the conceptual foundation for 'adaptive resource ecology', covering changes in diet composition, daily time allocation, foraging movements, metabolic rate, digestive capacity and fat stores. The functional response relating food intake rate to food availability changes its form depending on the temporal scale. To link resource variability in time and space to population dynamics, the intake response needs to be transformed into a biomass or energy gain response over a seasonal time frame. Foraging models based on rate averaging can be misleading, while challenges in applying dynamic optimisation models need to be surmounted. Models assuming equilibrium relationships between resource supplies and population growth are inappropriate for coupling resource gains to population dynamics.
Keywords. adaptive responses; diet selection; digestion processes; functional response; time budgets; trait plasticity

INTRODUCTION

Environments change in fundamental ways affecting resource availability for consumers over daily, seasonal and multi-annual time frames. To cope with such temporal variability, animals and plants must adapt phenotypically in physiology, behaviour, morphology and life history or, in other words, show 'trait plasticity' (Abrams 1995). For populations to persist, the individual organisms constituting these populations must be able to sustain their resource gains, relative to the expenditures involved in acquiring them, despite the ups and downs, troughs and pitfalls encountered while traversing a continually shifting fitness landscape

H.H.T. Prins and F. van Langevelde (eds.), Resource Ecology: Spatial and Temporal Dynamics of Foraging, 159-181.

(Figure 8.1). Models dealing only with average, normal or equilibrium conditions fail to capture what most crucially governs ecological success or failure in growing, surviving, reproducing and ultimately in transmitting genes over evolutionary time frames. Large mammalian herbivores in particular face wide seasonal contrasts in food quality, and substantial differences between years in the timing of vegetation growth and dormancy and amount of available forage produced. In the summer or wet season, there is a vast superabundance of relatively nutritious forage. During the winter or dry season, vegetation becomes largely a non-renewing resource, of much-reduced nutritional value. In this chapter, I consider how large mammalian herbivores respond to temporal variability in an adaptive way, at various temporal scales, and also how such consumers depend on spatial heterogeneity to cope with it. I outline the consequences for the functional response linking food intake to resource availability over different temporal scales, and evaluate the applicability of alternative foraging models. Finally I note the inappropriateness of equilibrium concepts in models linking population dynamics to resource supplies.

Large herbivores respond to temporal variability, at various scales, and depend on spatial heterogeneity to cope with it

Behavioural ecology considers specifically how organisms respond to spatio-temporal variability in their interactions with resources, physical conditions, conspecifics, predators, parasites and competitors (Sibly and Smith 1985; Krebs and Davies 1997; Houston and McNamara 1999). Recently the importance of environmental heterogeneity has become more widely recognised in ecological literature (e.g., Kolasa and Pickett 1991; Tilman and Kareiva 1997; Turchin 2003). The conceptual foundations for an 'adaptive resource ecology', linking foraging behaviour with its consequences for population and community ecology in spatio-temporally variable environments, was developed in Owen-Smith (2002a), with large herbivores specifically in mind. Crossing hierarchical levels, both temporal and spatial scales expand to encompass additional environmental influences affecting resource gains (Table 2.1).

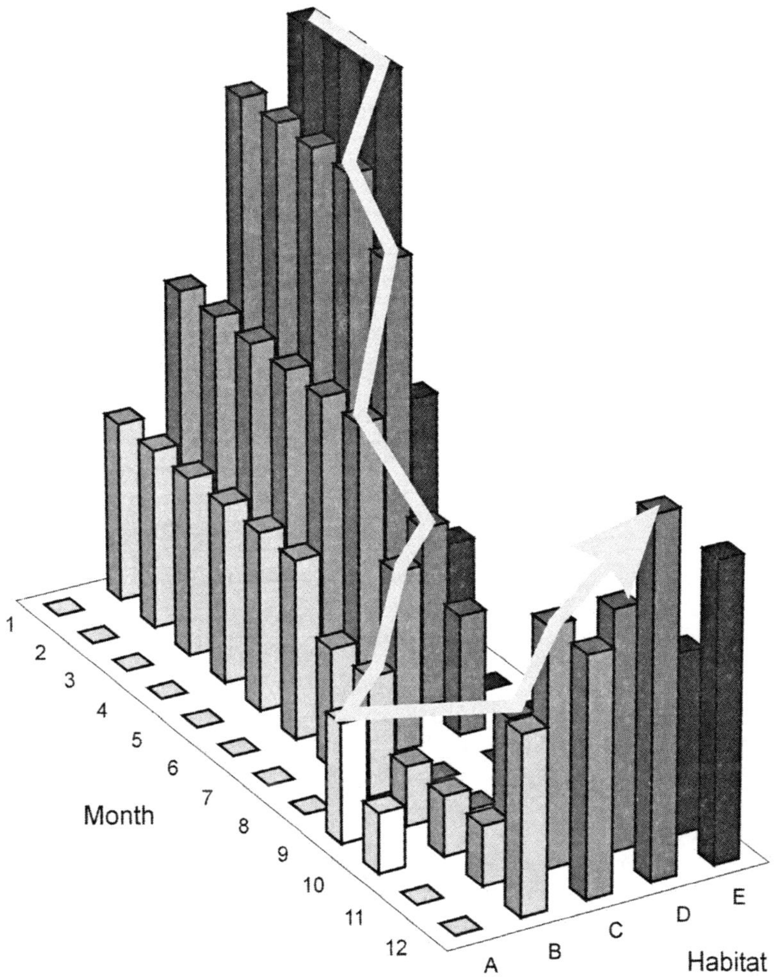

Figure 8.1. *The broken fitness landscape, with regular troughs and irregular pitfalls that organisms must traverse over time in order to survive and reproduce (from Owen-Smith 2002a)*

Short-term aspects of foraging behaviour including diet choice, intake rate and patch use are covered in earlier chapters (see Laca, Chapter 5, and Fryxell, Chapter 6). Subsequent chapters cover large-scale movements with landscapes and the consequences for population dynamics. My role in this chapter is to bridge between these realms, including not only changes in diet composition but also foraging-time allocation, searching movements, and associated physiological and morphological adjustments. The temporal scope extends from foraging spells of a few hours through daily and seasonal cycles to between-year differences in conditions. Specific questions addressed are:

- How do animals choose what to eat, when what is available to them is uncertain and changes over time?
- How do animals adjust their foraging behaviour at night when food is less visible and perhaps less nutritious, and the risk of predation and hypothermia higher than during the day?
- How do animals balance their foraging behaviour between days when conditions are favourable and other days when conditions are adverse because of extreme weather?
- How do animals cope with the adverse season when food availability as well as its nutritional quality is greatly reduced?
- How do animals counteract the extreme conditions that arise in some years, in food availability or weather?
- How does the functional response change its form across temporal scales?
- How well do alternative foraging models accommodate spatio-temporal variability?
- Are equilibrium conditions between consumers and their resource supplies ever reached?

VARIABILITY AMONG FORAGING SPELLS

Foraging spells of large mammalian herbivores typically extend over one or more hours, and encompass a sequence of feeding bouts interrupted by movements between food patches and other activities. They are separated by resting spells, or on occasions by extended travel to new foraging areas, water sources or other places (see also Bailey and Provenza, Chapter 2). Foraging spells take place within a foraging site defined by the speed of searching movements and the tortuousness of the foraging pathway. Successive foraging spells may take place within the same general area, or in different localities.

Food availability and thermal conditions differ between foraging spells, and animals adjust their diet choice accordingly

Diet selection models are concerned with how an animal selects what to eat given information about the range of food types (plant species and parts) available within some loosely defined area (Box 8.1). For herbivores the list of available plant species can be large, especially in or near the tropics. However, during foraging spells animals encounter only a limited sub-set of these food types. Rarer plant species, and more nutritious parts like new leaves or fruits, may not be contacted. How then should a hungry herbivore adjust its diets? Should animals keep searching for the best food types, expecting that these will eventually be found (even if only in a later foraging spell)? Or should they broaden their diet composition to encompass lower-value food types in the meanwhile?

***Box 8.1.** Diet-breadth model*

The classical diet-breadth (or contingency) model of optimal-foraging theory addresses food selection by a consumer confronted with a set of potential food items intermingled in the same region. The solution depends on ranking the food types from best to worst, based on their effective nutritional yield, determined by the ingestion rate (or the inverse, handling time) and digestible energy or nutrient content obtained. The optimal solution is given by the dietary combination maximising the rate of gain of the target nutrient. Food types are added to the diet in order of their effective value; hence the model indicates the optimal diet in terms of the range or breadth of food types included. Whether a food type should be incorporated in the diet is assessed by comparing the value that it would yield if eaten against the rate of nutrient gain obtained from the set of food types currently included in the diet, i.e., whether $E/T_h > \check{E}/(\check{T}_h+\check{T}_s)$, where E is energy or nutrient yield, T_h handling time, T_s search time and ˇ indicates values averaged over the food types in the diet. Hence a particular food type should be included in or excluded from the diet on an all-or-nothing basis (i.e., 1/0 acceptance). However, the model is idealised in only considering average nutrient yields, rates and times, and not their variability. Variation in nutrient contents, bite size or eating time within a food type, and in encounter rates with different foods indexed by the search time, could lead to departures from 1/0 acceptance, or so-called 'partial preferences'. Another problem arises when different factors affect the nutrient yields at different times, e.g., digestive processing time for herbivores often overrides the effects of handling time involved in food ingestion. For further reading, see Stephens and Krebs (1986), and for an application to large mammalian herbivores, see Owen-Smith and Novellie (1982).

The problem of dietary adjustments to stochastic variation in encounter rates with different food types has not been formally considered in the literature, at least not for large mammalian herbivores, so far as I am aware. Confronting it helps explain discrepancies between the predictions of simple rate-averaging models and observed diet choices. Herbivores generally select a broader diet than would seem optimal, and show partial rather than all-or-nothing acceptances of many food types (Westoby 1978; McNamara and Houston 1987a). A 'partial preference' means that animals eat certain food types on some encounters but not on others (Box 8.1).

The solution can be conceptualised using an elementary graphical model, incorporating a fundamental principle of environmental variability. Figure 8.2 illustrates a typical resource gain function for herbivores, and also many other consumers: a pattern of diminishing rate of gain towards higher resource levels. Accordingly, when by chance an animal finds itself in a foraging area richer than average in resources, it gains only a little more than it would at the mean resource level. When in a poorer-than-average area, it loses a lot more. The principle has become known as 'Jensen's inequality' (Ruel and Ayres 1999), and applies whenever gains are a saturating or decelerating function (convex upwards) of resource availability. The consequence of Jensen's inequality is that the overall gain over a sequence of foraging spells is enhanced if animals expand their dietary acceptance range to reduce deficits in poor localities. Hence partial acceptances should be shown for some less rewarding food types, eaten only when better food types are not encountered over some time period. The dietary range is consequently broader than predicted by a model treating diet choice only on the basis of average availability. This expectation has two underlying assumptions: (1) food types occur in patches containing limited sub-sets of the full range of food types available, rather than being dispersed in a fine-grained mix; and (2) foraging spells have a finite, relatively short time horizon, in relation to the resource area potentially available. Other hypotheses have been advanced to explain partial preferences (McNamara and Houston1987a; Prins and Van Langevelde, Chapter 7), but stochastic variability in encounter rates seems most fundamental.

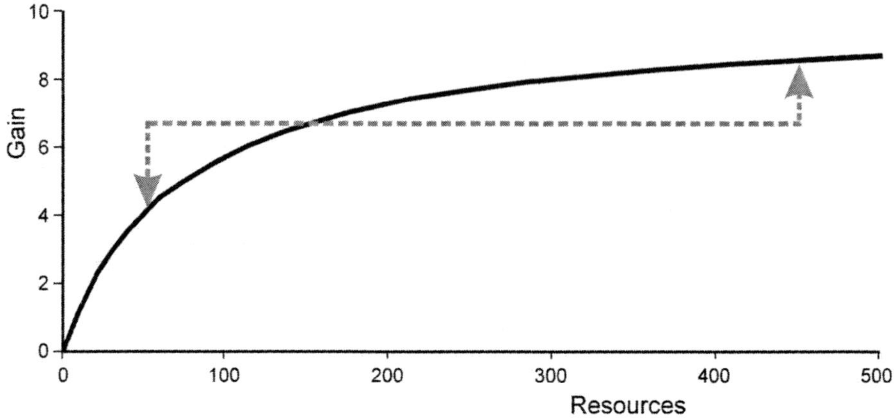

Figure 8.2. *Resource gain function illustrating Jensen's inequality (from Owen-Smith 2002a)*

Foraging spells are curtailed most basically because of heat build-up from muscular activity, with the heat load affected additionally by diurnal variation in ambient temperature, wind and solar radiation. I have in mind warm tropical environments, or summer conditions in higher latitudes, with activity levels of herbivores depressed by elevated midday temperatures. During winter, or even on relatively cold, windy days in the tropics, heat loss from exposure while foraging can restrict foraging activity (Parker 1988). Hence variation in thermal conditions over the diel cycle affects the duration of foraging spells, hence the time available to top up the rumen, and potentially also the diet composition within these spells. For most African ungulates, morning foraging spells are ended by rising heat towards midday, an exception being the African buffalo (Beekman and Prins 1989), while afternoon foraging spells can be prolonged into the cool of the evening. Buffalo in equatorial Tanzania even forage during the night (Prins and Iason 1989). Attenuated foraging periods should shift selection towards more rapidly ingested, less digestible food types. Over longer spells, animals have more time to seek out better-quality food types, even if these give reduced intake rates. For kudus, I observed such a pattern, with morning diets including relatively more woody browse, and afternoon diets more forbs (Owen-Smith 1993). A similar pattern has been reported for goats, in this case with more grass consumed in the morning and more woody browse in the afternoon (Solanki 1994).

DAY–NIGHT VARIATION

Over the 24-hour cycle, conditions differ between day and night not only in illumination and ambient temperature, but also in risk of mortality from predators relying on concealment. In Africa, the primarily nocturnal hunters for large herbivores include not only lions and leopards, but also spotted hyenas. The diurnal predators are cheetahs and African wild dogs, plus eagles for small antelope. Forage quality differs over the 24-hour cycle. Leaves are potentially less nutritious in soluble carbohydrates at night when they are respiring than during the day when they are actively photosynthesising.

A contrast between strategies of energy maximisation versus time minimisation was drawn early in the development of foraging theory (Schoener 1971; Fryxell, Chapter 6). The outcome is expressed through differences in the time allocated to foraging relative to other activities, because in either case the net food gain should be maximised during foraging activity. For energy maximisers, fitness increases monotonically with additional food, supporting enhanced survival and reproduction, hence foraging time should be the maximum permitted by environmental conditions. Time minimisation is expected when animals incur substantial predation risks or other fitness costs while foraging, compared with periods of immobility, so that as a consequence foraging should not be prolonged much longer than the time required to secure the maintenance metabolic requirement. For herbivores, a complication arises because the long-term food intake is usually constrained by digestive capacity rather than by the food intake rate while foraging (Owen-Smith and Novellie 1982). The trade-off between predation risk and foraging behaviour may also restrict the habitats selected for foraging at times when predation risk is high (Brown et al. 1988), but lead to risk-prone foraging when food is in short supply (Hik 1995; Sinclair and Arcese 1995b). McNamara and Houston (1987b) explicitly modelled such fitness tradeoffs.

Contrasts in illumination, temperature and risk of predation between day-time and night-time affect foraging-time allocation and food selection

Most African ungulates of medium-large size devote relatively more time to foraging during daylight than at night (Owen-Smith 1988). Warthog, which are highly vulnerable to predation, restrict their foraging entirely to daylight hours, while certain solitary antelope of medium-small size forage more at night than in the day, probably because they depend on concealment to escape predation (Novellie et al. 1984; Roberts and Dunbar 1991). Buffalo, which rely upon herd protection rather than concealment, and mega-herbivores like elephants and rhinos, with low vulnerability to predation as adults, forage equally day and night (Owen-Smith 1988; Beekman and Prins 1989; Prins and Iason 1989; Prins 1996; Winterbach and Bothma 1997). The habitat areas used for foraging may differ between day and night (Waser 1975), and animals may move much less while foraging at night than during the day (Jarman and Jarman 1973).

As a consequence of contrasts in foraging time and habitat use, differences in food selection might be expected between day-time and night-time. Animals could function effectively as energy maximisers during the day (subject to thermal limitations), and as time minimisers at night (see also Beekman and Prins 1989). However, because of the impracticality of observing feeding at night, no relevant data seem to exist.

The day–night contrast in conditions can have ramifying effects ultimately affecting population dynamics. In arid environments, ungulates such as oryx may shift their foraging activity into the cool of the night at times when surface water is not readily available, lessening evaporative water losses (Taylor 1969). This exposes animals to greater predation risks in places where nocturnally hunting predators are common, and hence to their exclusion from such habitats. The vulnerability of roan antelope to predation, which has evidently led to their population collapse in northern Kruger (after a zebra influx, promoted by augmented water supplies and followed by a lion increase; Harrington et al. 1999), could possibly be related to their need to do more foraging at night under hot conditions on account of their large size. A small number of roans placed in a fenced enclosure keeping out lions has thrived and expanded, despite the effects of drought conditions on the vegetation.

DAY-TO-DAY VARIABILITY

While foraging, herbivores can potentially ingest food much faster than they can digest it (Owen-Smith and Novellie 1982). Over the course of a day, food intake is limited more by digestive processing capacity than by the food intake rate obtained while foraging. In such circumstances, the food intake rate should be held below its potential maximum so as to keep the gut filled close to its capacity, within the available foraging time, as discussed above in relation to foraging spells. However, available foraging time also varies from day to day, dependent on daily differences in temperature and other ambient conditions. On hot days, animals must restrict foraging time to avoid over-heating. On cold or windy days, they may need to seek seclusion, also restricting the time available for foraging and the food types encountered.

Thermal conditions vary between days with consequent differences in foraging activity and dietary intake

For kudus, the proportion of time spent active (mainly foraging) showed wide day-to-day variability, ranging between 50 and 80% of daylight hours (Owen-Smith 1998). Only when the maximum daily temperature exceeded 36°C during the summer wet season, activity was restricted below the mean level of 66%, because kudus compensated partly for high midday temperature by foraging longer in the afternoon. In the winter dry season the temperature threshold appeared lower, about 30°C, probably because of the thicker hair coat animals possessed then. Weather records showed that 15% of days exceeded these temperature limits in both seasons, suggesting that the foraging activity of kudus was restricted by high temperature levels on only about one day in seven. As a consequence of their tolerance for high temperature levels, kudus are sensitive to mortality during extreme cold spells (defined by a midday temperature ≤ 14°C) that occur in the late dry season when their body reserves are depleted (Owen-Smith 2000). Observations on other African antelope species indicate that activity levels were reduced when midday temperatures exceeded about 32°C (Lewis 1977; Leuthold and Leuthold 1978; Klein and Fairall 1986). Northern ungulates are much less tolerant of high-temperature conditions, because of their greater insulation to cope with cold. Summer activity levels of elk and deer in North America decreased from 70% of the 24-hour day at a mean temperature of 10°C to about 40% at a mean temperature of 20°C (Belovsky and Slade 1986). During winter, deer showed reduced activity when the ambient temperature rose above -1°C (Schmitz 1991).

Herbivores may adjust their foraging behaviour, and potentially also their diet composition, in response to such weather variability. Under cold and wet conditions, hungry sheep increased bite size and biting rate to achieve almost double the rate of intake recorded a day later under more normal weather (O'Reagain et al. 1996). Sheep that were fasted overnight also showed a higher food intake rate than non-fasted animals, largely through taking large bites, and additionally foraged for longer during the day (Iason et al. 1999). Day-to-day variation in foraging time, and also perhaps in diet composition (such as observed for kudus, Owen-Smith 1993), suggests that dietary optimisation by herbivores takes place over periods somewhat longer than a day.

SEASONAL VARIATION AFFECTING FORAGING BEHAVIOUR

Environmental conditions vary seasonally in temperature and in precipitation, whether in the form of rain or snow. Close to the equator, annual temperature variation is slight. Beyond about 20° N or S, the summer – winter alternation becomes meaningful. In the tropics and subtropics, rainfall exerts a seasonal control on plant growth, and hence on food availability for herbivores. In high latitudes, precipitation commonly accumulates as snow during winter, making herbage somewhat inaccessible especially when a temporary melt leads to crusting of the snow. In northern regions, the spring growth of vegetation is supported largely by snowmelt, with the growth rate and duration of the growing season dependent on temperature conditions.

For kudus, the foliage biomass available within mouth reach on trees, shrubs and forbs declines, mostly due to leaf loss from deciduous species, to a minimum of around 8 g m^{-2} by the late dry season, amounting to about 10% of the wet-season peak (Owen-Smith and Cooper 1989; Owen-Smith 1994). For grazing ungulates, the seasonal change in available grass biomass is generally less marked, unless removed by fire or locally heavy grazing, but the proportion constituted by green leaves can be quite minute by the late dry season (e.g., Prins 1988; Prins and Beekman 1989). Hence grazers face a limitation in quality rather than quantity, because crude protein levels in dry grass commonly fall below 5% of dry mass, representing the minimum maintenance level for livestock (Owen-Smith 1982; Prins and Beekman 1989). I am not aware of comparable measurements of seasonal fluctuations in forage biomass for northern ungulates, but expect that it would be more extreme. The dependence on rainfall means that seasonal variation in food quantity and quality is somewhat more erratic in tropical and subtropical Africa and Australia than in northern latitudes where temperature is a more dependable influence.

During the adverse season, food availability and quality are reduced; herbivores alter their food selection, foraging activity and range use accordingly

During the summer or wet-season months, when there is a superabundance of food, herbivores can afford to be narrowly selective for the best-quality plant species and plant parts. During the winter or dry season, when food resources become progressively depleted through dieback and decay as well as consumption, herbivores must expand their diets to include lower-quality food types in order to maintain an adequate intake of food (Owen-Smith and Novellie 1982; Prins and Beekman 1989; Owen-Smith 2002a). Supporting adjustments also take place in daily foraging time, in the proportion of foraging time spent feeding, and in feeding durations in food patches (Beekman and Prins 1989; Owen-Smith 1994).

Figure 8.3 depicts how kudus expanded their diet over the course of the seasonal progression, from the mid-wet season through the dry season back into the early wet season, in terms of broad plant categories. The contribution of the staple deciduous trees and shrubs and favoured forb types declined progressively as these became less available during the dry season. Consumption of woody plant species with evergreen foliage was restricted mostly to the dry season. Less palatable deciduous species made their contribution during the early spring growing season from September on. Peaks in the consumption of fruits and flowers were evident when these plant parts became available. The daily food intake was elevated through most of the dry season, to compensate for the reduced nutritional quality of the food types included in the expanded diet. Only in September, when little green foliage remained, did daily food intake drop markedly.

The dietary expansion of the kudus was underlaid by seasonal changes in the frequency with which plants of particular types were eaten when encountered during foraging spells (Owen-Smith and Cooper 1987). The most favoured woody species remained highly acceptable throughout the year, as long as they retained leaves. These plant species were classed as palatable, i.e., whatever secondary chemicals they contained did not deter kudus from feeding on them. Relatively palatable evergreens showed a sharp increase in their acceptability when they became incorporated into the diet in the early dry season, after the favoured deciduous species had started shedding their leaves. Unpalatable deciduous species also showed marked changes in acceptance when they were added to the diet, either in the late dry season when they still retained some leaves, or when they flushed new leaves in spring. Finally, even the most unpalatable among the evergreens became eaten at the end of the dry season when little other food remained, but were consumed in restricted amounts.

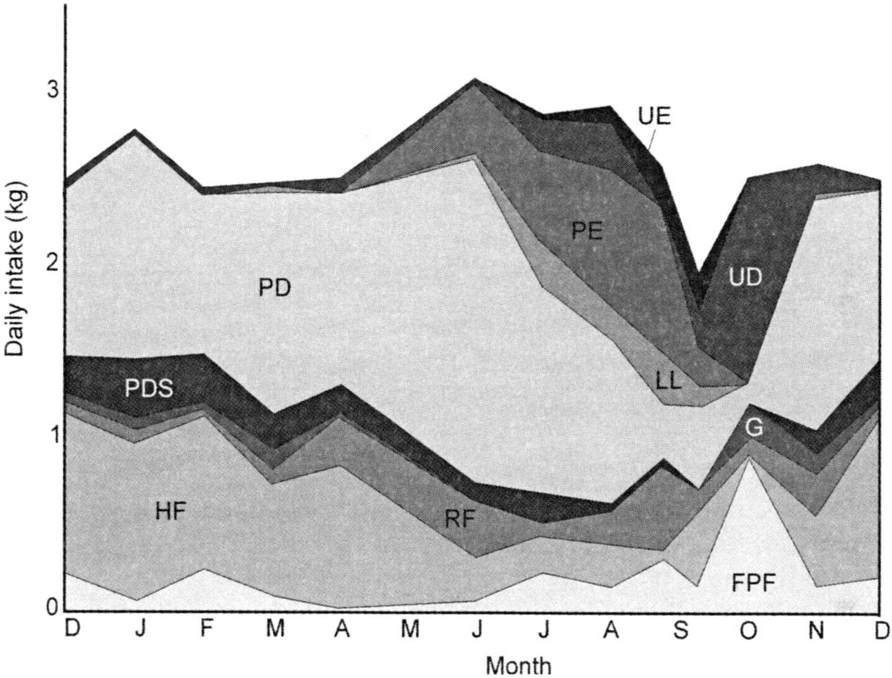

Figure 8.3. Changing diet composition of kudus over the seasonal cycle, in terms of broad plant types: FPF = fruits, pods and flowers, HF = herbaceous forbs, RF = robust forbs, G = grass, PDS = palatable deciduous spinescent browse, PD = palatable deciduous unarmed browse, LL = leaf litter, PE = palatable evergreen browse, UD = unpalatable deciduous browse, UE = unpalatable evergreen browse (from Owen-Smith and Cooper 1989)

Comparable shifts in the use of different grass species were recorded for free-ranging African buffalo through the dry season (Macandza et al. 2004). However, grasses favoured during the wet season can show decreased acceptability during the dry season, depending on phenological changes in green-leaf retention and in leaf:stem ratio (O'Reagain et al. 1996).

Dietary shifts make the measurement of food availability for herbivores problematic. During the wet season, kudus consumed only 3-6% of the potentially edible foliage biomass that they encountered within neck reach along their foraging pathway (excluding grass). By the late dry season, the fraction of plant biomass removed along the foraging pathway had increased to over 25% (Owen-Smith 1994). Although almost all plant species retaining foliage were accepted for feeding, animals still consumed only a portion of the potential forage offered by individual plants before moving on.

Herbivore species or feeding types differ in how they adjust their daily foraging time in response to seasonally changing food availability (Beekman and Prins 1989). Browsing kudus progressively increased the proportion of the daylight hours spent foraging as the dry season advanced, thereby partially compensating for reduced forage quality (Figure 8.4). A similar pattern was shown by impala and springbok, which although mixed feeders concentrated increasingly upon woody plants during the dry season (Jarman and Jarman 1973; Davies and Skinner 1986). Among grazers, blesbok, buffalo and white rhinos all showed reduced feeding time in the dry season, probably because of the lengthened digestion time required by the mature dry grass (Novellie 1978; Winterbach and Bothma 1997; Owen-Smith 1998). Nevertheless, all three of these grazers increased their daily foraging time in the transition period when green grass started reappearing in restricted amounts, especially in previously burnt grassland. In contrast, horses, which are also grazers but non-ruminants, expanded their daily foraging time from summer through winter in the Camargue region of France (Duncan 1985). Muskoxen, which are mixed feeders, showed a reduction in daily foraging time in mid-winter, apparently as an energy-conserving strategy, and a peak in spring (Forschhammer 1995). Browsing moose in Alberta, Canada, showed no change in daily foraging time between summer and winter, but a peak in spring (Renecker and Hudson 1989), while in Alaska the foraging time of moose was extremely low in mid-winter (Risenhoover 1986). This indicates responses to seasonally changing food quantity and quality. Temperature and day length can be subtler than a naive contrast between time minimisation and energy maximisation suggests, but a theoretical synthesis is still lacking.

Herbivores may also show contrasting responses in their foraging range to seasonal changes in food resources. Black rhinos and kudus, both of which are browsers, showed a contraction in their home-range extent during the dry season (Goddard 1967 and personal observations), whereas white rhinos, which are grazers, expanded the area they covered during the dry season (Owen-Smith 1975). This makes sense, because the woody plant species retaining leaves which the browsers depend on are localised in their occurrence and dependable in their phenology, while grazers could find areas where chance rain-showers or other local variability had promoted some green grass growth.

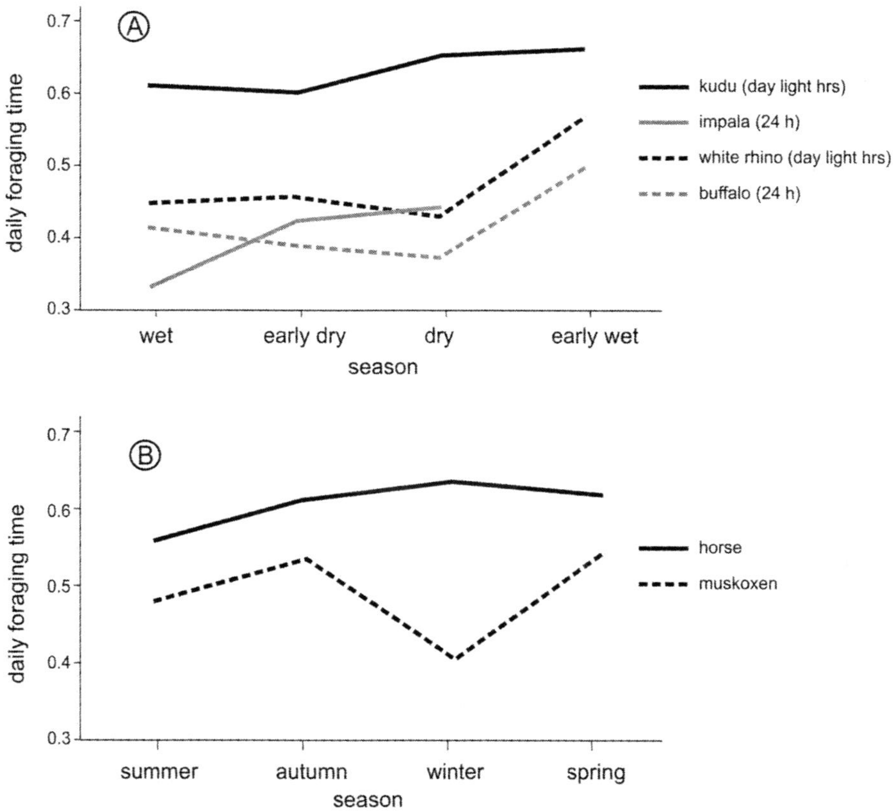

***Figure 8.4**. Seasonal changes in the proportion of the day spent foraging by representative grazers and browsers; (a) tropical or subtropical species, (b) northern species*

SEASONAL VARIATION AND PHYSIOLOGICAL AND MORPHOLOGICAL ADAPTATIONS

Additional physiological and morphological adaptations may aid herbivores in meeting their nutritional requirements through the adverse season. Northern deer may show substantial fluctuations in metabolic rate between summer and winter (Silver et al. 1969; Weiner 1977; Regelin et al. 1985), although partly as a consequence of differences in food intake or activity (Mautz et al. 1992). Red deer exhibit a dramatic reduction in heart rate indicative of reduced metabolism during winter nights, associated with peripheral body cooling (Arnold et al. 2004). Many ungulates show changes in pelage between summer and winter. Digestive capacity can be expanded above the fill levels

Herbivores seasonally adjust their metabolic rate, gut capacity, fat stores and allocation to maintenance, growth or reproduction

apparent in summer (Baker and Hobbs 1987; Boomker 1987, cited by Owen-Smith 1994), and the internal morphology of the rumen controlling the surface area for absorption of digestive products may alter seasonally (Hofmann 1973).

Fat stores help animals survive through periods when food gains are inadequate to meet requirements. However, storing such reserves is costly, not only from the extra foraging time required to build them and the energetics of carrying the extra mass, but also potentially from the heightened predation risk associated with diminished mobility. Fat stores carried by tropical or subtropical ungulates amount to no more than 5-10% of body mass (Ledger 1968; Smith 1970), compared with 15-30% for many northern ungulates (Tyler 1987; Parker et al. 1993). Moreover, the fat carried by African antelope is associated internally with the kidneys and mesenteries, rather than being subcutaneous. This suggests that tropical ungulates avoid thermoregulatory problems, while ungulates living in high latitudes benefit from the insulation provided by surface fat. Because of the costs of carrying fat stores, fat is generally laid down towards the end of the benign season, and should theoretically amount to little more than is needed to carry animals through to when foraging conditions improve again (Owen-Smith 2002a).

The additional demands of reproduction raise the daily energy requirements of mothers almost twofold, and that for protein more than twofold, through late pregnancy and early lactation (Oftedal 1984; Prins and Beekman 1989; Chan-McLeod et al. 1994; Prins and Van Langevelde, Chapter 7). Free-ranging cattle in the Netherlands increased their daily food intake by almost 60% relative to the expected mean for the food quality, when they were supporting newborn calves in autumn (Van Wieren 1992). Dairy cattle can increase their effective rumen capacity by up to 40% in response to such demands (Campling 1970). The daily food intake may nevertheless be insufficient to support reproductive demands, forcing mothers to draw upon stored reserves of fat and body protein (for buffalo see, e.g., Prins 1989a). Accordingly, births generally occur at the time of the year when nutritional conditions are most favourable, most narrowly at high latitudes where the seasonal fluctuation in resources is greatest.

For young, growing kudus, foraging gains as indicated by daily energy intake exceeded estimated metabolic expenditures for tissue maintenance and activity for most of the year, and fell below the maintenance requirement by less than 10% even at the end of the dry season in September (Figure 8.5). Young black-tailed deer in Alaska showed somewhat greater seasonal variation, with the daily energy intake dropping to about one third of the summer maximum by late winter, well below the maintenance requirement (Parker et al. 1999). This was because the deep snow cover made food largely inaccessible, and movement costly, so that the deer showed no compensatory adjustment in daily foraging time (restricted also by the brief period of daylight). These deer used stored body reserves to carry them through the period of deficits. Young growing cattle in Colorado showed a more than twofold decline in daily energy intake seasonally, but compensated during late winter by reducing daily energy expenditures (Senft et al. 1987b).

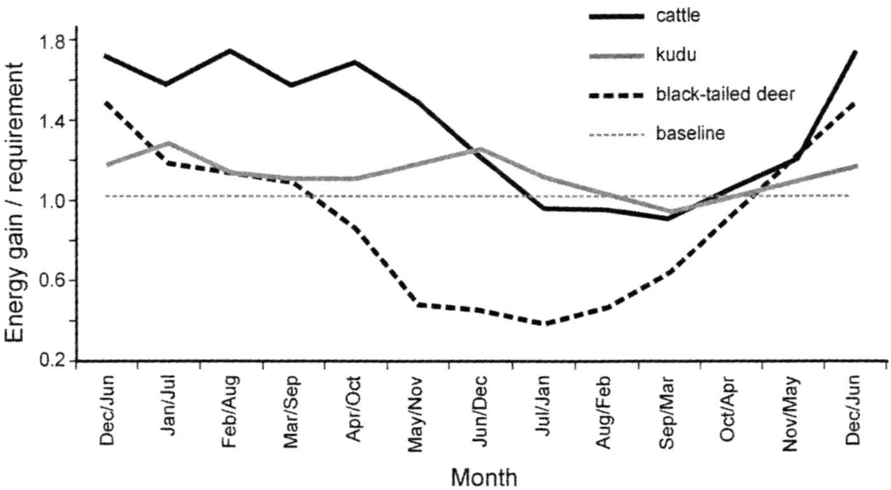

Figure 8.5*. Seasonal changes in daily energy intake relative to maintenance requirements for kudus in South Africa (from Owen-Smith and Cooper 1989), black-tailed deer in southern Alaska (from Parker et al. 1999), and beef cattle in Colorado (from Senft et al. 1987b). Alternative months represent corresponding seasonal stages in the northern and southern hemispheres*

Food resources differ in their contribution towards supporting the nutritional requirements of herbivores at different stages in the seasonal cycle (Owen-Smith and Cooper 1987; Owen-Smith 2002a). The following functional categories can be distinguished: (1) high-quality food types supporting peak reproductive outputs, (2) staple food types providing the bulk of the diet for much of the year, (3) reserve food types consumed when staple foods become depleted, (4) buffer food types consumed during critical periods, slowing the rate of starvation (Owen-Smith 2002a; see also Prins and Beekman 1989). Some food types can also serve as bridging resources during periods when little else is available, e.g., the deciduous trees with generally unpalatable foliage that leafed out ahead of the rains in early spring for kudus (Figure 8.3). Food types that are high in quality, but ingested at restricted rates, can be another functionally distinct category. They contribute towards enhanced diet quality, but need to be complemented by other food types that can be consumed more rapidly for animals to achieve an adequate daily food intake. Thorny acacia species fell into this category for kudus.

Food resources can be subdivided among functional categories in terms of their dietary contributions in different seasons

BETWEEN-YEAR VARIABILITY

While the basic seasonal cycle is fairly predictable in many regions, much variation can still occur between years in the specific conditions encountered at different times. Body reserves get progressively depleted through the adverse season, and whether the starvation threshold is reached depends on how prolonged the period of sub-maintenance diets lasts. Hence in tropical savannas, the timing of the early rains can be more important than how severe the food deficit was during late dry season. Kudus in Kruger Park died of hypothermia when cold spells occurred in September or October, in the transition period between winter and spring, although they tolerated colder temperatures during mid-winter (Owen-Smith 2000). On windy days, both kudus and white rhinos retreated to localities where dense bush provided some shield against the wind. In high northern latitudes, there is annual variation in the occurrence of extreme weather in the form of blizzards, snow depth and formation of ice crusts. Animals may survive through a few days of a blizzard by sacrificing feeding time to seek shelter, but if the bad weather is prolonged too long they die of hypothermia (Schaller and Junrang 1988). This may lead to episodic severe mortality, e.g., in Soay sheep associated with March gales (Grenfell et al. 1998).

Refuge resources or habitats are needed to cope with the extreme conditions that can arise in some years

Whether a population of a particular species persists in a region may depend crucially on how well animals are adapted to cope with the extreme conditions that occur infrequently, but in the long term inevitably. A population may thrive through nine years out of ten, but when the crunch conditions occur survival depends on having refuge resources or habitats available. No ungulate hibernates. The closest to hibernation is the congregation of white-tailed deer in 'deer yards' for a few weeks at the end of winter, while feeding little (Schmitz 1991). Stored body reserves can probably support animals for at most a month without food. Plant types not normally eaten can become crucial to bridge emergency periods. Fallen tree leaves may perform this role even for grazers like buffalo after virtually all grass has been consumed during severe droughts, although not for long.

TRANSFORMING THE FUNCTIONAL RESPONSE ACROSS TIME FRAMES

The relationship between the food intake rate obtained by a consumer and food abundance, commonly called the 'functional response', is of fundamental importance in theoretical ecology. For herbivores it is conventionally measured over short time periods while animals graze down available forage within an experimental enclosure. The classical 'Holling Type II' was originally conceptualised as an outcome of the trade-off between search time and handling time, but for large herbivores search time largely overlaps with handling time, so that changes in bite size exert the main control over short-term intake rates

(Spalinger and Hobbs 1992; Drescher 2003). Furthermore, over the daily cycle digestive handling time becomes the overriding constraint (except for domestic herbivores fed high-quality food) with bite sizes reduced below the maximum possible, leading to a truncated intake response (Figure 8.6a; Owen-Smith 2002a). If more abundant food is also less digestible, the daily intake response to changing food availability may become humped rather than asymptotically saturating in form (Figure 8.6b).

Functional response changes its form across time frames and should be transformed into biomass/energy gain for linking with population dynamics

The factors constraining daily food intake change over the seasonal cycle. When food is most abundant, digestive capacity may be limiting, but towards the end of winter or the dry season so little foliage may remain that the food intake rate becomes the main limitation. When forage quality is highest in spring, metabolic satiation may restrict the daily intake, i.e., animals eat as much as they need for maintenance and activity needs, and gain no further benefit from storing more fat than they already have. Metabolic satiation would not be a factor for females supporting growing foetuses or nursing offspring (see Prins and Van Langevelde, Chapter 7).

Adaptive changes in diet breadth also affect the form of the daily-intake response. Diet expansion plus additional behavioural compensation may result in the daily food intake remaining fairly constant, or even increasing, with seasonally diminishing food abundance, e.g., the daily food intake obtained by sub-adult kudus declined precipitously only when less than 2-3 g m^{-2} of accessible foliage remained on trees and shrubs (Figure 8.6c). Grazing ungulates may show a decline in daily food intake at somewhat higher levels of standing forage biomass than browsers, because forage biomass is determined largely by grass height, which also restricts bite sizes below some threshold height (Distel et al. 1995). Furthermore, diminishing grass quality lowers digestive processing capacity.

For linking with population dynamics, the gross food intake must be transformed into the consequent gain in consumer biomass or its energetic equivalent. Moreover, it is the seasonally changing food abundance that is most relevant. Net energy or nutrient gains generally decline over the adverse season, because the added food types do not replace the nutritional yields of the preferred but depleting foods. Hence the effective consumer gain response deviates from the food intake over the course of the seasonal cycle. For kudus, although daily food intake peaked at intermediate levels of food abundance in the early dry season, the daily energy intake declined progressively from the wet season through the dry season (Figure 8.6c). Small proportional differences in daily energy gains transform into huge differences when integrated over the annual cycle, e.g., a 1% increase daily results in a 38-fold compounded gain over the course of a year.

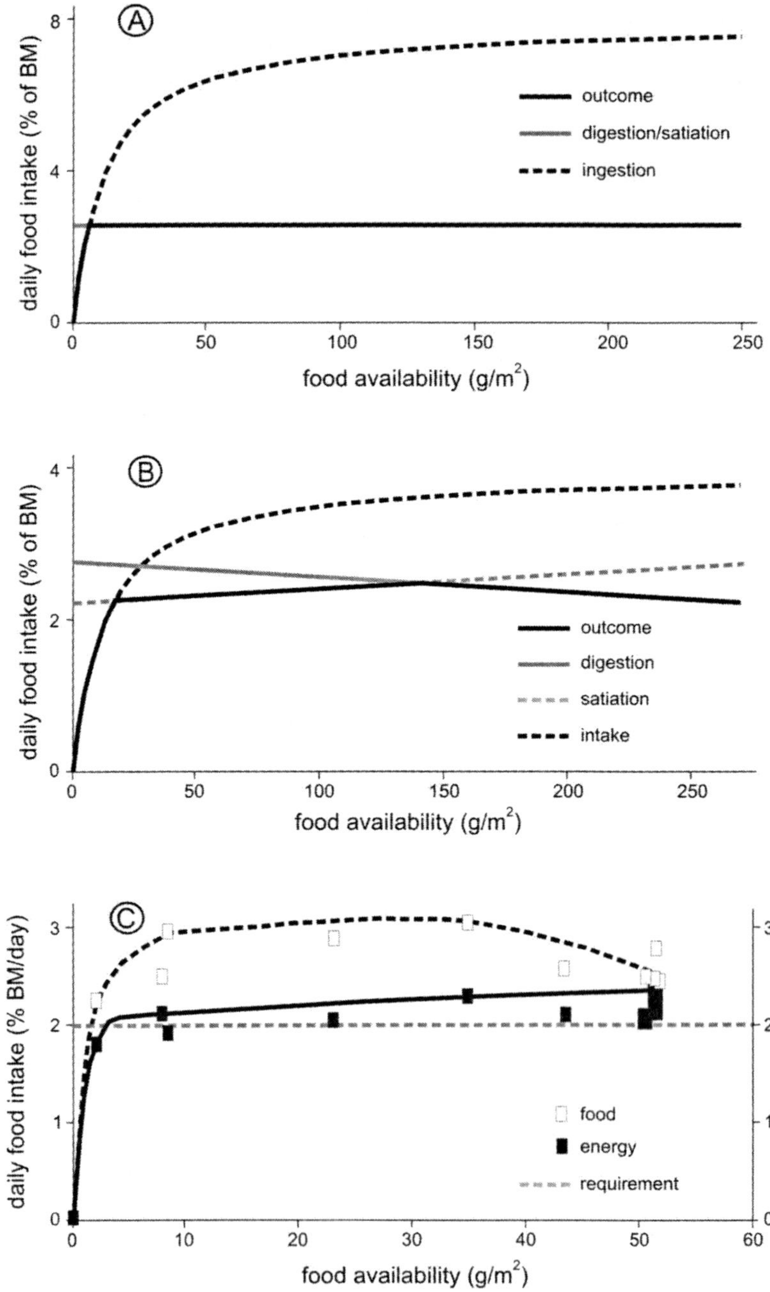

Figure 8.6. *Food intake and energy-gain responses of kudus to seasonally changing food availability, both in terms of gross food intake and rate of gain of digestible energy (from Owen-Smith 2002a)*

ACCOMMODATING ENVIRONMENTAL VARIABILITY IN FORAGING MODELS

Simple foraging models aimed at predicting average diets from average measurements on vegetation and other factors can be misleading. A good example is presented by the linear programming model (LPM, Belovksy 1986), which seemed amazingly accurate in predicting the broad diet composition, in terms of plant types like grass versus browse, for almost all herbivore species to which it was applied. This model proposes that herbivores trade the nutritional benefits of abundant food offering high intake rates but restricted digestibility against those of more nutritious foods available in lower quantities, given the constraints of daily foraging time and digestive capacity. The underlying assumptions are reasonable and well supported, as documented above. Yet, given the practical difficulties in measuring all of the relevant factors influencing food gains, it seemed surprising that the predictions of LPM should repeatedly be so close to reality (Hobbs 1990; Huggard 1994). I was initially surprised to find a close match between predicted and observed diets when applying LPM to the data that I had collected for kudus (Owen-Smith 1993). I doubted that kudus made such a crude distinction between broad plant categories (woody browse versus forbs), given the wide variability in nutritional quality within these classes.

Rate-averaging models fail to accommodate environmental variability while dynamic state-variable models remain to be overcome

Hence I investigated further whether kudus actually responded to variation in the parameters determining the optimal diet in LPM between foraging sessions and days. They did not adjust their diet composition as predicted. This meant that the putative constraints of digestive capacity and daily foraging time were not actually effective. The apparent predictive success arose from a logical circularity in the way the model was being applied. Assuming that digestive capacity and foraging time were constraining, their upper limits were assessed from average observed values of digestive contents and daily foraging time. These values were then used to make predictions, which of course confirmed the model. Even if the potential digestive capacity (or foraging time) was actually greater than measured, the intersection of the average observed settings of the supposed constraints must correspond with the average observed diet, if measurements are accurate (Owen-Smith 1993, 1996).

Box 8.2. *Dynamic state-variable models*

Dynamic optimisation models differ from rate-averaging models by assessing optimality in terms of the state (of the gut, or body condition, or whatever) achieved as a result of decisions made over some extended time period. This state has an expected fitness associated with it, e.g., an animal that has ample fat stores is more likely to survive and reproduce than one lacking body reserves at the end of the summer season. However, animals that are too fat could have reduced fitness, through being more likely to be predated. While it might be quite easy to define the optimal state to aim at, how to get there is more problematic. The trick in solving a dynamic optimisation problem is to work backwards from the end time to the start time. The optimal solution depends on being in the optimal state for future fitness at each stage, and hence is evaluated using a state-dependent fitness function. Solving a dynamic optimisation problem analytically constitutes a huge challenge, but it can be made more tractable by dividing time into discrete steps. Such models can also take into account uncertainty in the situation confronted at each time step, e.g., a predator may or may not be encountered, and the food type sought may or may not be found at that time. This leads into a procedure called 'stochastic dynamic programming'. The limitation for such models is that each additional choice doubles the number of computations that must be performed at each time step, to consider all the options and their consequences for fitness, an obstacle known as the 'curse of dimensionality'.

Another approach to optimal decision-making in complex and changing environments uses concepts from neural networks to establish the weights to be given to various factors influencing the decision through intensive computation to explore the outcomes. This can be taken further using genetic algorithms to compute which sets of behavioural responses are more likely to persist in a population than others.

Mangel and Clark (1988) introduced the concept of stochastic dynamic programming to behavioural ecologists, while Clark and Mangel (2000) and Houston and McNamara (1999) present more comprehensive treatments. Anderson (1995) gives an introduction to neural networks, while Goldberg (1989) describes how genetic algorithms can be used.

More fundamentally, in variable environments one would expect animals to have some reserve capacity to cope with the extreme conditions that they have to face at times. Hence, under average or benign conditions they should appear somewhat slack in their foraging behaviour. How much reserve capacity should animals have, in digestive space, temperature tolerance or any of the other factors affecting foraging efficiency? I noted above that the daily foraging time of kudus was limited by high midday temperatures on about one day in seven. Over what period of the year is the maximum digestive capacity actually filled at the end of a foraging session, allowing for possible seasonal adjustments in the physical capacity?

Models projecting equilibrium outcomes are inappropriate for representing effects of resource variability on population growth

Dynamic state-variable models projecting the trade-offs between current decisions and future states seem to offer the most appropriate framework for accommodating environmental variability (Box 8.2). However, the obstacle to be overcome is the 'curse of dimensionality'. Each additional choice doubles the number of computations, and once the number of options exceeds three or four, computer memory and computing time begin to become a restriction. At a plant species level, the number of food types available to kudus was well over 100. Nevertheless, most of these plant species are rare, and for the diet breadth model I reduced the effective number of food types to around seven (Owen-Smith 1993). At any point in time, the alternate decisions are simply either widening the diet by one food type, or eliminating a food type from the diet. Hence I believe that a dynamic diet-choice model is feasible, if structured around a restricted set of choices at each time step. Neural network models (Anderson 1995) offer an alternative approach towards establishing the optimal choice in complex environments. They have been applied to habitat selection of, for example, panda bears (Liu 2001; Liu et al. 2002), but have yet to be applied to foraging behaviour.

The challenges of dynamic optimisation are more readily overcome considering the allocation decisions that herbivores must make for the surplus resources they have acquired: to grow bigger, to grow fatter, or to grow babies (Owen-Smith 2002a). If storing fat is costly for survival, animals should store fat as late as possible, and just enough for their needs to survive the adverse season, plus whatever additional amount is needed to ensure successful reproduction in spring. Hence during times of the year when fat reserves are adequate, animals may appear somewhat slack in their foraging behaviour.

For most models in ecology, equilibrium solutions are sought and identified analytically. For example, the environmental 'carrying capacity' is designated as the equilibrium population that can be maintained by the balance between density-dependent birth and death rates, dependent in some undefined way on resource availability. This zero-growth density may be asymptotically stable despite environmental perturbations to its level, and the disruption of lagged density feedbacks (Turchin 2003). An alternative approach links the population dynamics interactively to the growth potential enabled by resources consumed, less background mortality losses (Caughley 1976). It can generate either an asymptotic approach to an equilibrium density, or oscillations generated by the delayed effects of consumption on resource production. Both approaches overlook the enormous fluctuation in the vegetation resources supporting herbivore populations during the course of a year (e.g., Sinclair 1977; Prins and Beekman 1989; Prins 1996). Any equilibrium between population growth and resource supplies is no more than transient. In the benign summer or wet season, there is more food available than herbivores can possibly use, while during the adverse winter or dry season remaining resources do little more than alleviate starvation. The population level sustained is the emergent outcome of the counterbalancing of the changing gains and losses at different stages of the seasonal cycle. Storage buffers like body-fat reserves help dampen the seasonal fluctuations that might otherwise occur, but when thresholds are surpassed herbivore populations can crash (Walker et al. 1987).

To have predictive value, population models must incorporate the adaptive responses of consumers to the changing conditions that they face daily, seasonally and between years. The foundations for such a modelling approach, integrating foraging behaviour into population and community dynamics, are laid in Owen-Smith (2002a). Consumer-resource models incorporating functional heterogeneity in resources coupled with adaptive responses by consumers generate radically different dynamics to those assuming uniform, unchanging environments (Owen-Smith 2002b, 2002c).

SYNTHESIS

The foundations for theoretical resource ecology are taking shape, below the ecological sterility of classical population dynamics, which links consumers with resources through a nebulous 'carrying capacity', and the mechanistic vagueness of food-web analysis at the community level. Animal ecology needs to look upwards towards higher-level processes taking place beyond the time frame of bites and steps, and encompass phenotypic adaptation in physiology, morphology and life-history events as well as behaviour. Population ecology needs to explore the mechanisms linking the survival and reproductive rates of consumer to resource variability. The impacts of parasites and predators on populations operate to a large extent within the context of the resource status affecting the vulnerability of consumers to such amplifying influences on mortality (Hik 1995; Sinclair and Arcese 1995b; Prins 1996).

Box 8.3. *Testable hypotheses for future research*

Hypothesis 1. Changes in foraging behaviour are adapted more to reduce losses during adverse periods than to maximise gains during good times.
Hypothesis 2. Consumers have surplus capacity or tolerance to cope with adverse extremes that occur no more frequently than once in seven days, or other appropriate period of environmental variation.

This chapter has explored some of those links, moving upwards across temporal scales from periods within days through the diel and seasonal cycles to variability between years. It has encompassed not merely the direct consequences of foraging behaviour in terms of diet composition, but also the additional behavioural responses involved in food procurement, i.e., daily time allocation, searching movements, plus the phenotypic adjustments associated with processing and allocating the food gains. In the light of these responses, simplistic notions of the 'functional response' need to be modified, and alternative optimisation approaches accommodating environmental variability explored. Some hypotheses for future research are formulated in Box 8.3. Wide seasonal fluctuations in food availability expose the non-linearities inherent in functional relationships, and emphasise how the adverse extremes override the more prevalent benign conditions. Population models incorporating equilibrium 'carrying capacities' or average conditions are inappropriate and mechanistically misleading.

In a previous exploration of foraging theory, a 'dumb' ungulate, with inflexible food selection, was contrasted against a 'clever' ungulate, adjusting its diet selection to maximise its immediate rate of nutrient gain (Owen-Smith and Novellie 1982). The more far-sighted strategies of a 'wise' ungulate remain to be defined. The need to recognise 'individual trait plasticity' in both population and community ecology has become increasingly widely recognised (Schmitz et al. 2003). The importance of the key resources supporting animals during crucial periods of the year and in crunch years was emphasised by Illius and O'Connor (1999). Landscapes retaining functional heterogeneity in the resources supporting herbivore populations could avert the roller-coaster dynamics to which these species are prone (Owen-Smith

2004). The chapters of the book contribute substantially towards expanding the foundations for rigorous resource ecology, addressing consequences of the spatial and temporal variability that is a basic feature of the real-world environments that large herbivores, and indeed most other organisms, occupy.

ACKNOWLEDGEMENTS

I thank Sander Oom, Frank van Langevelde and Herbert Prins for their helpful suggestions for improving this chapter, and all the participants of the workshop in Wageningen that helped shape its content.

CHAPTER 8B

COMMENTS ON "EFFECTS OF TEMPORAL VARIABILITY IN RESOURCES ON FORAGING BEHAVIOUR"

WILLEM F. DE BOER AND AREND M.H. BRUNSTING

Resource Ecology Group, Wageningen University, P.O. Box 47, 6700 AA Wageningen, The Netherlands
E-mail: fred.deboer@wur.nl

Owen-Smith (Chapter 8) describes the current state-of-the-art of the foraging models in relation to different temporal scales. He argues that the static, equilibrium approach of the traditional optimal foraging models does not seem to hold for real-life foraging studies, as trait plasticity is needed to cope with changes in foraging conditions over time. Addressing different temporal scales, from the variability on a certain day to day-to-day changes, and seasonal fluctuations of foraging conditions, it becomes clear that present-day models must be improved in order to be able to accommodate the variability in environmental conditions. Apparently, current models cannot yet fully cope with the importance of scale in foraging models. Indeed, the effects of both temporal and spatial scales on foraging behaviour need to be incorporated in the available models. An important gap is that few studies have been carried out that implicitly study the impact of these scale issues on foraging theory, let alone the hierarchy of different scales. There is an urgent need for studies that address the effect of scale on foraging behaviour.

One of the major problems in these foraging studies is that the thresholds (e.g., the marginal value, or thresholds used in the diet breadth analyses) or optima are dynamic as well; there is probably no fixed optimal intake rate or fixed optimal diet composition. They will change, depending on the changing local conditions, and fluctuate over time. Stephens and Krebs (1986) showed that the variation of the thresholds or of the expected intake rate can have important consequences for foraging choices, thereby explaining partial preferences. Moreover, Bailey and Provenza (Chapter 2, but see also Bryant et al. 1991) highlight the importance of a

H.H.T. Prins and F. van Langevelde (eds.), Resource Ecology: Spatial and Temporal Dynamics of Foraging, 183-185.

balanced diet, the satiety hypothesis, as an alternative explanation for partial preferences. In the real world, probably all these processes influence foraging choices simultaneously, so we have to develop elegant studies that are able to disentangle the effect of these different variables. Due to the complexity, experiments under controlled conditions are certainly required. One should be careful with testing new ideas by searching for evidence in studies that were not aimed at addressing these multiple scales, such as done in Chapter 8 in the rhino home-range contraction example.

In contrast to Owen-Smith we would like to argue that the equilibrium concept in the description of systems is very important. Models without an equilibrium cannot persist over long time whether or not accompanied by variation around the equilibrium. Therefore, animal numbers never reach equilibrium. The equilibrium-modelling approach can therefore be of limited use and could miss the patterns we want to understand. On the other hand, the concepts of equilibrium and stability should not be dismissed. Even though the equilibria themselves might not be reached, the system nonetheless tracks these equilibria, even when varying over time. Savanna systems are highly dynamic. It is important to identify a stable point to which a system is inclined to move, the point of attraction, irrespective of the starting conditions. There exists an intriguing degree of constancy in nature; a good example is the constancy in the bison population that lived for over 5 million years in America under highly variable conditions. Sometimes individual species seem to fluctuate erratically, but the biomass of the whole grazer community appears to be remarkably constant (e.g., the Ngorongoro crater or Manyara National Park; Prins and Douglas-Hamilton (1990)). The question rises whether constancy necessarily implies stability.

A dynamic optimisation approach could be a useful alternative for the classic optimal foraging models, or neural networks or genetic algorithms could be used. These are probably only part of the alternative modelling approaches; others include multiple criteria or multiple objective optimisation, or approaches that include conflicting demand (Schmitz et al. 1997) or evolutionary modelling. Moreover, the concept of satisficing (Ward 1992), i.e., maximising the probability of exceeding a moving target (Parlar and Weng 2003) could be useful in modelling the diet choices of herbivores that have so many objectives and constraints. Realism urges us to consider the effect of imperfect information of the animal, or decaying spatial memory and risk aversion strategies as well. The latter approach seems essential; decreasing the probability of poor decisions might be an evolutionarily more promising strategy than just nutrient optimisation that could include some poor decisions under imperfect information or just from the stochasticity in resource availability. However, it might be true that these approaches may not give more insight into diet selection and optimisation of intake. These new models might be able to detect optimal solutions under a set of local conditions, but not be able to describe the mechanisms behind the underlying selection process. A first step would be to test this, and Owen-Smith clearly shows that we lack studies addressing different temporal scales simultaneously, studying the variability of the resource availability over time, and addressing the hierarchy of these different scales for the foraging decisions taken at a specific moment in time. These studies are essential to parameterise and test new models. However, building in more and more biology and realism in the models does not necessarily lead to more insight and to more reliable diet predictions. Conclusions about partial food preferences maybe slightly altered, but might not be fundamentally different from what one could infer using more simple models. Moreover, there is the problem of the enormous number of parameters needed and dimensionality. A simulation model incorporating even some simple biologically sound rules may already generate enormous complexity. Incorporation of all relevant biological variables yields results as inconceivable as reality itself, as Owen-Smith also clearly underlines. Simulation hides the applied optimisation rules and might results in nothing more than what was already stated by the formulation of the model rules; much more work should therefore be done on the objective function. Even more so, the first priority should be to develop a theoretical framework that combines scale issues with foraging theory. This would certainly be a prerequisite if one wants to extrapolate field study results for the understanding of population dynamics.

One of the testable hypotheses put forward in this paper is that, due to Jensen's inequality, adverse periods are more important in shaping foraging behaviour than times of plenty. This is an elegant, attractive theory, but how useful is it when studying foraging behaviour? Can we test this? We think that these studies seem feasible, but also here controlled intake experiments are being called for. Moreover, the classical approach of presenting foods to animals and recording selection and intake, could be extended to address the effect of multiples scales, another topic of Owen-Smith's chapter. These experiments seem also the most appropriate to disentangle the effect of spatial and temporal scales, which in field studies are so often confounded.

Owen-Smith's chapter is very valuable in that it clearly pinpoints the shortcomings of the classical foraging theories. It stimulates our creativity, and urges us to start planning to test new ideas, in order to be able to incorporate scale in foraging theory.

ACKNOWLEDGEMENTS

We would like to thanks all participants at the workshop for sharing their opinions and ideas with us. This chapter therefore not only synthesizes our own ideas, but also the output of a very fruitful discussion with our colleagues to whom we owe the credits.

CHAPTER 9A

LARGE-SCALE MOVEMENTS OF LARGE HERBIVORES

Livestock following changes in seasonal forage supply

RANDALL B. BOONE, SHAUNA B. BURNSILVER, JEFFREY S. WORDEN, KATHLEEN A. GALVIN AND N. THOMPSON HOBBS

Natural Resource Ecology Laboratory, Colorado State University, 1499 Campus Delivery, Fort Collins, CO 80523-1499, USA
E-mail: rboone@nrel.colostate.edu

Abstract. Large-scale movements allow large herbivores to cope with changes in seasonal forage supply. Pastoralists use mobility to convert low-value ephemeral forage into high-value livestock. Transhumant pastoralists may move livestock less than ten to hundreds of kilometres. In semi-arid tropical sites, water and forage shortages in the dry season cause pastoral livestock to move to water or key resource areas. In temperate summers, livestock may be moved to higher-elevation snow-free meadows. In winters, animals may be moved lower to warmer sites, or to mountain valleys protected from steppe winds. Despite the recognised value of mobility, pastoral mobility is being reduced around the world. Changes in the mobility of three pastoral groups are reviewed, the Aymara of the South-American highlands, Mongolians, and the Maasai of Kenya and Tanzania, for which quantitative results are given. The Maasai of Kajiado District, Kenya are subdividing some group ranches into individually owned parcels. In subdivided Osilalei Group Ranch, herders moved an average of 5.6 km per day, whereas in undivided northern Imbirikani, herders moved 12.5 km per day. Residents of northern Imbirikani accessed more green vegetation the more they moved, whereas those in subdivided southern Imbirikani did not. Maasai selected areas with more heterogeneous vegetation during the dry season than found at their permanent households. In modelling, subdividing to 100-ha parcels allowed Eselengei Group Ranch to support 25% fewer livestock by mass, even though the area remained the same. For any pastoralist, the costs of mobility must be weighed against benefits, but pastoralists have demonstrated flexibility in their mobility, if constraints such as human population growth and limitations in land access are not too great. We show that pastoralists have successfully evolved methods of herding livestock to access adequate forage in areas of variable climate.
Keywords. Aymara; fragmentation; Kenya; Maasai; Mongolia; pastoralism; subdivision

H.H.T. Prins and F. van Langevelde (eds.), Resource Ecology: Spatial and Temporal Dynamics of Foraging, 187-206.

INTRODUCTION

Semi-arid and arid rangelands that are generally too dry to support rain-fed agriculture but have vegetation comprise about 25% of the landscapes of the world, excluding Antarctica (reviewed in Groombridge 1992). Twenty million or more households make their living as pastoralists on these lands, and ten times as many obtain a significant source of income from raising livestock (De Haan et al. 1997). Some form of pastoralism is practiced in every continent, excluding Australia and Antarctica, and a diversity of pastoral cultures and subcultures have evolved, especially in Africa, the Near East and West Asia, and the Indian region (FAO 2001). Most of these groups must contend with rainfall that is more variable within years, between years and across space than in more mesic regions (Ellis 1994). At its most basic, pastoralists have had to develop means of converting a spatially and temporally variable resource of little intrinsic value (grass) into a more stable, mobile resource of greater nutritional, economic and social value (livestock) (Swift 1977; Goldschmidt 1979). Adaptations allowing pastoralists to use areas with spatially and temporally variable rainfall are varied, but a central adaptation is through movements of livestock to make use of ephemeral forage resources. Livestock herders move their animals to different degrees (Box 9.1). This chapter focuses on transhumance and the effects of seasonal movements on livestock.

Movement of livestock is a crucial adaptation allowing pastoralists to use areas with spatially and temporally variable rainfall

***Box 9.1.** Livestock and pastoral movements*

Livestock herders move their animals in ways that may be broadly categorised into three classes (FAO 2001), although a continuum exists. Some movements are nomadic, using a given foraging resource, then moving on to other pastures following variable rainfall, with movement patterns notably different from year to year. Other movements are transhumant, where animals and people move between locations where forage is available seasonally. Movements may be short (< 10 km) or long (hundreds of km), and may be absent in years of very good rainfall (Kavoori 1999) or extreme in years of severe drought (Bekure et al. 1991), but movements in years of typical rainfall follow a predictable pattern. Agropastoralism is practiced by those that cultivate lands and raise livestock. Their livestock movements tend to be short, allowing family members to remain close-by and to work their agricultural plots.

In rangelands around the world, the mobility of pastoralists has been, or is being, reduced. Reductions are due to exogenous sources, such as increased transportation costs, land subdivision and changing government policies, as well as endogenous sources reflecting the pastoralists' desires, such as to be near schools, hospitals and other services, or to work agricultural plots. The literature of the past 15 years includes pleas for the mobility and land access of pastoral peoples to be maintained (e.g., Behnke and Scoones 1993; Scoones 1995; Niamir-Fuller 1999; Chatty and Colchester 2002). However, mobility has been reduced, as evident in the case studies we present. Today an important research focus is on quantifying the effects of sedentarisation and on adaptive strategies pastoralists may invent or adopt that allow them to lessen the negative effects of sedentarisation and improve decision making in the face of uncertainty.

We briefly discuss some general principals in transhumant pastoralism. We then seek to introduce transhumance patterns, but transhumance is as variable as the pastoralists the literature describes (Dyson-Hudson and Dyson-Hudson 1980). To limit our contribution, we review traditional patterns of livestock movements in three groups inhabiting three continents, selected to represent short-, long- and medium-range seasonal movements: the Aymara of the South American Andes, the Mongols of Mongolia, and the Maasai of Kenya and Tanzania, where we focus upon southern Kajiado District, Kenya. Some effects of fragmentation and other interventions on the seasonal movements of Mongolian and Aymara pastoralists are briefly cited, and the status of Maasai transhumance in Kajiado is reviewed. We then present quantitative effects of declining access of livestock to a diversity of forage patches due to profound land tenure changes in Maasailand. Maasai herders' selection of seasonally available green forage patches is quantified, and modelling results quantify the effect of declining parcel size on livestock production and human welfare. We conclude by reviewing some effects of fragmentation and emphasise the flexibility of pastoralists to adapt to stressors, if limitations are not too extreme.

Most pastoral systems have evolved responses that entail moving livestock seasonally

SEASONAL MOVEMENTS OF LIVESTOCK

Most pastoral systems have strong seasonality, with extremes in temperature (summer and winter), precipitation (dry season and wet season) or both. Forage quality and quantity vary through time in any pasture, but in semi-arid and arid lands, seasonal changes in forage quality and production can be extreme. In many regions, pastures cannot support livestock throughout the year, and water may be unavailable for portions of the year. Access to forage may be limited (e.g., because of snow depth), production may be inadequate, or the nutrient content of forage may be low (Bokdam and WallisDeVries 1992; Turner 1998; Schareika 2001; Kerven 2002; Mishra et al. 2003; Mishra et al. 2004). Pastoralists have evolved responses that entail moving livestock seasonally, so that the aggregate access to forage is the sum of access to forage 'pulses' within grazed landscape patches (Pickup and Stafford Smith 1993). Rotational movements also allow grazed pastures to rest between uses, reduce the likelihood that diseases or pests will become a severe problem (Kavoori 1999), and can help maintain biodiversity in some pastures (Zervas 1998). Pastures may be burned to reduce insects and encourage new growth (e.g., Bassett and Koli Bi 1999; Van de Vijver et al. 1999). Large herds are apt to be moved longer distances, in part because large herds require more forage, and in part because the costs per animal are too high when moving small herds (Humphrey and

Sneath 1999; Kerven 2002). Herds may be moved several times throughout the year, or even throughout a season, and if labour is available herds are split into groups (Evans-Pritchard 1940; Stenning 1959; Swift 1986), such as young and milking animals that stay near the households and heifers and steers that are taken far afield. Transhumance should not be viewed solely as livestock and people moving about on foot. Transhumance can be improved or made possible using vehicles, concrete loading bays, watering troughs, etc. (Chang 1993; Karoovi 1999; Kerven et al. 2003); we provide an example from Mongolia. In tropical systems, high-quality forage and water are generally most plentiful during the wet season (e.g., Prins 1989a; Prins and Beekman 1989). Areas where permanent water is not available are often used in the wet season, leaving areas with year-round water as reserves for use in the dry season (Bernus 1979; Galaty 1980). Livestock may be moved closer to temporary or permanent households to reduce travel costs and allow families easier access to lactating livestock. In temperate systems, summer months are times of plenty. In mountainous areas, livestock are typically moved to higher elevations, to make use of snow-free high mountain meadows and to prolong milk production (Chang 1993; Jina 1999; Mishra et al. 2003).

In the tropical dry season, forage availability and quality, rainfall, humidity and water availability decline (Stenning 1957). Some livestock species (e.g., goats, camels) may remain on landscape patches used in the wet season, relying upon woody vegetation. However, typically livestock are moved to areas where forage remains greenest and water is available (e.g., Evans-Pritchard 1940). Dry season sites may be highland slopes that receive more rainfall, areas with soil properties that lead to better plant growth (Schareika 2001), grazing reserves intentionally avoided other times of the year, areas free from insect pests (Stenning 1957, 1959), heavily grazed areas around water sources that are the only remaining options (e.g., Schareika 2001), or drainages, wetlands or other key resource areas (Box 9.2) that provide forage even in dry periods. Crop residue can be a key resource for some pastoralists (Jina 1999; Kavoori 1999; Turner 2003). Livestock may be trekked long distances to forage on residues left after crops have been harvested, often in formal arrangements that benefit both the pastoralist (access to residue and perhaps payment) and agriculturalist (manure and urine deposited on the cultivated plot) (Heasley and Delehanty 1996).

In temperate regions, winters bring cold, snow, and reduced quality and access to forage. Grasses may be covered by accumulated snow, and crusted snows make access to forage difficult or impossible. Herders move animals to sheltered valleys and lower elevations with less snow and higher temperatures (Jina 1999; Mishra 2003), although access to areas swept free of snow by wind may be valued. In some systems, such as those of Inner Asia and the Andes, pastoralists move from their autumn pastures to higher elevations, locating sites in mountain valleys that are protected from strong winds on the steppe or plains (Humphrey and Sneath 1999).

***Box 9.2.** Key resource areas*

Many semi-arid and arid grazing areas are expanses of rangelands with low or episodic vegetative productivity, with smaller areas of higher, more reliable, or extended primary production. Drawing from his work in Zimbabwe and examples from elsewhere, Scoones (1991) documented the importance of small wetlands in livestock management, and coined the term key resources, or key resource areas, for the wetlands. Primary production within key resources may limit the number of ungulates that can occur in an area (Illius and O'Connor 1999, 2000), and may be limiting factors (Blackman 1905), but are generally small and can be delineated from the surrounding landscape. Examples of key resources include wetlands, lake and river floodplains, and high-elevation grasslands that stay green longer than lowland rangelands (see also Scholte and Brouwer, Chapter 10).

Key resource areas in semi-arid and arid areas throughout Africa are threatened due to land-use intensification and human population growth. For example, in Kajiado District, Kenya, the margins of the swamps outside Amboseli National Park are being converted to cultivated plots by agropastoralists, with water from the swamps used in irrigation and their livestock grazed nearby year-round (BurnSilver et al. 2004; Worden et al. 2003). Stakeholders are concerned about the effects that loss of swamp area, access to water and continuous grazing by livestock have on Maasai food security and area wildlife.

Seasonal movements of livestock are not solely associated with forage quality or quantity or water availability. Traditionally, movements were constrained or altered by social, tenural, labour or political restrictions, military or other security threats, large rivers or disease (e.g., Stenning 1957; Dahl and Hjort 1976; Frantz 1978; ILCA 1979; Turner 1999b), and those constraints and others exist today. Increasingly, it is a combination of socioeconomic and political factors that strongly influence the ability of pastoralists to continue using mobility as an adaptation to seasonal resource heterogeneity in dry rangelands.

TRADITIONAL RESPONSES TO SEASONAL FORAGE AVAILABILITY

The Aymara of the highlands of South America provide an example of short-range seasonal movements of livestock in a temperate system. The Aymara raise llamas and sheep for meat and alpaca for wool, with other species (e.g., cattle, horses, pigs) less common (Orlove 1977). The environment is extreme; the Bolivian area studied by Buttolph and Coppock (2001) was 3,900 m in elevation, with 260 days with frost in an average year and large swings in diurnal temperature. During the summer wet season, families move to houses at lower elevations to make use of productive grasses and herbs. In the dry winter, families return to the highlands, which remain relatively moist (Orlove 1977). Households use designated landscape patches within lands owned communally. These landscapes include *bofedales*, which are natural or man-made high-elevation peatlands with more than 60 species of perennial grasses, herbs and sedges (Moreau et al. 2003). *Bofedales* are an important grazing resource in the dry season. Llamas tend to graze in upland habitats during the wet season, but alpaca and sheep are regularly moved between upland habitats and *bofedales*, conserving forage in the *bofedales* while maintaining an adequate nutritional state for the livestock. In the dry winters, alpaca and sheep predominately use *bofedales* for forage and water, and llamas use these resources somewhat more than in the wet season as well (Buttolph and Coppock 2001).

Mongolians have adopted complex movement patterns in response to extreme climatic conditions (Enkhtuvshin and Tumujav 2002). This system provides an example of some groups making short seasonal movements, and some very long movements. Mongolian growing seasons are brief, with most of the annual rainfall in the summer, with rainfall totals of less than 300 mm annually, except for the northern zones. Sheep, camels, goats, cattle, horses and yaks are herded for meat, milk, wool and transport – mostly indigenous breeds that can withstand the low winter temperatures without housing and restore body condition quickly during the short growing season. Seasonal movements are made to access some or all of the desert, desert steppe, mid-altitude steppe, mountain steppe and forest steppe (Mearns and Swift 1995). Movements span from 10-km shifts two to four times a year between protected valleys used in winter to nearby summer pastures, to 300-km treks between open mountain passes used in the summer, autumn in lowlands, with a return to mountain passes in winter (Fernandez-Gimenez and Allen-Diaz 1999; Enkhtuvshin and Tumujav 2002), seeking snow as a water source for livestock and shelter for livestock and people from strong steppe winds (Suttie 2000). In general, four seasonal grazing areas are used (Fernandez-Gimenez and Allen-Diaz 1999; Enkh-Amgalan 2002). Winter and spring pastures are most important to the survival of livestock and are in limited supply, whereas summer and autumn pastures are often understocked (Suttie 2000).

Pastoralists have adopted complex movement patterns in response to extreme climatic conditions

The final example of seasonal movements is for our focal pastoral group, the Maasai of southern Kajiado District, Kenya. Within the district, and elsewhere in Maasailand, herders make medium range movements throughout the seasons to access green forage. Traditionally, Kajiado herders used lands communally, and movements were subject to complex use rights, within large Maasai sections (Figure 9.1a) (Galaty 1980). The short and long wet seasons brought highly nutritious forage that was readily available (Bekure et al. 1991), and many Maasai grazed their cattle, goats and sheep near their permanent households. Others moved their herds to temporary households within wet-season grazing areas. As forage was consumed or dried, livestock were moved farther away from the permanent settlement areas, to nearby areas of remaining green forage, with herds ultimately occupying dry-season zones. The timing of return to permanent households for those that migrated to dry-season grazing areas was often determined by water shortages, as well as by forage availability.

RESPONSE TO SEASONAL FORAGE AVAILABILITY UNDER INTERVENTIONS

Buttolph and Coppock (2001) provide an example of a negative effect of intervention on behalf of transhumant Aymara people – in this case, the pastoralists' own production association. In 1993, Project Alpaca was begun by the Asociación Integral de Granaderos en Camélidos de los Andes Altos, comprised of Aymara herders. They sought to improve alpaca wool production. Among the interventions made, the association provided credit and barbed-wire fencing for herders to fence the *bofedales* that they used, so that grazing could be controlled seasonally and land-use conflicts reduced. There were some benefits to fencing *bofedales*, but the effect of interest was that *bofedales* that were once managed communally were fenced for private use. In one site, about half of the accessible *bofedales* were fenced within two years. Instead of land-use conflicts being reduced, they had been intensified. More importantly, this magnitude of loss of access to key resources does not bode well for the Aymara in drought (Buttolph and Coppock 2001).

In Mongolia, intervention came in the form of profound political change. Under socialism, the livestock sector was collectivised in 1950, although some stock remained privately held. Families were required to raise single-species herds under relatively intense management, including increased hay and fodder production and use of government-provided mechanised transport, boreholes and simple livestock shelters. Unlike in the past, households were associated with management units, called *negdels*, which restricted their opportunities to move to access forage relative to their historic seasonal movements (Suttie 2000). Livestock were using pastures for longer periods than under the traditional system. The centralised government also attempted to avoid overstocking, although stocking rates were elevated and degradation did occur (reviewed in Kerven (2002) for areas to the west). In 1990, the centralised system of government ended and efforts were put in place to create a market economy (Mearns and Swift 1995). Much of the subsidised support for livestock production ceased, including most hay production and mechanised transport, and many wells failed. In the years since, the degree to which traditional transhumance patterns have re-emerged is mixed. Many families owned too few animals to maintain a transhumant, or even pastoral, lifestyle – in 1995, more than 40% of households had fewer than 50 head of livestock, which is the poverty line (Suttie 2000). In some areas, land-use rules are now absent or weak, with new (ex-urban) or displaced herders using lands not traditionally theirs to use (Mearns and Swift 1995). Many pastoral families have re-established mixed-species herds and have resumed some seasonal movements, although distances travelled are shorter than what was traditional (Humphrey and Sneath 1999). In general, although control of grazing and movements have persisted or has re-emerged among family units, such control is at a spatial scale too small for efficient management of the variable and extensive grazing resources of the region (Humphrey and Sneath 1999; Suttie 2000).

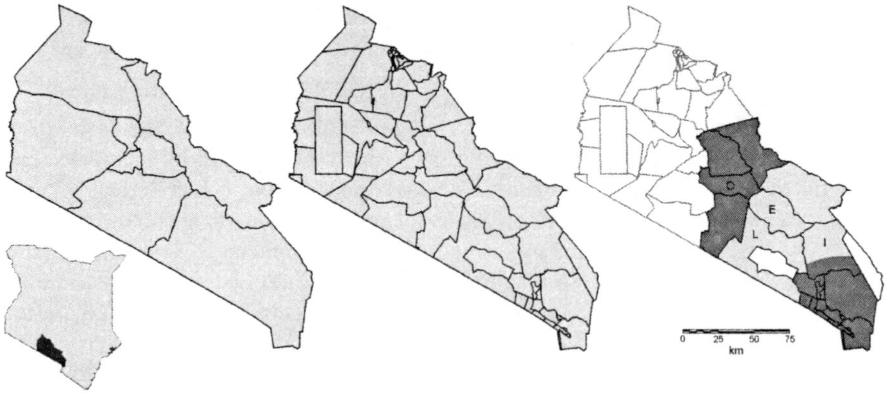

Figure 9.1. *(a) Maasai sections within Kajiado District, south-western Kenya (Ole Katampoi et al. 1990), (b) Kajiado group ranches and (c) the state of subdivision of ranches within the area we modelled are shown. Areas in (c) that are dark grey are subdivided, those that are light grey remain communally held, Amboseli National Park in the southwest is white, and West Chyulu Game Conservation Area to the east is in white. Group ranches cited include Imbirikani ("I" in c), Eselengei ("E"), Olgulului/Lolarashi ("L"), and Osilalei ("O") (Group ranch boundaries are ill defined; approximate boundaries are shown)*

In Kajiado District, Kenya, land tenure has changed markedly in the last 30 years after a series of economic and political interventions instituted for the most part from outside the pastoral system. The Kenyan government, in cooperation with the World Bank, began dividing Maasai sections (Figure 9.1a) into group ranches (Figure 9.1b) in the late 1960s and early 1970s (Galaty 1980). Group ranches were formed to improve livestock production, ease the provision of services, and secure land ownership. In general, group ranch formation failed to meet its original goals (Galaty 1994; Heath 2000), although ranch formation has allowed lands to stay largely in Maasai hands. From 1965 to 1975, Kajiado District was adjudicated, and the district was divided into 52 ranches (Figure 9.1b) that are used somewhat exclusively by group ranch members (Kimani and Pickard 1998). Members graze their livestock within their own ranches throughout the year, but in years of drought, agreements can allow herders to move between group ranches. In 1983, the government sanctioned subdivision of ranches (Kristjanson et al. 2002), and today subdivision continues, with group ranches being further subdivided into parcels held by individual herders or families (Figure 9.1c). There have been many social and institutional effects of subdivision in Kajiado (e.g., Galaty 1980; Bekure et al. 1991; Rutten 1992; Galaty 1994; Kristjanson et al. 2002; BurnSilver et al. 2004); in the next section, we focus upon effects on livestock and household status.

REDUCED ACCESS TO FORAGE HETEROGENEITY

We turn to qualitative and quantitative analyses that reflect the importance of access to heterogeneous forage patches in livestock production and the associated costs of fragmentation for herders under land subdivision. In analyses relating to biocomplexity, we are assessing the effects of fragmentation on ungulates and human welfare. Theoretically, we hypothesise a humped-shaped relationship between the importance of landscape fragmentation to livestock and system productivity (Box 9.3). Here we focus upon the middle portion of that hypothesised curve where loss of access to heterogeneous forage patches can reduce herbivore capacity (Figure 9.2), the range of productivity represented by southern Kajiado, Kenya. We use satellite images in analyses to represent the strength of selection for green vegetation by Maasai, or alternatively, the cost of sedentarisation due to landscape fragmentation. Process-based ecosystem modelling is used to quantify the effects of landscape fragmentation on livestock stocking rates.

Reduced access to heterogeneous forage patches in livestock production under land subdivision results in costs for herders

Box 9.3. *Primary production and effects of fragmentation*

We hypothesise a quadratic (humped-shaped) relationship between the importance of landscape fragmentation to livestock and system productivity. Very arid systems with low primary productivity and low stocking rates, where livestock travel costs cannot be increased and primary and secondary productivity are weakly linked (Ellis and Swift 1988), may be insensitive to fragmentation at broader scales. Conversely, at exceedingly productive sites, forage production may be adequate to supply livestock their needs, and stocking is limited by other factors (e.g., behavioural restrictions because of crowding, disease transmission risks, etc.); fragmenting the landscape into small units may have little effect upon livestock production. In turn, livestock inhabiting homogeneous pastures are less sensitive to fragmentation than those inhabiting heterogeneous pastures. For heterogeneous pastures in a moderately productive system (i.e., near the top of our humped-shaped curve), fragmentation can reduce the foraging choices available to livestock.

Images and modelling tools

Satellite images have often been used to represent vegetation greenness. Ratios of the near-infrared and infrared bands are termed Normalised Difference Vegetation Indices (NDVI), and reflect vegetation biomass and vigour. The images cannot represent all relevant aspects of semi-arid lands, such as the prevalence of unpalatable or exotic vegetation, but NDVI values are correlated with ungulate stocking rates (Oesterheld et al. 1992; 1998; Ottichilo et al. 2000b), and have been used in research extensively (e.g., Tucker et al. 1985; Eklundh 1998; Boone et al. 2000; Skidmore and Ferwerda, Chapter 4). Satellite images were acquired from the SPOT program, Earth Observation System, which was developed by the Centre National d'Etudes Spatiales of France, with cooperation from the governments of Sweden and Belgium. Recent SPOT satellites have included a vegetation sensor, which has a coarse resolution (1.15 km square pixels). Vegetation NDVI images are freely available at full resolution (1 km pixel) for entire continents (VITO 2003), with the earliest images from April 1998. These are 10-day (i.e., decadal) composite images, where the best NDVI value available (based on sun and sensor angles, etc.) is selected for the 10-day period. We acquired the NDVI images for Africa from 1999 and 2000.

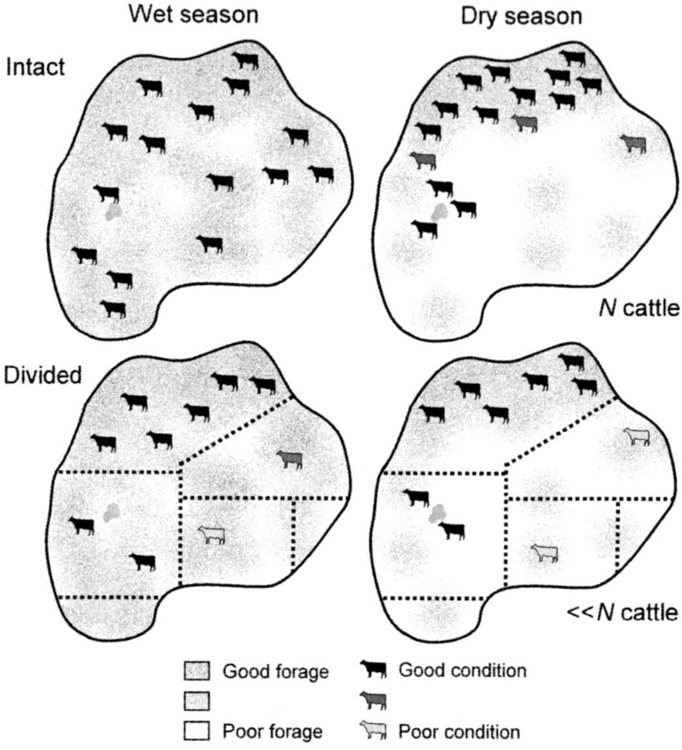

Figure 9.2. *A schematised view of movements by livestock in an area of heterogeneous forage patches of moderate productivity. Animals move about freely (a) in an intact landscape, shifting to high-elevation grazing areas and a key resource in the dry season. If the landscape is divided into parcels (b), some parcels contain adequate forage through the dry season and livestock prosper, but other parcels cannot support livestock or support animals in poorer condition*

Two models were used in the analyses, the SAVANNA ecosystem model and a pastoral-household decision model called PHEWS (Pastoral Household Economic Welfare Simulator). A full description of these models is beyond the scope of this review, but more detail is available (Ellis and Coughenour 1998; Boone 2000; Boone et al. 2002; Thornton et al. 2003). In general, SAVANNA is a series of interconnected computer programs that model primary ecosystem interactions in arid and semi-arid landscapes, simulating functional groups for plants and animals. SAVANNA is spatially explicit and represents landscapes by dividing them into a system of square cells that have spatial data associated with them. The model predicts water and nitrogen availability to plants using rainfall and soil properties, for each of the cells. Based upon water, light and nutrient availability, products of photosynthesis are calculated for plant functional groups, using process-based methods. The carbohydrates are distributed to leaves, stems and roots using plant allometrics, yielding estimates of primary production and from that, plant populations. A habitat suitability index is calculated for each cell in the landscape, at weekly intervals and for each animal functional group, based upon forage quality and quantity and physical attributes of the cell. Individuals in the population are distributed in the landscape based upon these indices. Animals will feed upon the available vegetation, and energy gains and losses are tracked, as well as changes in populations. Summaries of the status of vegetation, herbivores and climate are produced at monthly intervals.

The PHEWS model simulates decision making in Maasai households (Thornton et al. 2003). A series of rules that reflect decision making in Kajiado were incorporated, determined from interviews and published sources. Families seek to meet their caloric needs, while simultaneously seeking to build livestock and monetary holdings. Calories are gained from milk, tea with sugar, and livestock slaughtered due to disease or for occasional ceremonies. More calories are needed, so available maize and other crops are eaten. If there remains a caloric deficit and the family has money or animals to sell, grains and other crops are purchased. Finally, if a deficit remains and livestock cannot be sold, the families' needs are met through supplemental food. The PHEWS model is tightly linked to SAVANNA. For example, SAVANNA reports to PHEWS livestock populations, and PHEWS reports back to SAVANNA the numbers of livestock sold so that population dynamics may be tracked.

Grazing-area analyses

Based on survey results, we have shown that the daily pathways (i.e., grazing orbits) were shorter in a completely subdivided group ranch than in unsubdivided ranches (BurnSilver et al. 2004). BurnSilver and Worden conducted surveys in 6 communities within four group ranches: Imbirikani, Olgulului/Lolarashi, Eselengei and Osilalei Group Ranch (Figure 9.1c). Osilalei Group Ranch is fully subdivided, with ranch members each owning individual parcels of approximately 40.5 ha (100 ac). The other group ranches are not subdivided, but the wetlands of southern Imbirikani Group Ranch are being subdivided for cultivation. 61 daily grazing pathways from 32 herds were recorded during a wet and dry season, using global-positioning technology. Herders in subdivided Osilalei moved 5.6 km per day in the wet season, whereas herders in communally held northern Imbirikani moved 12.5 km. These movements may be additionally affected by differences in vegetation productivity between the ranches (Osilalei is more productive than Imbirikani), but the differences in distances travelled are large, and 86% (18 of 21 people surveyed) stayed on their own parcel in Osilalei during 1999.

As part of the surveys, BurnSilver and Worden asked Maasai herd owners to identify areas where they had grazed livestock. For 46 herds, locations were recorded each month for a calendar year of average rainfall (1999) and an extremely dry year (2000); here we present results from the average year. The approximate centres of the seasonal grazing areas were later identified using geographic-positioning instruments, and grazing areas are assumed to be circular, with radii defined using summaries from daily grazing orbits cited above. Whether each location was associated with a permanent or temporary household was noted. Two spatial data sets were generated from these lists: the locations of herds as they moved in reality, and the location of herds if they remained near their permanent households, emulating sedentarisation through subdivision. The differences between mobile and simulated-sedentarised herd locations in communal northern Imbirikani Group Ranch were large, whereas the differences were small for southern Imbirikani, where most pastoralists are engaged in agriculture and typically do not move their animals long distances.

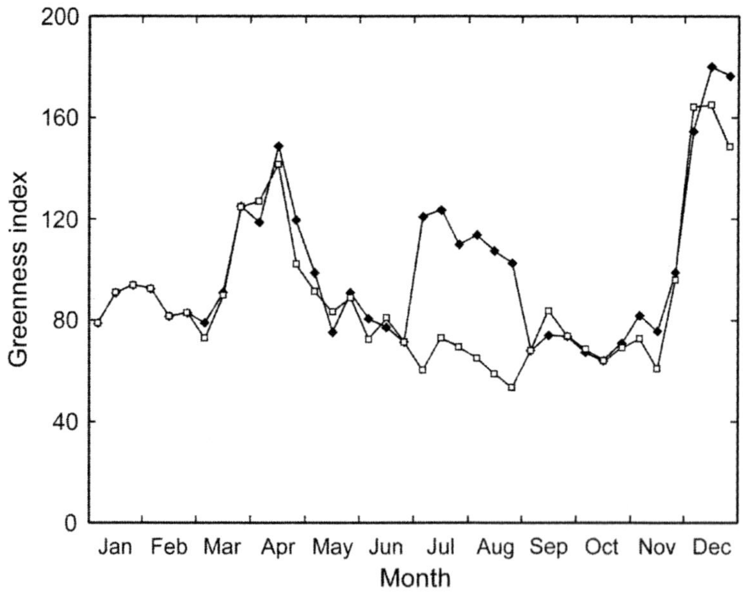

Figure 9.3. *Greenness indices, from 1999 NDVI, tracked by pastoralist "98" in Imbirikani Group Ranch in reality (solid diamonds) and if forced to graze only near permanent settlements (open boxes). In 1999, the herd was moved 6 times, and access to green forage improved in the stressful long dry season, relative to if the herder used only the permanent settlement area. Here the difference in access to green forage was large (380 units), but on average integrated greenness accessed by Maasai herders of northern Imbirikani that moved was 61 NDVI units higher than when those same herders were simulated to be sedentary. NDVI indices are based on greenness measured by satellite images, and are a good indication of primary productivity (e.g., Tucker et al. 1985; Paruelo et al. 1997)*

Integrated (i.e., summed) greenness profiles were generated from NDVI based upon seasonal movements, and compared to greenness profiles based on the locations of permanent settlements. Pastoralists improved their herd's access to green forage by moving, compared to if they remained around their permanent settlements (e.g., Figure 9.3). Access to greenness improved (Figure 9.4) as inhabitants of northern Imbirikani moved more ($R^2 = 0.59$, $P < 0.001$, with one herd that left Imbirikani Group Ranch excluded) with up to 6 movements made, whereas inhabitants of subdivided south Imbirikani did not move more than three times and showed no improvement in access to green forage as movements increased ($P > 0.1$).

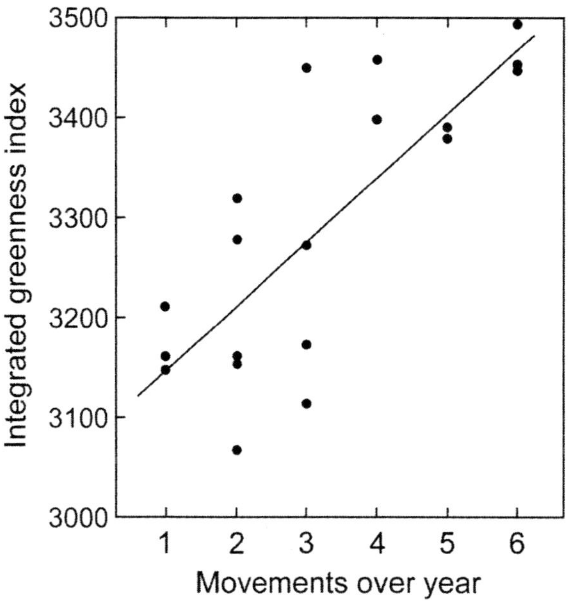

Figure 9.4. *Pastoralists in northern Imbirikani Group Ranch accessed more green forage if they moved, as reflected in integrated greenness indices from NDVI. NDVI indices reflect primary productivity, as cited in the legend of Figure 9.3*

We hypothesised that areas used for seasonal grazing by Maasai would be more temporally and spatially variable in vegetation greenness than areas around their permanent settlements. Measures of vegetation heterogeneity were created by calculating standard deviations in changes in greenness across images within wet seasons (combined short and long seasons, i.e., last image in October, November, December, March, April, May) and dry seasons (i.e., January, February, June, July, August, September and the first two images of October). Standard deviations were used rather than coefficients of variation to avoid standardising the variation by the mean; a 50-g increase in forage production is of similar value to livestock whether in a pasture with 100 or with 350 g standing biomass. The mean of the standard deviations within a 2-km moving window around each pixel in the image was then calculated using Arc/Info (Environmental Systems Research Institute, Redlands, California, USA). This created heterogeneity indices (Figure 9.5) similar to those in BurnSilver et al. (2004), except that elevation and soils were not incorporated. Monthly grazing areas used by Maasai in Imbirikani Group Ranch and neighbouring northern Chyulu (Figures 9.1 and 9.5) were overlaid upon the seasonal heterogeneity indices and mean heterogeneity indices calculated.

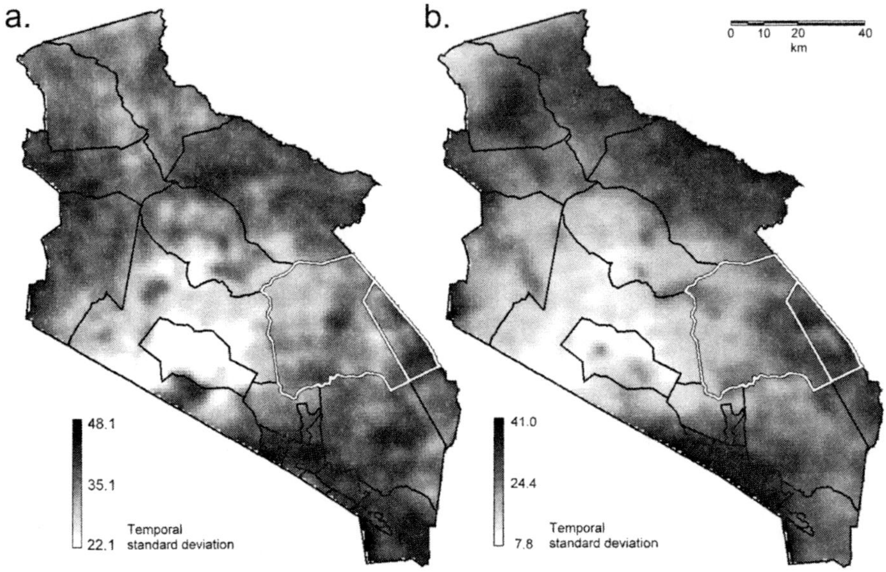

Figure 9.5. *Vegetation heterogeneity indices in the (a) wet seasons and (b) dry seasons. Indices reflect the standard deviation in NDVI across time, smoothed across space. The areas bounded in black and white are Imbirikani Group Ranch and the northern portion of Chyulu Hills, the area used in the analyses*

Maasai in communal northern Imbirikani Group Ranch selected more heterogeneous landscape patches in the dry season (Figure 9.6) than when we simulated herders remaining at their home settlement year-round. Mobility allowed more access to heterogeneity in the wet season as well, although the differences were small. Maasai in southern Imbirikani rarely move between seasons, and if simulated to be entirely sedentary, showed no difference in selection for vegetation heterogeneity (Figure 9.6). Maasai of southern Imbirikani occupy swamp margins – subdivided key resources that are highly heterogeneous, yielding large indices for their permanent settlements.

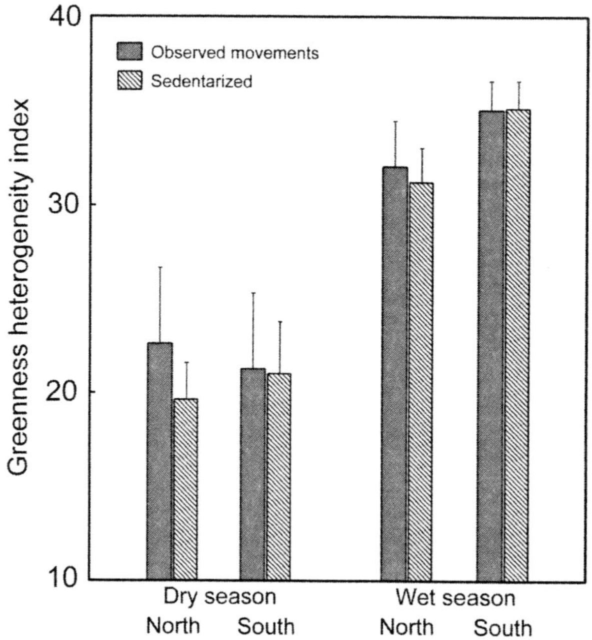

Figure 9.6. *Heterogeneity indices of areas grazed by Maasai herders (grey bars) in the wet and dry seasons in northern and southern Imbirikani Group Ranch, and indices if herders used only their permanent households (hatched bars). Northern Imbirikani residents selected heterogeneity differently in the dry seasons ($N = 126$ movements; $P < 0.001$) and the wet seasons ($N = 89$; $P = 0.012$). Heterogeneity indices were calculated from NDVI images, which reflect primary productivity, as cited in the legend of Figure 9.3*

SAVANNA / PHEWS *modelling*

In a theoretical setting emulating a semi-arid ecosystem, the SAVANNA model was adapted to include only cattle in a 300-km^2 landscape, and to disregard effects of water supply (Boone and Hobbs in press). Simulations were then run for each parcel with the block fragmented into two 150-km^2 parcels, three 100-km^2 parcels, ..., fifteen 20-km^2 parcels and thirty 10-km^2 parcels (Figure 9.7a). Fragmenting the system into 10-km^2 parcels caused a significant decline in the livestock population that could be supported across the entire block of land (Figure 9.7b); 19% fewer animals could be supported when entirely fragmented.

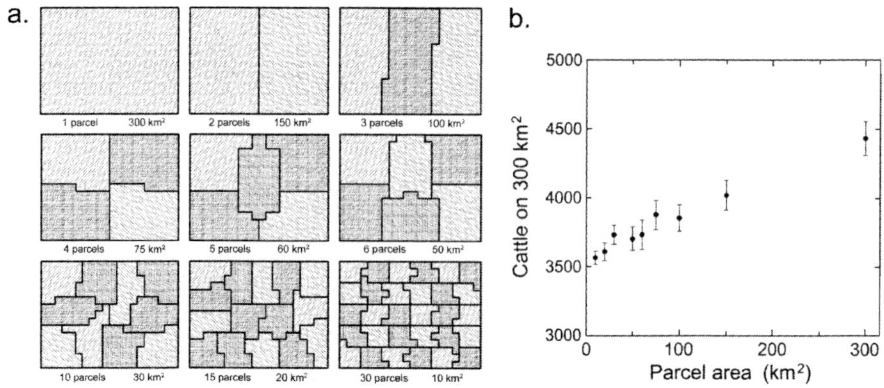

Figure 9.7. *Parcels of different areas (a) were used in SAVANNA simulations in a hypothetical landscape. Cattle that could be supported over the entire 300 km^2 block over the long term (b) declined by 19% when fragmented to 10 km^2 parcels*

We may ask what the effect of subdivision into small parcels in Kajiado may be on livestock and human welfare, or alternatively, what level of external inputs will be required to maintain human welfare under subdivision. The joined SAVANNA/PHEWS model was adapted to southern Kajiado District, and included three livestock and eight wildlife populations (Boone et al. in review). Using replicated simulations, the effects of fragmentation on livestock and household welfare were quantified. For Eselengei Group Ranch, livestock populations for the entire ranch declined by 25% as the ranch was fragmented into 1-km^2 (250 ac) parcels (Figure 9.8) (Boone et al. in review). Incidental to boding poorly for Maasai food security, these results highlight the inappropriateness of assigning a 'carrying capacity' to an entire landscape, regardless of patch size. When simulated with PHEWS, effects of these losses on human welfare were extreme. As livestock holdings declined and food security lessened, Maasai sold animals to purchase grain, which further reduced food security and led to the sale of more animals. In analyses, Maasai households at their current density that were forced to graze their animals on 196-km^2 parcels in Eselengei Group Ranch could not persist without massive economic support from outside the system or dramatic changes in pastoral economic strategies.

Fragmenting the system into small parcels caused a significant decline in the livestock population that could be supported across the entire block of land

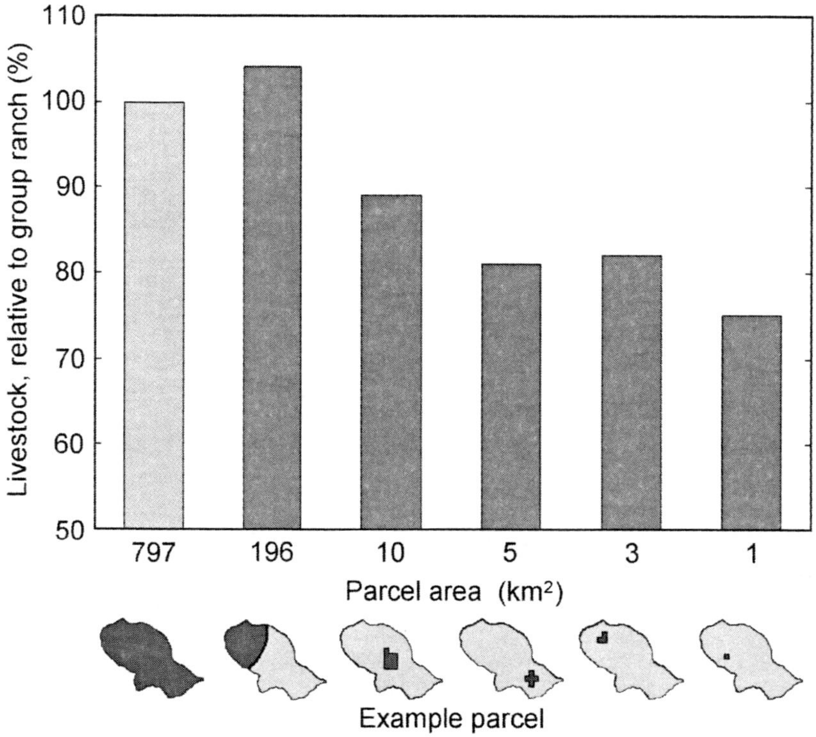

Figure 9.8. *In a SAVANNA/PHEWS application to southern Kajiado District, livestock (in tropical livestock units: TLU = 250 kg) declined significantly as Eselengei Group Ranch was fragmented from its full area (797 km²) to 1 km²*

SYNTHESIS

Lane and Moorehead (1994, p. 123) put it plainly, that "settlement of nomadic pastoralists is the greatest single transformation of pastoralism as both a production system and a way of life". Sedentarisation has been pursued as a specific goal of policy reforms, a secondary outcome of governmental administration or neglect, and as a philanthropic goal of non-governmental organisations to ease the provisions of services (Niamir-Fuller and Turner 1999). But what was once a laudable goal and remains a frequent outcome of fragmentation is now discouraged, as one of the three hard-earned lessons Sandford (1994, p. 179, emphasis added) cites "My personal opinion is that we social scientists have not yet structured our views rigorously enough to have any clear message for policy makers and practitioners except that everything is very complex, that Hardin (1968) was wrong *and that livestock mobility is to be encouraged*". Dramatic examples of improved survival in herds that moved relative to sedentary herds have been reported (e.g., Scoones 1992; Kavoori

1999). We have cited negative effects of reduced mobility in three areas on three continents, quantified the positive effects of greater mobility in Maasailand, and provided modelling results that quantify losses under increasing fragmentation.

That said, the costs of moving livestock, especially for small herds, must be weighed against the benefits (Dyson-Hudson and Dyson-Hudson 1980; Schareika 2001; Kerven et al. 2003), and the benefits of other adaptations to fragmentation, such as changes in herd structures, production systems or intensification (e.g., Dahl and Hjort 1976; Swift 1977; Dyson-Hudson 1980). There are real costs associated with movement, such as transportation and labour costs, plus costs associated with the maintenance of complex social networks. Transhumant pastoralists cannot simply move about seeking the greenest pastures, irrespective of social constraints (Evans-Pritchard 1940; Stenning 1959). Areas may be set aside as grazing reserves, to provide late-season forage or to rest the vegetation. Beyond that, societal relationships can be very complex, changing throughout the season, affected by social norms, religious views and politics (Dyson-Hudson and Dyson-Hudson 1980; Lane and Moorehead 1994; Sylla 1994; Niamir-Fuller and Turner 1999; Turner 1999b), and the social networks to which pastoralists must appeal have sometimes themselves been fragmented, such as by emigration, diversification or disease (e.g., HIV/AIDS). Competition for land is high, especially for key resources and access points (Prins 1987b; Turner 2003) – indeed, some key resource areas are individually controlled and are no longer in competition, such as the fenced *bofedales* of the Aymara (Buttolph and Coppock 2001) or the riverine trees privately owned by Turkana families or controlled by well-armed rivals (Lind and Sheikh 2001; Mbogo 2003), a de facto privatisation. Areas used by livestock that are marginal for agriculture are now being converted to cultivation, and herds are in closer proximity to cultivated lands (Ottichilo et al. 2000b; Turner 2003). In regions such as West Africa, areas used seasonally by transhumant pastoralists are favoured for cultivation, because of the build-up of manure (Heasley and Delehanty 1996), although livestock make use of some cultivated lands (Kavoori 1999).

The importance of maintaining transhumant patterns has gained acceptance, but benefits gained from transhumance may be outweighed by changes in land tenure systems – private property and intensification as a foundation of investment and economic growth pervade economic policy (Stenning 1959; Oxby 1982) – and rapidly expanding human populations. The Aymara studied by Buttolph and Coppock (2001) had not emigrated to pursue non-pastoralist lifestyles, and the population was high, increasing rates of trespass. In Mongolia, even though livestock numbers had been fairly stable from 1950 to 1996, the numbers of livestock per person had dropped by two-thirds due to human population growth (Suttie 2000). Similarly, in the well-studied Maasai system of Ngorongoro (Kijazi et al. 1997; NCAA 2000; see Prins 1992), livestock biomass has been relatively stable for 40 years, but livestock-to-person ratios have declined dramatically (Figure 9.9), a pattern that is similar for Kenya (Ottichilo et al. 2000b) and much of semi-arid East Africa. Land subdivision within Kajiado District may be inevitable, given the value of subdivided land in securing loans and maintaining control of group ranch resources. We do, however, encourage those holding lands individually within group ranches to avoid fencing their properties, and retain open access.

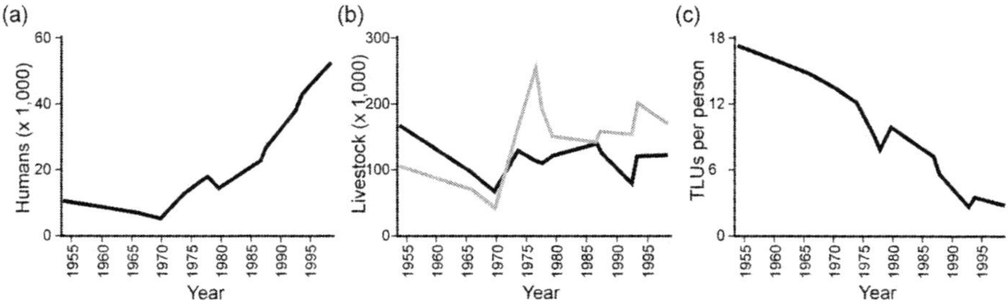

Figure 9.9. *(a) Human and (b) livestock (cattle – black line; small stock – grey line) populations in Ngorongoro Conservation Area, Tanzania. Livestock populations have been relatively constant (b), but increasing human population has led to a dramatic decline in Tropical Livestock Units (TLUs) per person (c), a unit of standardised livestock biomass, where cattle is 180 kg, goats and sheep 18 kg, and 1 TLU = 250 kg*

Pastoralists have successfully evolved methods of herding livestock to access adequate forage in areas of variable climate. Environmental, political, demographic and socioeconomic relationships are altering these long-term movement patterns, necessitating further adaptations or leading to insecurity, and spawning new research questions (Box 9.4). From local changes such as fencing of parcels to regional changes in climatic variability due to global climate change (Fowler and Hennessy 1995; Mason et al. 1999), pastoralists will have to adapt to new conditions. Calls for returns to historic patterns of transhumance are likely unrealistic because of human population growth and socioeconomic changes, and indeed may be detrimental to pastoral well-being – there are advantages to moving families shorter distances, such as access to hospitals and schools. That said, evidence to-date reflects well on mobility as a strategy allowing pastoralists to find new and creative ways to adapt to changing conditions, provided that political and socioeconomic restrictions on flexibility are not extreme.

Pastoralists have successfully evolved methods of herding livestock to access adequate forage in areas of variable climate

Box 9.4. *Testable hypotheses for future research*

Many questions remain about livestock seasonal movements, and management and policies that apply to the resources the livestock use.

Hypothesis 1. Sedentarisation of families and reduced mobility of livestock herds will cause declines in livestock productivity, but these may be offset by external inputs. At some point, declining mobility in moderately productive areas will cause populations to collapse. The area available to herbivores at that point of collapse should be related to measures of vegetative heterogeneity, from simple counts of land-cover types to more complex heterogeneity indices. If heterogeneity is related to minimum viable herd sizes, it will have important implications for stakeholders and policy makers, as well as implications under global change.

Hypothesis 2. Theory and model simulations have demonstrated that key resource areas can influence the number of livestock an area can support to such a degree that livestock populations may not be related to primary productivity in areas outside key resources (Illius and O'Connor 1999, 2000). Such a response would emulate non-equilibrium dynamics relative to the region, but in reality would represent equilibrium dynamics relative to the key resource (Illius and O'Connor 2000; Cowling 2000), although others disagree (see Sullivan and Rohde 2002). The simulation results have not been demonstrated in reality. An assessment of the theory would be an important contribution to a continuing debate (Briske et al. 2003).

Hypothesis 3. As mobility is reduced for livestock and human populations increase, a research focus has been on diversification of pastoral people, as they cultivate, start small businesses and work as wage labour. Recent research results (BurnSilver unpublished data) suggest that for Kenya, although diversification is occurring, economic returns are variable, and intensification is a dominant change in the system – livestock continue to bring the vast majority of income to Kajiado Maasaii. The relative importance of intensification to diversification is not well known elsewhere.

ACKNOWLEDGMENTS

We extend our appreciation to the pastoralists of southern Kenya Maasailand for allowing us to collect movement information and conduct surveys. Analyses by Philip Thornton quantified effects of fragmentation on Maasai households using PHEWS. The views and comments of the editors and other participants of the workshop on which this volume is based are appreciated. Their comments helped greatly in improving the chapter. The International Livestock Research Institute, Nairobi, Kenya, kindly provided institutional support. Original research reported here was sponsored by the Global Livestock Collaborative Research Support Program under Grant No. PCE-G-98-00036-00 from the US Agency for International Development, and by the US National Science Foundation under Grant No. DEB-0119618.

CHAPTER 9B

COMMENTS ON "LARGE-SCALE MOVEMENTS OF LARGE HERBIVORES: LIVESTOCK FOLLOWING CHANGES IN SEASONAL FORAGE SUPPLY"

WILLEM F. DE BOER[#], THOMAS A. GROEN[#], IGNAS M.A. HEITKÖNIG[#] AND KOEN KRAMER [##]

[#] *Resource Ecology Group, Wageningen University, P.O. Box 47, 6700 AA Wageningen, The Netherlands*
[##] *Alterra, Wageningen University and Research Centre, P.O. Box 47, 6700 AA Wageningen, The Netherlands*
E-mail: fred.deboer@wur.nl

Mobility is indeed a perfect tool to optimise exploitation through tracking changing resources, as shown in Boone et al. (Chapter 9) by the example of transhumance systems in various parts of the world. Fragmentation and private access can limit mobility of herds so that key resources can no longer be used, decreasing the overall productivity of these livestock systems. The chapter recommends therefore the re-instalments of mobility wherever possible, and indicates the risks associated with a reduced mobility of herds.

A pivotal question is whether key resources really exist. What are key resources (see also Scholte and Brouwer, Chapter 10)? How large does the landscape heterogeneity have to be in order to affect overall pastoral-productivity levels? Intuitively, the idea makes sense, but it would be good to identify the thresholds that produce significant effects on herd production.

Assuming that these heterogeneous key resources exist, the question remains whether an increased mobility or return to former transhumant systems is feasible. An important threat for transhumance is the privatising of lands. Privatisation can have an autocatalytic effect. Once a small group of pastoralists starts to privatise certain areas, others feel urged to do so as well. Conversion from privatised lands into communal lands, enabling transhumance is a difficult, if not impossible road with few examples. Fragmentation and privatisation are not easily stopped.

H.H.T. Prins and F. van Langevelde (eds.), Resource Ecology: Spatial and Temporal Dynamics of Foraging, 207-210.

Factors that hamper the re-instalment of transhumance are the increase in the human-population size, public services that are more accustomed to sedentary lifestyles, economic constraints or social changes. One of the problems is a rapid deterioration of vegetation due to overstocking on private grounds. So, not only access to pastures has changed but also pasture quality. An additional problem is that it is profitable to subdiverse land for future sales. Furthermore, banks do not provide loans without a clear ownership. Unfenced land with free-ranging cattle will not be accepted for loans. Therefore, the re-installation of transhumance often fails.

The realism of re-installation of transhumance is therefore low. A first lesson to be learned though is that if transhumance in an area still exists, and access is open to different herd owners, one should try to avoid fragmentation, privatisation, or other processes that lead to a reduced access to key resources. Secondly, a more important issue to be solved is: what is the best mobility strategy in a fragmented landscape? How can herd mobility, but also stocking rates or other aspects of the pastoral production system contribute to improved herd productivity in a privately owned, smaller fragment?

Moreover, fragmentation leads to a reduced overall productivity, and thereby results in increased prices. This reduces the overall gain but locally the gains are variable: some win, some lose. Fragmentation and privatisation of a heterogeneous landscape create owners possessing high-quality pastures or other key resources. The profits for those privileged owners are probably larger than the ones generated by a communal system, lowering the maximal profits per herd. Some capitalist herd owners might therefore see new opportunities, stimulating privatisation and limiting common access.

The chapter focuses on the negative effects of fragmentation on livestock production, but numerous positive effects have also been documented. Restricted access can provide stimuli for investments, improving resource quality through pasture management, decreasing the chance of overgrazing, or improving water availability. Small-scale investments of smallholders are often only possible when land can be used as security for the financier. In fact, the graphs presented in the chapter indicate that with relatively stable livestock numbers, and an increasing human population, livestock production efficiency could have gone up, apparently able to sustain a higher number of people. This brings us to an important question that is not addressed by the paper: what are the underlying causal mechanisms responsible for the decrease in livestock numbers or herd productivity with increasing fragmentation? The understanding of the mechanisms is instrumental when one wants to initiate mitigating measurements.

Another topic that needs urgent attention is the goal function of the herd owners. How important are risk minimisation strategies in shaping herd mobility? Owen-Smith's paper (Chapter 8) used Jensen's inequality principle to illustrate differences in foraging behaviour. This principle certainly also applies to transhumant livestock owners. Do owners minimise deficits, or maximise profits? How do pastoralists accommodate for uncertainty in their decision-making? A different goal function would influence herd productivity, change benefits, and therefore change mobility patterns in relation to the fragmentation level. The relationship between herd productivity and landscape fragmentation, as proposed by Boone et al., must be able to incorporate these different strategies in order to increase realism, thereby leaving the focus on maximum herd productivity. Studies that tried to optimise pastoral production have received, rightly so, a lot of criticism, as risk minimisation, social status, herd diversification, dowry and other factors all determine production decisions. It is therefore a scientific challenge to try to model these different factors, and compare the different scenarios, starting with risk minimisation strategies. The approach of Boone et al. is a very important first step, but in order to be able to use their modelling results, we need a more realistic goal function.

Transhumance was in the past a good system to maximise productivity. Herd mobility through agricultural areas is sometimes only possible for wealthy herd owners (e.g., in Mongolia) who can use trucks for the transportation of their cattle to seasonal pastures. The costs for livestock transport in other areas are sometimes very low (Africa). Return to transhumance is therefore no longer attractive, as herd owners sometimes use modern transport facilities to track fluctuating resources.

A possible solution seems to be the formation of grazing associations or cooperations. An option to overcome most problems may be to try to convince private landholders not to fence their private lands, thereby enabling transhumance in fragmented landscapes. However, this is probably only feasible where there is a (social or economic) compensation for opening key resources for others to use. Grazing associations seem to be able to supply a framework for this. The new government in Kenya froze the transfer of land. The trend towards fragmentation can be rolled back, not so much in ownership, but by not having the areas fenced. Individual contracts in this sense already occur (Chapter 9).

Another option discussed to solve the problem of overstocking and losing transhumance in systems was the option to create game farms (Prins et al. 2000). However, in some countries, legislation prevents this. For instance game and trophy hunting is not allowed any more by Kenyan law, so a reduction in cattle and increase in game is not feasible. However, ecotourism is occurring and this also sometimes leads to a reduction of fragmentation, as landowners join larger management units in order to optimise management with neighbouring owners who share similar interests, such as happened in and around the Klaserie area near the Kruger National Park.

Social fragmentation was not included in the paper, and might also have effects on land degradation and transhumance. Fragmentation could also have positive effects on ecosystems, e.g., when manure is collected from the corals where animals are kept during the night. Moreover, the link of transhumant pastoralists with sedentary agriculturalist is fundamental in their production strategy. Fertilisation of agricultural fields, guarantees food supplies in the form of millet or maize. How do we value these future benefits?

In general, the paper clearly illustrated that movement of a consumer is instrumental in optimising fluctuating resources in time and space. Resource access (e.g., through territories or ownership) has large implications for the overall benefits derived from these resources, so social organisation and access cannot be neglected when studying resource exploitation.

ACKNOWLEDGEMENTS

We would like to thank all workshop participants for sharing their opinions and ideas with us. This chapter therefore not only synthesizes our own ideas, but also the ideas of our colleagues to whom we owe the credits.

CHAPTER 10A

RELEVANCE OF KEY RESOURCE AREAS FOR LARGE-SCALE MOVEMENTS OF LIVESTOCK

PAUL SCHOLTE[#] AND JOOST BROUWER[##]

[#] NC-IUCN, Leiden University and Independent Consultant, Nieuwe Teertuinen 12C, 1013 LV Amsterdam, The Netherlands
[##] Resource Ecology Group, Wageningen University, P.O. Box 47, 6700 AA Wageningen, The Netherlands
E-mail: scholkerst@cs.com

Abstract. Semi-arid rangelands show much spatial heterogeneity, with some parts producing more and better quality food for herbivores. The concepts of 'Key Resource' and 'Key Resource Area' have been developed to describe a resource that 'provides good-quality forage' and that 'reduces (inter-)annual variation in forage supply'. Illius and O'Connor (1999) formalised these concepts, arguing that in key resource areas herbivores experience a density-dependency relation with food resources, generally during the dry season. In other areas, generally during the wet season, non-equilibrium conditions govern the relation between herbivores and their food resources. They further argued that it is implicit that key resources show lower inter-annual variability than occurs on the (alternative) dry-season range, buffering livestock densities from climatic conditions. Key resource and outlying areas must further operate in a source–sink manner. In this chapter, we discuss the various assumptions and conclusions regarding key resources and key resource areas, using the floodplains of the Sahel, especially those of Waza-Logone in Cameroon, as examples. Sahelian floodplain grasslands are intensively exploited during the dry season, with cattle densities on a year-round basis about five times as high as in surrounding drylands. We come to the conclusion that the inter-annual variability in the *quantity* of the forage production of the Sahelian floodplains is not less, but often greater than that of surrounding areas. Forage *quality*, however, may be more constant. The model of Illius and O'Connor would be more realistic if it included intra-annual variability in forage availability, variability in accessibility of that forage, and associated differences therein between the dry-season range and the wet-season range. The importance of a resource varies from year to year, depending among other things on inter-annual variability in rainfall in the wet-season grazing range and in (the catchment upstream of) the dry-season grazing range. When it is of great importance, it may be considered a 'key resource', but in another context the same resource is not necessarily a key resource. Because of this spatial and temporal variability in rainfall and forage availability, there is no unequivocal source–sink relationship between the Sahelian floodplains and the associated wet-season grazing ranges. Forage in a key resource area does not necessarily provide the only key resource in the grazing system. Water, for instance, can be important as well. We end by discussing what our findings mean for the key resource area concept of Illius and O'Connor, and by presenting a new definition of key resource area which is also relevant to other trophic systems.
Keywords. floodplain; key resource area; livestock; population control; Sahel

H.H.T. Prins and F. van Langevelde (eds.), Resource Ecology: Spatial and Temporal Dynamics of Foraging, 211-232.

INTRODUCTION

Semi-arid rangelands form the habitat of many large herbivores, and have been the domain of cattle herders for hundreds of years (e.g., Prins 2000). Spatial heterogeneity in these semi-arid rangelands has been attracting the attention of ecologists for about half a century (Macfadyan 1950; Coughenour 1989; Hary et al. 1996). Some of the most distinctive features associated with this heterogeneity are drainage lines and wetlands and their accompanying vegetation. Examples are *Acacia* woodlands along dry rivers in Eastern Africa (Wuant and Ellis 1990), wet 'dambo' depressions in Southern Africa (Scoones 1995), and seasonally flooded grasslands in Sahelian Africa (Hiernaux and Diarra 1983; Howell 1988). At such features, water and nutrients are concentrated, primary and secondary production potential are greater, and production risk is often less than in the surrounding rangelands. Higher-altitude areas, with associated higher rainfall and more humid vegetation, may also constitute a distinctive feature in otherwise arid and semi-arid regions (Prins and Loth 1988; Hary et al. 1996). Because of the temporal heterogeneity found in semi-arid areas, in particular seasonal and inter-annual periods of drought, this spatial heterogeneity can be crucial for the survival of local herbivores. It allows herbivores to alternate between food sources (Drent and Prins 1987), according to their needs and the time of year. To this purpose long-distance migrations are often undertaken, by wild as well as domestic herbivores (Breman and De Wit 1983).

Spatial heterogeneity in (semi-)arid regions can be crucial for the survival of herbivores as they migrate

Herbivores shift to these high-production areas because they provide so-called 'Key Resources' (Scoones 1995). The term 'Key Resource' has been applied in a range of disciplines, varying from management science to anthropology and ecology, to stress the importance of a relatively limited resource for the survival of an individual or a population. In rangeland science, key resources have been mentioned since the early 1990s, although often in passing only, to stress their importance for the survival of herbivore populations during prolonged dry seasons (Drent and Prins 1987; Bayer and Waters-Bayer 1994; Scoones 1994; 1995; Hary et al. 1996). In relation to herbivores, two characteristics of key resources have received particular attention: they provide "good quality forage" and they "reduce (inter-)annual variation in forage supply" (Bayer and Waters-Bayer 1994). No reference is made to the minimum quantity of forage production, however.

'Key resource' has been applied in a range of disciplines to stress the importance of a limited resource for survival of individuals or populations

Until now the use of the term key resource seems to have been limited to forage resources, excluding other factors that may also be crucial to the survival of herbivores. In the African Sahel, for instance, as in most semi-arid areas, grazing patterns depend essentially on water availability. Dry-season grazing of cattle, but also of most wild herbivores, generally occurs within a distance of 20 km of sources of water (e.g., Le Houerou 1989; Fryxell and Sinclair 1988).

The term 'Key Resource Areas' was conceptualised by Illius and O'Connor (1999). They formalised the distinction between key resource-producing areas, where herbivores experience a density-dependency relation with food resources, generally during the dry season, and areas where non-equilibrium conditions are dominant, mostly in the herbivores' wet-season ranges (Box 10.1). In key resource areas, animals should further have a sufficient impact on the vegetation to experience intraspecific competition. Illius and O'Connor (1999, 2000) further argued that "it is implicit that these resources show lower inter-annual variability than occurs on dry-season range", buffering livestock densities from climatic conditions. Key resource and outlying areas must further operate in a source–sink manner, with key resource areas maintaining by definition a higher level of herbivory in outlying areas than these could support on their own (Illius and O'Connor 1999).

Box 10.1. *Historical development of the 'Key Resource Area' concept*

A central concept in range ecology is carrying capacity. Herbivore numbers are controlled by the availability of forage, and the availability of forage is controlled by animal numbers. This pattern of negative feedback eventually produces a stable equilibrium between animal and plant populations (Behnke and Scoones 1993). This equilibrium concept was largely based on the then reigning Clementsonian climax model of vegetation change. Its inadequacy in especially rangelands dominated by annual grasses was highlighted for arid Australia in the early 1980s (e.g., Westoby 1980). For such circumstances the alternative 'state and transition' model was postulated, which contested simple linear vegetation change (Westoby et al. 1989). Increasingly based also on examples from Africa, it is argued that plant production in highly variable climates is largely determined by rainfall and unaffected by herbivore population densities (e.g., Ellis and Swift 1988). The new paradigm of 'Rangeland at Disequilibrium' came to the fore in the early 1990s, and challenged the prevailing rangeland management practices (Behnke et al. 1993; Scoones 1995). In a reaction to this, Illius and O'Connor (1999) developed a somewhat different view of African rangeland functions. They argued that (1) herbivore numbers are regulated in a density-dependent manner by the limited forage available in so-called 'Key Resource Areas', utilised during the dry season mainly; (2) strong equilibrium forces exist over this limited part of the grazing system, while the animal population is virtually uncoupled from resources elsewhere in the system; (3) the wet-season grazing range is more heavily utilised by animal populations sustained by key resource areas than would apply in the absence of the key resource areas; and (4) the uncoupling of the animal population from the wet-season grazing range vegetation, in systems containing a key resource area, carries the risk of increased degradation of vegetation resources in the wet-season grazing range. In addition, they assumed in their associated modelling study (Illius and O'Connor 2000), that (a) potential primary production in the key resource area is dependent on rainfall only and was therefore the same in the key resource area as in the remainder of the grazing system; and (b) rainfall in the key resource area and in the remainder of the grazing system varies synchronously, but with a lower Coefficient of Variation (CV) in the key resource area than in the remainder of the grazing system. This was achieved by setting the deviation from mean annual rainfall in the key resource area in a particular year at a fraction of the deviation from mean annual rainfall in the remainder of the grazing system. Illius and O'Connor also looked at (c) different area ratios between key resource areas and wet-season grazing ranges. The greater that ratio, the greater the positive effect of the key resource area on livestock numbers.

Table 10.1A. *Cattle densities in Sahelian seasonally flooded grasslands*

	Waza-Logone (Scholte et al. 2006)	**Logone Floodplain** (Schrader 1986)	**Chad, Lake Fitri** (DHV/Labo 1994)	**Mali, Inner Delta** (Wilson et al. 1983)		**Floodplains theoretical** (De Bie 1991)
Period	Entire dry season 1993-99	Dry season March–June 1985 drought	Feb. 1993	Flooding season Oct 1980	Dry season Febr. 1980 / March 1981	
Area (km^2)	500	± 740	1,600	11,400		
Cattle (km^2)						
during obs. period	27 → 69[1]	30[2]	61[3]	10	75/78	80-137[4]
on 12-month basis	13 → 34		31			
Calculated min. number of cattle involved[5]	*34,000*	*22,000*	*98,000*	*114,000*	*890,000*	

1. Cattle pressure recalculated to average density during the six months that the floodplain is accessible to livestock (see text); does not take into account sedentary herds
2. Including sedentary herds (approximately 10% of total)
3. Half of the surveyed Fitri area has a density lower than 5 cattle km^{-2}, i.e., open water or land far from the lake, motivating the presented doubling of the recorded densities
4. Based on theoretical calculations, considering soils with high nutrient status. Recalculated to average densities over 6 months as under 1
5. Indicated to appreciate the importance of the area; does not take into account differences in peak densities and averaged densities.

In relation to the concept of 'Key Resource Areas', the seasonally flooded grasslands, or 'floodplains', of the African Sahel deserve further investigation. These grasslands are intensively exploited during the dry season, with livestock densities up to 60-100 cattle km^{-2}, ten times as high as cattle densities in dry-season ranges at the same time of the year. On a year-round basis, cattle density on floodplains is only about 3-6 times as high as in surrounding drylands, because high water levels make the floodplains inaccessible to cattle for up to six months each year (compare Tables 10.1A and 10.1B). Box 10.2 provides a brief general description of the seasonally flooded grasslands of the Sahel.

This important concentration of livestock motivates the consideration of these large seasonally flooded grasslands as key resource areas. But do they behave according to the above-cited properties? Are they really 'Key Resource Areas' *sensu* Illius and O'Connor (1999)? Should key resource areas perhaps be defined slightly differently? Or should different types of key resource areas be recognised? We address these matters by discussing the following questions. The examples used to illustrate that discussion mostly come from the Waza-Logone Floodplains grazing system, of which details are provided in Box 10.2 and Box 10.3.

Table 10.1B. *Cattle densities in Sahelian wet and dry season ranges (drylands)*

	Chad Basin				**Niger Basin**							**Theoretical** (De Bie 1991)[2]
	Chad, Centre (DHV/Labo 1994)		Nigeria / Borno (ERGO 1990)[1]		Mali, Inner Niger Delta (Milligan et al. 1982)[1]		Mali, Gourma (Milligan 1983; RIM 1985)		Niger, SW (Milligan 1982)			
Rainfall (mm yr^{-1})	150-550	150-800	400-800		400-600		300-500		200-500			200-800
Season	rainy	dry	rainy	dry	early dry	late dry	late rainy	dry	late rainy	early dry	dry	year-round average
Large wetlands	yes		yes		yes		no		no			
Area (km^{-2})	60,000	117,000	117,000		36,000		81,000		81,600			
Cattle (km^{-2})	10.1	7.8	21.4	25.3	22.5	22.5	5.7	4.3	4.1	4.6	3.5	
TLU (km^{-2})	9.5	7.1	21	23.2			5.6	4.3	5.9	6.6	5.0	8-16
Calculated min. number of cattle involved[3]	*600,000*	*910,000*	*2,500,000*	*2,960,000*	*810,000*	*810,000*	*460,000*	*350,000*	*335,000*	*375,000*	*285,000*	

[1] High (peak) density because of the proximity of floodplains
[2] Based on theoretical calculations, considering soils with low nutrient status
[3] Indicated to appreciate the importance of the area; does not take into account differences in peak densities and averaged densities.

Temporal variability

- Is the inter-annual variability in food production of the Sahelian floodplains less than that of alternative dry-season grazing lands?
- Does the food production of the dry-season ranges vary in synchrony with that of the wet-season ranges, and is it important whether it does?
- Does the role of Sahelian floodplains in the associated grazing systems vary between wet years, normal years and dry years?

Spatial variability and density dependence

- Do Sahelian floodplains offer better-quality food in greater quantity than do alternative dry-season ranges?
- Do the Sahelian floodplains regulate, in a density-dependent way, the number of livestock grazing the associated wet-season dryland grazing areas?
- Related to that, do the Sahelian floodplains act as a source for livestock numbers, and the wet-season grazing areas (that is, the dryland) as a sink?
- Further: does the presence of the Sahelian floodplains occasionally lead to an increased degradation of the wet-season grazing range?

Other assumptions by Illius and O'Connor (2000)

- Does it matter whether the potential primary production in a key resource area, on a per-hectare basis, is assumed to be the same as in the associated wet-season grazing area?
- Does it matter whether plant growth in dry-season grazing areas, or key resource areas, is assumed to vary in synchrony with plant growth in wet-season grazing areas?
- Does it matter whether dry season and wet season are both set at six months of the year, instead of another ratio?
- Are potential effects of key resource areas on surrounding wet-season grazing areas, including on species composition, perhaps related to food accessibility as much as to food availability?

Additional aspects

- Would the effects of a number of small key resource areas forming one grazing unit, such as a number of adjacent isolated wetlands, differ from the effects of one large key resource area?
- Does it matter whether, in relation to key resource areas, the focus is so generally on food instead of on water or some other factor?
- The above questions implicate that herbivore production is increased by the presence of key resource areas, but does this also hold for the individual animals?

We end by discussing what our findings mean for the key resource area ideas of Illius and O'Connor and by presenting a new definition of key resource area which is also relevant to other trophic systems.

Box 10.2. *African seasonally flooded grasslands and their utilisation by pastoralists*

Flooding and primary production

The major African floodplains are associated with rivers that have strong seasonal differences in volume (Denny 1993). These include the Zambezi, Nile and Niger Rivers, as well as the rivers flowing into Lake Chad and their tributaries (Figure 10.1). Spilling of the river water over the levees onto the associated floodplains can take place from once to several times a year. The regularity of flooding and its depth and duration obviously influence what plant, and animal, species are present (Denny 1993).

Maximum flood depth also determines aboveground biomass production of perennial-grass communities in African seasonally flooded grasslands (Scholte 2005). Under deeply inundated circumstances, i.e., 2-3 m, aboveground standing herbaceous biomass may reach 30 tons DM/ha (Hiernaux and Diarra 1983), up to ten times as high as in surrounding dryland areas (Le Houerou 1989; Prins 1996). The forage quality on floodplains, when characterised by its protein content, is generally negatively correlated with aboveground biomass. At the end of the flooding season, these floodplains are covered with a large quantity of grasses of below maintenance quality (Breman and De Wit 1983; Hiernaux and Diarra 1983; Howell et al. 1988; Prins and Olff 1998; Olff et al. 2002). The main grazing asset of the seasonally flooded grasslands is regrowth, which is of much higher quality. This regrowth is triggered by burning and grazing, and gradually becomes available during the dry season (Hiernaux and Diarra 1983; Howell et al. 1988; Scholte 2005). On the Inner Niger Delta floodplains in Mali, regrowth biomass was found to be a linear function of previous aboveground biomass (Breman and De Ridder 1991), and thus indirectly a function of maximum depth of the preceding flood. Regrowth assessments in Logone, Cameroon, suggest a regrowth production threshold at ±50-100cm maximum flood depth, corresponding to an aboveground biomass of ±10 tons DM ha^{-1}. With a lower flood depth hardly any regrowth is produced because of the lack of moisture stored in the soil in those parts of the landscape (Scholte 2005).

Pastoral exploitation

Similar to most other seasonally flooded grasslands, the almost featureless floodplains of Mali's Inner Niger Delta, the Sudd (Sudan) and the Lake Chad wetlands are home to more than a million cattle when the floods have receded. The same can be said of the floodplains of the Zambezi, and of other major rivers in those regions of Africa where disease does not preclude the grazing of, in particular, cattle. The herders let their cattle graze these floodplains in a complicated system established over perhaps centuries, with various traditional rights of grazing, passage, water access and management, now threatened by a host of new developments. By and large the herders use the floodplains when they can, i.e., once the floods have receded and there is sufficient grass and herb growth of sufficient quality available. Water availability generally does not pose problems. Labour needed to water animals is an important factor in the selection of areas (i.e., floodplains), yet production is generally not influenced by this factor. As mentioned above, good-quality regrowth can be triggered by setting fire to the old, poor quality growth (e.g., Van de Vijver 1999; Van de Vijver et al. 1999; Van Langevelde et al. 2003). When access to the floodplains becomes impossible, for example, because of the lack of water or because of the availability of better forage in the surrounding dryland areas, the herders move their cattle to those uplands. How and when this happens depends on, among other things, the timing of the flooding of the river plains relative to the falling of the rains in the drylands.

TEMPORAL VARIABILITY

Is the inter-annual variability in food production of the Sahelian floodplains less than that of the wet-season range?

Relatively low inter-annual variability in food production is considered to be an important characteristic of key resource areas (Illius and O'Connor 1999, 2000). Key resource areas may depend on groundwater, infiltrated surface water or permanently available surface water to buffer them against rainfall variation and maintain high production levels with relatively low inter-annual variability. Examples include the afore-mentioned small, ephemeral drainage lines (Scholte 1992), (semi-)permanent drainage lines (Wuant and Ellis 1990), lake shores (Loth and Prins 1986; Prins 1996), and higher-elevation areas with higher rainfall than the surrounding (semi-)arid lowland areas (Hary et al. 1996). Such areas occur in the Sahel (Le Houerou 1989) as well as in Eastern Africa (Prins and Loth 1988). In all these cases local hydrological processes appear to dominate the concentration of water in the key resource area, leading to greater hydrological and production security than in the surrounding drylands. But need this always be the case?

Water resources in floodplains are a combination of local rainfall and flooding. In the Lake Chad basin the discharge of feeding rivers is responsible for 50% (Waza-Logone floodplains) to more than 90% (Lake Chad itself) of the volume of the water resources (Naah 1992; Olivry et al. 1996). The Coefficient of Variation (CV) of the annual discharges of these rivers range from 26 to 98%, higher than the CV of annual rainfall of 26-43% in surrounding Dry-Season Ranges (Table 10.2). The situation in the Inner Niger Delta floodplain (Mali) is quite similar (Table 10.2). Inter-annual variability in the extent of the area inundated is greater still, sometimes more than 100%.

Inter-annual variability in food quantity of the Sahelian floodplains is not less than that of the wet-season range

There are a number of reasons for this large variability in river discharge and extent of flooding. (1) Not all rainfall events cause run-off and an increase in river flow. Only rainfall events above a certain intensity and duration threshold will do so, and the occurrence of such extreme events is more variable than annual rainfall. Run-off and river flow are also influenced by the distribution of rainfall through the rainy season, which again is more variable than annual rainfall totals. (2) Topography of the floodplains also influences the extent of flooding. If a certain flood level is reached, a whole new basin may be flooded, which would not have happened if the flood level had remained fractionally lower. This, too, can lead to an increase in the variability of the extent of flooding. (3) Human intervention of course also has an effect. Within a floodplain humans may try to influence the flooding of certain areas to further pastoral, agricultural or fishing aims. At a higher scale, through the operation of dams, a more or less constant, large volume of river flow is diverted or lost to evaporation each year. This reduces the average flow downstream, but the absolute variations, and thus the standard deviation, of the flow are influenced less. As the CV equals the standard deviation divided by the mean, regular annual water takeoffs via dams increase the CV.

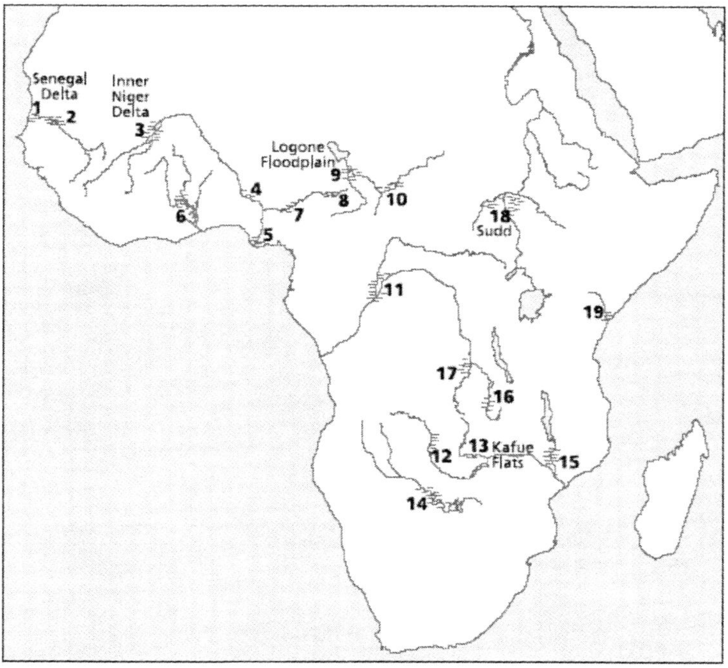

No and name	Country	Area at peak flood (km^2)	Floodplain type
1 Senegal Delta	Senegal	8 000	Coastal delta
2 Senegal Valley	Senegal	5 000	Fringing floodplain
3 Inner Niger Delta	Mali	30 000	Internal delta
4 Niger fringing plains	Nigeria	5 000	Fringing floodplain
5 Niger Delta	Nigeria	36 000	Coastal delta
6 Volta River	Ghana	8 500	Fringing floodplain
7 Benoué River	Nigeria	3 100	Fringing floodplain
8 Benoué River	Cameroon	1 000	Fringing floodplain
9 Logone Floodplain	Cameroon	5 000	Fringing floodplain
10 Chari and Logone	Chad	63 000	Fringing floodplain
11 Congo River	Congo	?	Fringing floodplain
12 Barotse plain	Zambia	10 750	Fringing floodplain
13 Kafue Flats	Zambia	4 300	Fringing floodplain
14 Okavango	Botswana	17 000	Internal delta
15 Shire River	Malawi	1 000	Fringing floodplain
16 Kifakula Depression	Congo	1 500	Fringing floodplain
17 Kamulondo	Congo	12 000	Fringing floodplain
18 Sudd	Sudan	92 000	Fringing floodplain
19 Tana Delta	Kenya	1000	Coastal delta

Adapted from Drijver and Marchand 1986 and Welcomme 1979.

Figure 10.1. *The location of large (> 1000 km^2) seasonally flooded grasslands in Africa (Scholte 2005)*

Furthermore, the major floodplains in the African Sahel (Figure 10.1) are surrounded by medium-rainfall areas, with higher-rainfall areas located 100-500 km to the south. Rainfall CV decreases from an average 26% in the medium-rainfall areas to an average of less than 18% in the high-rainfall areas (Table 10.2). The forage quantity is relatively moderate in the medium-rainfall areas and high in the high-rainfall areas. Forage quality is generally low throughout, with exception of browse and regrowth in generally scarce lower-lying areas (Breman and De Wit 1983; Le Houerou 1989). On the floodplains themselves, forage production quantity is generally high, and quality varies (not so high immediately after flooding, high following burning and regrowth).

***Table 10.2**. Coefficients of variation of key parameters of wet- and dry-season ranges, including key resource areas (KRA) in the African Sahel*

		Area	Average rainfall (mm yr^{-1})	Rainfall CV (%)	CV (%) rainfall in main catchment areas	CV (%) Flood level[2]
NORTH ↔ SOUTH	Wet-season range	North Sahel	55	81[1]		
			153	43[1]		
	KRA flood plains	Lake Chad	284-(576)	(26)-43	11-26	35 Chari River discharge 51 Lake levels[3]
		Inner Niger Delta (Mali)	300-600	30	17-26	55 Water level Niger river 108 Area of Inner Niger Delta inundated
		Lake Fitri (Chad)	394	30	26-30	98[4] Batha River discharge at Ati
		Waza-Logone (Cameroon)	576	26	11-14	26 Logone River[5] 39 Secondary sources [6]
	Dry-season range	South Sahel (Ndjamena, Chad)	576	26		
	High-rainfall areas, Cameroon	Maroua	806	18		
		Guider	919	16		
		Garoua	972	16		
		Ngoundéré	1513	11		

[1] Lack of reliable rainfall data available, based on data set of, respectively, 11 and 9 stations in the Sahel (Le Houerou 1989)
[2] Maximum river discharge was considered the best available parameter to predict flooding levels in the Logone floodplain and Lake Chad (see Naah 1992; Mott Macdonald 1999), Inner Delta, both flood level and flooding area (Quensiere 1994; Zwarts 2002; Zwarts pers.comm.)
[3] Quasi-linearly correlated with surface area of lake (Olivry 1996). Data set does not include the dry 1980s, CV thus underestimated
[4] Total annual discharge volume, based on 1955-1989 data set only
[5] Based on 1933/1948–1997 data set
[6] Based on data set of 1970 and 1980s.

Box 10.3. *Pastoralist responses to inter-annual variability in Sahelian floodplains: two examples from Waza-Logone*

The key resource area characteristics described in Box 10.2 for Sahelian floodplains show that the discussion on population size regulation by density-dependent factors or by density-independent factors should not ignore inter-annual variability. We give here two examples of such variability for the Waza-Logone Floodplains during a period of six wet years (1993-1999) and during a severe drought (1985), to show the implications of this inter-annual variability for the exploitation of the available grazing resources by cattle.

1. Effects of reflooding of part of a floodplain previously dammed out

In 1993, the Waza-Logone project initiated a pilot reflooding by breaching an embankment that had closed off a small branch of the Logone river, triggering the annual reflooding of a downstream area of ±180 km^2 (Scholte et al. 2000; Scholte 2005). From 1993 till 1999, the project monitored the impact of this reflooding, thus imitating a period of six wet years following a prolonged period of dry years. Such a sequence was not exceptional during the last century.

During the study period, annual grasses were replaced by rhizomatous grasses, with an annual conversion rate of 7-10% of the 180 km^2 reflooded area (Scholte et al. 2000). Here, as well as in an additional area of ± 500 km^2, the maximum flood level was raised by about 20 cm, leading to a ±30% increase in aboveground biomass (Box 10.2). We monitored nomadic pastoralists' responses to these changes through interviews about their migration patterns in the reflooded area (Scholte et al. 2005; Scholte 2005). Grazing intensity of nomadic herds, expressed as cattle density averaged over the six months of dry season, increased 2.6-fold from 1993 till 1999 (Figure 10.2). In the first year, the increase in grazing pressure was caused by a longer stay of herds already present (Figure 10.3). In later years the reflooded area experienced especially an inflow of herds and herders.

The assumption that the monitored reflooding imitated a period of six wet years holds especially for the floodplain vegetation. As indicated, with a lag of one year annual grasses started to be replaced by perennial grasses, generally spreading from rhizomes. Contraction and expansion of rhizomatous grasses are normal phenomena during periods of long-term climatic and annual rainfall fluctuations. Flood depth – aboveground biomass relations also showed a lag in the full response, with the 1996 production per unit of flood depth higher than in 1994, the first year of reflooding.

The reaction of the pastoralists was influenced by developments elsewhere in the floodplain, or lack thereof. In the remainder of the floodplain, rainfall and flooding during the 1993-1999 study period were rather average, and thus not similar to 'wet years' as in the reflooded area. Most of the increase in pastoral camps was caused by the arrival of pastoralists from elsewhere in the floodplain, pulled in by the increasing availability of forage resources in the reflooded floodplain. In case of more widespread favourable rainfall and flooding conditions, they would have stayed where they were elsewhere in the floodplain. In addition, a small group of pastoralists had changed their migration completely and entered the Logone floodplain for the first time since years, settling in the reflooded area immediately following the reinstatement of flooding. The relative grazing pressure in the reflooded zone, compared to the entire floodplain, therefore increased till 1996 but subsequently stabilised, suggesting a saturation in cattle density in the reflooded zone after three 'wet years'.

Pastoralists' responses to floodplain rehabilitation were in line with the Ideal Free Distribution model, with any increase in forage production subject of increased consumption (Scholte et al. 2005; Scholte 2005). No overshoot in number of pastoralists and cattle has taken place. Initially feared for, such overshoot would have had an impact on perennial vegetation that however continued to expand during the study period. Nor were there any signs of territorial blocking of newcomers, except for the first-year delay in responses (Figure 10.2). Cattle herds should further be considered as one (meta-)population, with regular exchanges, resulting in rather uniform productivity levels.

(cont.)

Box 10.3. (cont.)

2. Dynamics during a period of drought

In the years 1983-1985, on the other hand, the entire Sahel experienced a period of severe drought. Rather strikingly, the 1985 cattle densities in the Waza-Logone floodplain, following the drought and six years after the cessation of annual flooding of parts of the floodplain, were very comparable to the densities prior to the reflooding in 1993, a year with rather average rainfall and flooding (Table 10.1A). The origin of the migrating cattle, however, was completely different. During the dry season in the drought year 1985 those pastoralists normally present in the floodplain had migrated further south into high-rainfall areas (Table 10.1A). Pastoralists present in the floodplain in 1985 came from further north, and had changed their migration, which used to be directed into the Lake Chad bed (Schrader 1986; Clanet 1996). Lake Chad flood sources show a higher CV in their annual discharge than the flood sources of the Waza-Logone floodplain (Table 10.2). At least in 1985, the Lake Chad grazing resources were apparently less dependable than those of the Waza-Logone floodplain. In the same year 1985, cattle losses due to diseases and exhaustion in the Waza-Logone floodplain were much higher than those recorded in 1997, estimated at, respectively, 10 and 0.9% of the total number present.

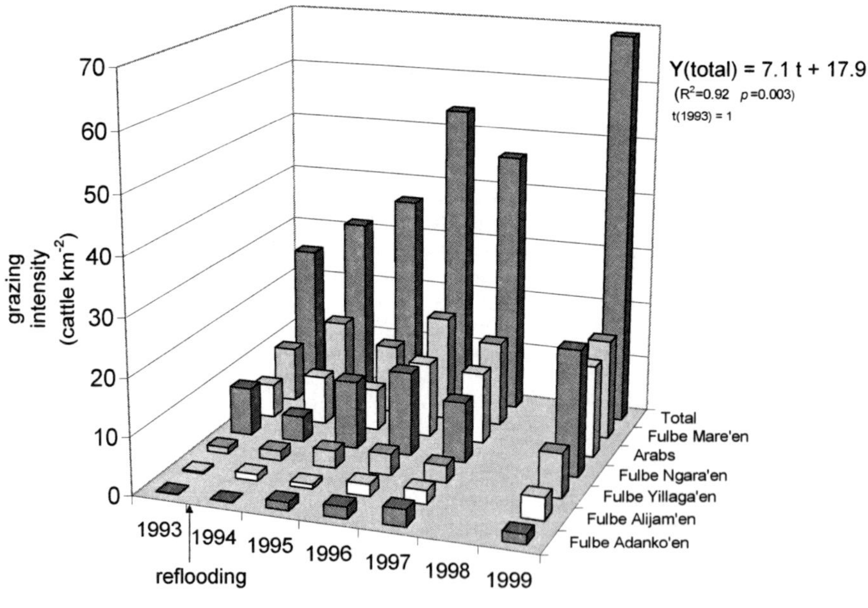

Figure 10.2. *Linear increase in grazing intensity, averaged over a dry season of 6 months, following reinstatement of old annual flooding regime and gradual recovery of the grassland vegetation in the Waza-Logone Key Resource Area (1993-1999). Cattle density given for each of the six pastoral clans and as total*

Based on the preceding we conclude that the inter-annual variability in the *quantity* of the food production of the Sahelian floodplains is not less, but often greater than than that of surrounding medium-rainfall areas. This does, however, not necessarily hold for the inter-annual variability in the *quality* of the food production of the Sahelian floodplains. Data on the inter-annual variation in food quality are, however, near-absent from the literature.

Does the forage production of the dry-season ranges vary in synchrony with that of the wet-season ranges, and is it important whether it does?

Illius and O'Connor (2000) only looked at total annual forage production during a year, not at intra-annual variation in forage availability. They then modelled three situations, with either no, partial or complete restriction in the seasonal accessibility of the two range areas. They assumed that the quality of dead forage biomass was equivalent to the quality of living forage biomass. Thus they excluded intra-annual temporal variability in forage availability from their modelling, and also differences between the wet-season range and the dry-season range in that intra-annual temporal variability in forage availability.

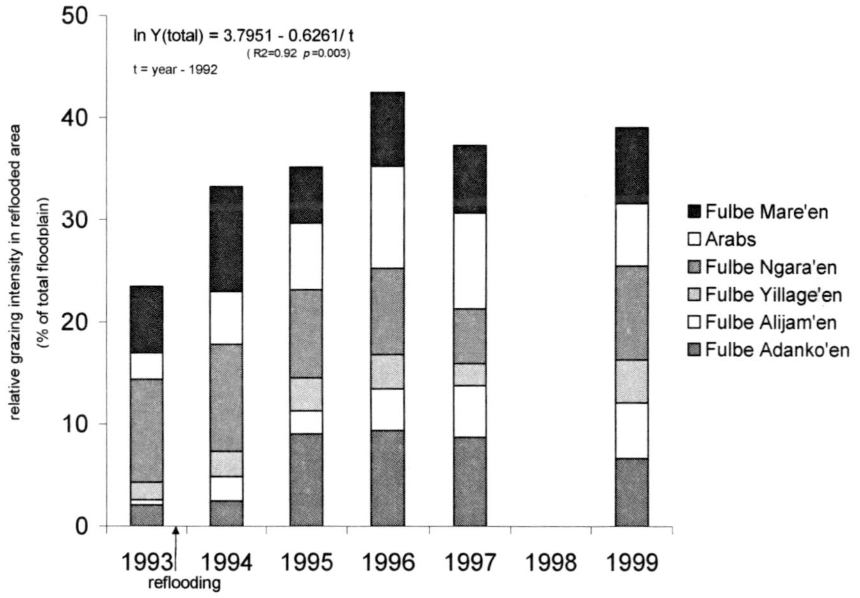

Figure 10.3. *Changes in grazing time by the six pastoral clans spent in the reflooding impact zone, as a percentage of the grazing time they spent in the Waza-Logone floodplain as a whole during each six-month dry season (Note that the y-axis scale refers to the totals for the six clans together, and not to contributions by individual clans)*

Whether forage production of the dry-season ranges varies in synchrony with that of the wet-season range will depend on where the dry-season range, in our case consisting of floodplains, gets its water from for primary production. If flooding is from local rainfall, then forage production will vary more or less in synchrony with wet-season ranges. If the flooding is caused by rainfall far away, then there will often be a delay in forage production on the floodplains compared to the surrounding drylands. Moreover, the storage of water in the soil during flooding allows primary production to resume following grazing or burning, long after the rains, and the flooding, have ceased. Primary production in the Okavango Delta (Botswana), for example, is nearly completely out of phase with that of the surrounding drylands.

What is also important, however, is the *accessibility* of the forage produced. In the wet-season ranges, there may still be some forage left during the early part of the dry season, but the drying-up of surface water for drinking can make it inaccessible to large grazers. Construction of permanent water points can make such forage more accessible, but during poor-rainfall years it can also increase the danger of overgrazing (Le Houerou 1989). In the floodplains of the dry-season range, on the other hand, the flooding that causes the forage to be produced can also limit the accessibility of that forage. What is important to the grazing system is not only that the floodplains produce more or better forage than the wet-season range, but that the forage is accessible to the grazers at a time when food left on the wet-season range is not accessible.

Including intra-annual temporal variability in forage availability and differences therein between the dry-season range and the wet-season range, would make the model of Illius and O'Connor more realistic. It would also show that the dry-season key resource area does not necessarily provide the only key resource in the grazing system (see the proposed new definition of a key resource at the end of this chapter).

Does the role of Sahelian floodplains in the associated grazing systems vary between wet years, normal years and dry years?

Illius and O'Connor did not look into this, but the role of the Sahelian floodplains certainly varies between years (Box 10.3). Again, this shows how the importance of the floodplains in a particular year depends on the circumstances that year in both the floodplains and the associated drylands. In other words, a key resource area does not fulfil a key role independently of its spatial or temporal context.

SPATIAL VARIABILITY AND DENSITY DEPENDENCE

Do Sahelian floodplains offer better quality food in greater quantity than does the wet season range?

Illius and O'Connor (2000) assume that the quality of the food in wet- and dry-season ranges is the same. They also assume that, during below-average rainfall years, rainfall and forage quantity are always better in the key resource area, i.e., the floodplain. By and large it depends on the time of year whether Sahelian floodplains offer better-quality food in greater quantity than do alternative dry-season ranges, also in below-average rainfall years. See Box 10.2, and the answers to the first two questions above.

Do the Sahelian floodplains regulate, in a density-dependent way, the number of livestock grazing the associated wet-season dryland grazing areas?

The concept of key resource areas, with their density-dependent effects on the associated herbivore populations, implies a coupling of the population dynamics to the key resource area's vegetation, at least during periods of drought (Illius and O'Connor 1999). Increased grazing intensity and intraspecific competition on high-quality forage did indeed take place during several years in the Logone floodplain. During such years pastoralists generally adjusted the length of stay in the floodplain based on the availability of forage sources (Figure 10.3).

However, in none of the relatively well-studied major seasonally flooded grasslands in Africa, an inter-annual impact of grazing on the vegetation was reported, not even after severe droughts (Ellenbroek 1987; Hiernaux and Diarra 1983; Howell et al. 1988; Scholte et al. 2000). Degradation of soil and vegetation was, however, reported from surrounding dryland grazing ranges (Howell et al. 1988). Characteristics that explain the resilience of the seasonally flooded grasslands include the high belowground biomass (±70% of total biomass) that is inaccessible to grazing animals because of the firm vertisols and the six months of seasonal protection due to rainfall pounding and subsequent flooding. Drent and Prins (1987) also argued that, because the herbivore is prisoner of its food supply, vegetation under 'natural conditions' is free from disturbance due to the herbivore. Ideal free distribution, observed in at least one Sahelian Floodplain (Box 10.3; Scholte 2005; Scholte et al. 2006) is certainly one of the mechanisms that make it rather unlikely that herbivores exercise long-term disturbance.

The floodplains are protected from lasting damage from overgrazing by their annual flooding. This makes it likely that herbivore numbers are controlled in these areas but not (density-dependent) regulated. The drylands directly surrounding the floodplains are not flooded, and the vegetation there is at times damaged accordingly: density-dependent regulation is more likely to occur in these particular drylands.

Do the Sahelian floodplains act as a source for livestock numbers, and the wet-season grazing areas (that is, the dry 'bush') as a sink?

The ecological theory of sources and sinks implies that the reproduction rate of individuals or the production level of the animal population in the source area determines population size in the sink areas, and not the other way around. Illius and O'Connor (2000) calculated that their key resource area, if large enough, does indeed positively influence livestock numbers in the associated wet-season range. In the case of the Sahelian floodplains, however, there is also a reverse effect of the wet-season range on the livestock numbers in the associated floodplains.

Seasonally flooded grasslands in the Sahel are indeed areas with high production potential for livestock (Box 10.3), and also for wild herbivores. The Waza-Logone floodplain harbours the only national park with substantial floodplain habitat in the African Sahel. Waza National Park used to have a wild-herbivore density of more than 2000 kg km^{-2}, well above the curve of wildlife densities against rainfall plotted for West and Central African reserves with only limited floodplain areas (De Bie 1992). With the cessation of annual flooding in Waza National Park in 1979, due to the construction of a dam upstream, wild-herbivore densities dropped to about 1000 kg km^{-2}, well within the rainfall–wild-herbivore curve just mentioned. This drop in the floodplain herbivore density also caused a drop in herbivore densities in surrounding upland areas, where herbivores migrate to during the rainy season (Scholte et al. 1996). It is concluded that annual flooding was a crucial factor in the high productivity of the floodplain, with an impact on herbivore populations beyond the floodplain.

Similarly, the Lake Chad Basin, characterised by the large ratio of its floodplains to its wet-season ranges, has amongst the highest livestock densities in the African Sahel (Table 10.1B). In the World Atlas of Degradation (UNEP 1992), the Lake Chad Basin was also considered to be less degraded than similar areas in the Niger basin. Illius and O'Connor (1999, 2000) indeed predicted high animal populations for areas with a large key resource area ratio to wet-season ranges. We speculate that the relatively high hydrological variability of the Lake Chad Basin is compensated for by the relatively low percentage of agricultural land, compared to areas surrounding the other major floodplains of West Africa. This has allowed longer migration routes and enhanced use of the Lake Chad Basin by large grazers (see also Van Keulen and Breman 1990). All this suggests that the floodplains of the Lake Chad Basin have a positive effect on the density of large grazers in the surrounding drylands.

The point is, however, that the drylands and floodplains in the Lake Chad Basin do not exist next to, and more or less separate from, each other. They are both an integral component in the same annual cycle of the associated large grazers. Poor conditions in the drylands during the time of year that the floodplains are inaccessible will reduce numbers returning to the floodplains, but good conditions may increase those numbers. As far as large grazers are concerned there is therefore not a simple source–sink relationship between the floodplains and the drylands of the Lake Chad Basin

Does the presence of the Sahelian floodplains occasionally lead to an increased degradation of the wet-season grazing range?

Illius and O'Connor hypothesised that this would be the case, contrary to what Behnke et al. (1993) and Scoones (1995) suggested. Our conclusions are that the presence of the Sahelian floodplains does indeed occasionally lead to greater degradation of the wet-season grazing range than would be the case without them. See the discussion under the preceding question and under the first point of 'Spatial variability and density-dependence' above.

OTHER ASSUMPTIONS BY ILLIUS AND O'CONNOR (2000)

Does it matter whether the potential primary production in a key resource area, on a per-hectare basis, is assumed to be the same as in the associated wet-season grazing area?

No, it does not. Illius and O'Connor varied the ratio between areas of the key resource area and the associated wet-season grazing range. Doubling that ratio at constant potential primary production per hectare is equivalent to doubling potential primary production per hectare at a constant area ratio.

Does it matter whether plant growth in dry-season grazing areas, or key resource areas, is assumed to vary in synchrony with plant growth in wet-season grazing areas?

Actually, Illius and O'Connor did not look at intra-annual variability in forage availability. Working only with total annual forage production, they implicitly assume that this food is available throughout the year unless already eaten. They also make the explicit assumption that living and dead forage biomass are of the same quality, which clearly is not normally true (see, e.g., Beekman and Prins 1989; Prins and Beekman 1989; Prins 1996). It really matters whether the plant growth in dry-season and in wet-season grazing areas is assumed to vary synchronously. First, because in real life biomass quantity and quality vary enormously throughout the year (Breman and De Wit 1983; Prins 1988, 1996). Second, because *accessibility* of the dry-season and wet-season grazing areas often does not vary synchronously (Box 10.2). If accessibility varies asynchronously, then it can be very 'useful' if quality varies asynchronously as well. That way good-quality forage may be available as well as accessible at different times of the year in the two areas, so that the dry-season and wet-season grazing ranges can be as complementary a pair of forage sources as possible.

Does it matter whether dry-season and wet-season length, and access to the associated grazing ranges, are both set at six months of the year, instead of another ratio?

Within the constraints that Illius and O'Connor set themselves, it does not matter. Changing the ratio of area or potential primary production has the same effect on the modelling outcome as changing the ratio of dry- and wet-season length.

Are potential effects of key resource areas on surrounding wet-season grazing areas, including on species composition, perhaps related to forage accessibility as much as to forage availability?

As the modelling by Illius and O'Connor (2000) shows, the positive effects of their key resource areas on grazer numbers are greater when there is more restriction in the seasonal accessibility of the range areas. When there is no restriction in accessibility the effect on grazer numbers was as good as zero. It follows from this that potential effects of key resource areas on surrounding wet-season grazing areas, including on species composition, must indeed be related to forage accessibility. If the forage is there in the key resource area, but the grazers cannot get to it because of, for example, flooding, the associated dryland or wet-season grazing areas will be searched for the required food, whether they can provide it or not.

ADDITIONAL ASPECTS

Would the effects of a number of small key resource areas forming one grazing unit, such as a number of adjacent isolated wetlands, differ from the effects of one large key resource area?

There are two important sides to this: accessibility and forage production reliability. If, for whatever reason, it is not possible for grazers to move from one small key resource area to another, then the grazers are obviously better off with one large key resource area than with a number of small ones. If there is no problem with access, then the forage production reliability becomes important. If forage production in all the small key resource areas depends on the same hydrological events, then all the small areas will react to hydrological events as though they were one large area. But if the small areas have a certain degree of hydrological independence from each other, then it becomes a different story.

If grazers cannot move from one small key resource area to another, one large key resource area is better than several small ones

Rainfall events in the Sahel are often rather local storms. Over a distance of 5-7 km it is possible for total annual rainfall to vary from 600 to 850 mm in the same year (Wallace et al. 1994; see also Prins and Loth 1988 for East Africa). Small isolated wetlands, each with their own catchment quite close to each other, can therefore vary considerably in the degree to which they flood in a particular year.

This spatial variability reduces the chances of all wetlands flooding poorly, and producing forage poorly, in the same year. In this way a fragmented resource can be a more reliable source of food than a similarly sized resource consisting of just one hydrological unit.

Does it matter whether, in relation to key resource areas, the focus is so generally on forage instead of on water or some other factor?

What do grazers need? Food, water, and places for shelter, feeding and reproduction, all in the right (optimal) amount, in the right place and at the right time. Grazers also need a relative absence of factors that prevent the grazers from making use of these five factors. Food is necessary for grazers, but not sufficient. A key resource may therefore just as well be water, or a safe place, as forage. It is, of course, no coincidence that forage has received most attention, as it is, within the limits of access to water, the bottleneck in Sahelian grazing systems (Breman and De Wit 1983; Le Houerou 1989).

Production characteristics of herbivores with access to Sahelian floodplains

Large-scale migration is reputed to enhance livestock production (Breman and De Wit 1983; Le Houerou 1989; Niamir-Fuller 1999). Whereas sedentary livestock produced an estimated 0.4 kg protein $ha^{-1}year^{-1}$, transhumant migration into medium-rainfall area produced 0.6 kg protein $ha^{-1}year^{-1}$ and transhumant migration into the Inner Niger Delta floodplain was found to produce up to 3.2 kg protein $ha^{-1}year^{-1}$ (Breman and De Wit 1983). These production characteristics are in line with observed differences in livestock densities (Table 10.1).

As expected with the observed ideal free distribution (Box 10.3; Scholte 2005; Scholte et al. 2006), available data indicate that Fulani cattle from herds that have access to Sahelian floodplains have comparable production characteristics, expressed per head of livestock, as (semi-)sedentary Fulani herds that remain outside floodplains (Table 10.3). If any changes can be detected between herds under otherwise comparable conditions (as indicated by average annual rainfall), they tend to show even a somewhat lower production of individual cattle having access to the floodplain compared to those that do not have such access (Table 10.3). We hypothesise that this lower individual production is caused by the lower 'risk' of cattle in the floodplain. Imagine a 100-km^2 floodplain with 100 herds (of 100 cattle each) and surrounding dryland with 10 herds in an area of 100 km^2. A 'marginal herder', joining with his single herd of 100 cattle, poses for the floodplain herders a cost of 1% 'competition', whereas the dryland herders face a 'competition cost' of 10%. Only with 9 herds joining the floodplain and 1 the drylands, the cost will be comparable. We postulate that the number of conflicts indicating these costs is higher in the drylands than in the floodplain. Indeed, mostly qualitative information from the Logone floodplain suggests a very limited number of conflicts inside the floodplain compared to outside (Moritz et al. 2002). Differences in herd management, such as low labour costs of providing drinking water in the floodplain compared to outside the floodplain, might alternatively explain the somewhat lower production inside the floodplain.

When considering Sahelian floodplains as key resource areas, it is clearly not the production ('fitness') of the individual herds or individual animals that characterises these areas. In contrast, herds that receive supplementary feeding, show higher production characteristics per individual animal (Table 10.3). The analogy of key resource areas with supplementary feeding as indirectly suggested by Illius and O'Connor (1999) does not hold, at least not for Sahelian floodplains.

Table 10.3. *Comparison of Fulani cattle production with and without access to floodplains*[1]

<table>
<tr><th rowspan="2"></th><th colspan="4">Cattle with access to the Inner Niger Delta floodplain</th><th colspan="4">Cattle without access to floodplains</th></tr>
<tr><td colspan="3">Herds lumped [2]</td><td>Years lumped</td><td>Mali[3]</td><td>Nigeria[4]</td><td>Nigeria[5]</td><td>Improved manage-ment [6]</td></tr>
<tr><td>Sample size</td><td>820</td><td>910</td><td>879</td><td>2550</td><td>734</td><td>?</td><td>1367</td><td>?</td></tr>
<tr><td>Year</td><td>1979</td><td>1980</td><td>1981</td><td>1979-81</td><td>1979-81?</td><td>?</td><td>?</td><td>?</td></tr>
<tr><td>Average annual rainfall (Table 10.2)</td><td colspan="4">± 500mm</td><td>± 500 mm</td><td>±850 mm</td><td>±1050 mm</td><td>±500mm</td></tr>
<tr><td>Cow viability (%)</td><td>0.95</td><td>0.92</td><td>0.94</td><td>0.92</td><td>0.97</td><td>0.97</td><td>0.96</td><td>0.97</td></tr>
<tr><td>Calving rate (%)</td><td>0.55</td><td>0.58</td><td>0.55</td><td>0.54</td><td>0.57</td><td>0.45</td><td>0.46</td><td>0.77</td></tr>
<tr><td>Calf viability (%)</td><td>0.79</td><td>0.50</td><td>0.62</td><td>0.64</td><td>0.75</td><td>0.86</td><td>0.88</td><td>0.69</td></tr>
<tr><td>Calf weight at 1 year (kg)</td><td>85</td><td>80.3</td><td>73.0</td><td>79.6</td><td>81</td><td>91.1</td><td>80</td><td>125</td></tr>
<tr><td>Milk off-take (kg)</td><td>266</td><td>205</td><td>185</td><td>218.6</td><td>193</td><td>286.4</td><td>234</td><td>522</td></tr>
<tr><td>Prod. index (cow/yr)</td><td>49.9</td><td>33.6</td><td>34.0</td><td>37.2</td><td>45.7</td><td>47.5</td><td>42.5</td><td>108.8</td></tr>
<tr><td>Adult cow weight (kg)</td><td colspan="4">215</td><td>242</td><td>268</td><td>245</td><td>302</td></tr>
<tr><td>Prod. index [7] (year/100 kg cow LW)</td><td>23.2</td><td>15.6</td><td>15.8</td><td>17.3</td><td>18.9</td><td>17.7</td><td>17.3</td><td>36</td></tr>
</table>

[1] Adapted from Wagenaar et al. (1986)
[2] Herd R and S (with 210 cattle each), following the same grazing orbit, and V (with 400 cattle) had a prod index (year/100 kg cow LW) of 20.7, 15.6 and 15.8, respectively (averaged over 1979-1981)
[3] Agropastoral system, Niono, Mali (Wilson 1983)
[4] Settled Fulani, Kaduna Plains, Nigeria
[5] Settled Fulani, Jos plateau, Nigeria
[6] Sudanese Fulani cattle under improved management, Niono, Mali
[7] [cow viability × calving rate × calving survival × calf weight at 1 year (kg)] + [cow viability × calving rate × milk offtake (kg) divided by 9)].

SYNTHESIS

What makes an area with resources in it a so-called 'Key Resource Area'? Illius and O'Connor (1999, 2000) have tried to give a quantitative definition, but according to that definition the floodplains of Waza-Logone are not key resource areas for the pastoralists and their livestock that use them. These floodplains diverge from that definition in that they are not the only resources that regulate grazer numbers in the associated grazing system. If, however, these floodplains were no longer available to those pastoralists, their grazing systems would change dramatically, and cattle numbers would probably be reduced substantially. In the case of the floodplains of Waza-Logone, the alternative would be for the pastoralists to migrate further south for the dry season. The costs of migration southward to areas with more reliable rainfall, and forage, including the time and energy needed to cover the relatively long distance, are high. Associated with these direct costs, there is an increased risk of exposure of the livestock to diseases that are more frequent in higher-rainfall zones, and the need to pass through agricultural areas with little available grazing, but also an increased risk of conflicts with farmers (Kari and Scholte 2001).

Important for KRAs is that the resource is accessible when a comparable resource is not accessible in sufficient quantity and/or quality elsewhere

For the floodplains of Waza-Logone and other floodplains in the Sahel, it is therefore not the previously assumed low inter-annual hydrological variability that explains their importance to the pastoralists that use them. Rather, it seems to be the availability of 'good'-quality forages in 'fair' quantity especially in the beginning of the dry season, combined with 'low'-quality forages in 'ample' quantity throughout the dry season, coupled to a relatively low risk of complications of resource accessibility and livestock diseases, which explain the relatively high livestock densities in the floodplains. Put differently, what the pastoralists involved most likely seek is not so much low variability as high 'assured' production and reduced risk. It is therefore not so important whether a particular area produces more and/or better forage, or another resource, than do other parts of the grazing system. What *is* important is that in such an area the resource is *accessible* to the grazers at a time of year, or in a year, when a comparable resource is not accessible in sufficient quantity and/or quality elsewhere in the grazing system.

Our alternative definition of a key resource within a grazing system is therefore "a spatially defined resource that allows a grazing system to maintain one or more populations of herbivores in disproportionately higher numbers than could be maintained without that resource"; a key resource area is the area where this key resource is found. Based on this definition, we postulate some hypotheses for future research (Box 10.4). Our definition, like the definition of Illius and O'Connor, implies that:

Key resource allows a grazing system to maintain herbivores in disproportionately higher numbers than could be maintained without that resource

a. Other parts of the grazing range are more heavily utilised by animal populations sustained by key resource areas, than would the case in the absence of the key resource areas.
b. The key resource area does not necessarily act as such every year: it may be the key to understanding population dynamics of a larger area only during extreme years.
However, this definition expands on the definition and its application by Illius and O'Connor (1999, 2000), and acknowledges that:
c. Herbivore numbers are not necessarily regulated in a density-dependent manner by the limited resource available in the key resource areas. Herbivore numbers may, as in the case of Sahelian floodplains, be merely controlled, without feedback with the grazing resources.
d. Temporal variability in the availability of the resource concerned need not be less than temporal variability in the remainder of the grazing system, and can even be greater. However, if such variability is greater than that of the remainder of the system, it is most likely important that it is also asynchronous with the variability of the remainder of the system.
e. The key resource need not be a dry-season resource nor necessarily be food, it can also be water, or a lack of predators or disturbance or pathogens, or another factor essential for grazer existence and reproduction.
f.. Although total production is increased through the presence of a key resource area, this does not necessarily hold for the individual herbivore and depends on the distribution model. A key resource should not be considered as a kind of supplementary feeding.

Box 10.4. *Testable hypotheses for future research*

Hypothesis 1. Sahelian floodplains are characterised by a high livestock density and resulting high animal production per unit area. We showed that it is not the previously assumed low inter-annual hydrological variability that explains their importance for pastoralists. We postulate that the availability of 'good'-quality forages in 'fair' quantities especially in the beginning of the dry season, combined with 'low'-quality forages in 'ample' quantity throughout the dry season, coupled to a relatively low risk of complications of resource accessibility and diseases, explains the high livestock density. What pastoralists seek is not so much low variability as high 'assured' production and reduced risk, which spatial and temporal variability in food availability and accessibility can help achieve.
Hypothesis 2. In a system with Ideal Free Distribution, the presence of high-quality forage ('supplementary feeding') may lead to increased herbivore densities and subsequent increased total animal production, but not to increased individual herbivore production.
Hypothesis 3. A key resource within a grazing system is a spatially defined resource that allows a grazing system to maintain one or more populations of herbivores in disproportionately high numbers compared to what could be maintained without that resource. A key resource area is the area where this key resource is found.
Hypothesis 4. A grazing resource fragmented over several hydrological units is a more reliable grazing area than a similarly sized resource located in just one hydrological unit.

CHAPTER 10B

COMMENTS ON "RELEVANCE OF KEY RESOURCE AREAS FOR LARGE-SCALE MOVEMENTS OF LIVESTOCK"

HERBERT H.T. PRINS, AREND M.H. BRUNSTING AND FRANK VAN LANGEVELDE

Resource Ecology Group, Wageningen University, P.O. Box 47, 6700 AA Wageningen, The Netherlands
E-mail: herbert.prins@wur.nl

This chapter deals with the issue of so-named 'key resource areas'. These particular localities within a landscape are endowed with resources that allow many more animals to live there than would have been expected on the basis of the 'general' features of that landscape or ecosystem. Scholte and Brouwer (Chapter 10) advocate the point that the resource under scrutiny in the 'key resource area' is not by necessity herbage; it can also be water. In deserts, oases have in fact been considered as such for millennia and people found them even worth defending at quite great cost. This is confusing, though, because indeed water is a resource for the vegetation and indeed water is a conditional necessity for most animal species; however, it is not a key resource in the sense of Illius and O'Connor (1999), because once the conditionality of the presence of water is sufficiently met there will be no further increase in herbivores. Yet, Scholte and Brouwer rightfully concentrate on floodplains and wetlands. Wetlands and especially their associated grassy floodplains have for hundreds of years played a key role in the economies of Fulani (Peul) and other cattle-herding societies. The same holds for those in southern Africa along, for example, the Zambezi, where Barotse have herded their cattle for generations, or along the Nile, where Nuer and Dinka have done the same (see, for instance, the work of Evans-Pritchard 1940). In East Africa, key resource areas have also been identified by anthropologists already in the 1940s: in areas where floodplains did not fulfil this function, mountains were catching higher amounts of monsoonal rainfall (Huntingford 1933, 1953a,b; Homewood and Rodgers 1991; McCabe 1994; Prins and Loth 1988; Sperling and Galaty 1990; Ruttan and

H.H.T. Prins and F. van Langevelde (eds.), Resource Ecology: Spatial and Temporal Dynamics of Foraging, 233-236.

Borgerhoff-Mulder 1999; Prins 1999). So, at first sight one might think that Illius and O'Connor's (1999, 2000) formalised approach to this phenomenon was unimportant. When they mulled over this age-old concept, they concentrated on herbage (and not water) in these key resource areas, and looked at the significance of a localised high resource concentration within a landscape with a much higher resource dilution. Moreover, even though water is of the essence for plants and plant-eaters, and every herdsman or grazier will have a vested interest in it, herbage is much more interesting a resource than water is from a theoretical (and management) perspective, because of the more complicated feedback relations between herbage and its consumers. Illius and O'Connor then basically asked two questions, namely, "What is the effect of such an uneven distribution for the population dynamics of herbivores making use of that resource?" and "What consequences does this have for utilization and the chances of over-utilization of the resources in these two different compartments of the ecosystems?". Especially this latter issue may be of great concern when floodplains are restored or cut off from a migratory system.

The strength of the chapter by Scholte and Brouwer is that they systematically discuss the strong and weak points of the concept. They do this at the hand of a particular floodplain, the one of the Logone River, within a semi-arid, larger ecosystem south of Lake Chad. Scholte has a deep and multifaceted knowledge of this particular area, and both he and Brouwer have a long familiarity with other West-African wetlands. In their systematic discussion of the different aspects they clearly draw on this understanding, and they try to sharpen up the concept of 'key resource areas'. They point out that the herbivore density in these key resource areas is much higher than expected on the basis of the general relation between rainfall and herbivore mass, as was also found for East-African areas with upwelling groundwater (Lake Manyara NP: Prins 1996; Lake Nakuru NP: Mwasi 2002).Their main insight comes from a paper by Scholte et al. (2006), in which they studied the distribution of cattle and of cattle herds with their owners across the floodplain (the key resource area) and the adjacent dry lands. Although the density was as expected, namely most herds in the key resource area, the individual herder is equally well off in the non-key resource areas. However, because of the lower available herbage biomass outside the floodplain, the cost of trying to make additional use of forage is higher for a marginal herder outside the key resource area than inside the key resource area; herders thus distribute themselves over the landscape according to an ideal free distribution. Supporting revealing information is presented by Scholte and Brouwer in their Table 10.3. This underscores their point that the average individual herbivore has the same production parameter values in a key resource area as in the surrounding dry lands. In other words, the conclusion of Scholte and Brouwer is that individual herbivores are distributed across the landscape according to an ideal free distribution, and the fitness of individuals is on average the same whether they live in a key resource area or not.

The weakness of the approach of Scholte and Brouwer is that they do not provide insight into the questions of population dynamics that are so central to Illius and O'Connor's papers: whether the Logone floodplain acts as a source and the surrounding dry lands as sink, does not become clear. Scholte and Brouwer focus on the point that in West Africa, areas with floodplains have a higher average livestock density than areas without. This indeed is covered in our previous paragraph. If the fitness of individuals in the different areas is equal, and if the costs of settling in the surrounding dry lands is, on a *per capita* basis, higher than in the floodplain, then it may be speculated that these floodplains do not act as source at all!

Another small weakness of this chapter is that floodplains are implicitly supposed to be areas with a high production of palatable grasses (and, thus, key resource areas). Indeed, if one thinks of *Leersia* meadows, then this might be true. However, many wetlands are covered with plants that are not limited by the availability of water; they invest heavily in tallness so as to outcompete other non-woody species. The *Typha* swamps in East Africa, the *Papyrus* beds along the Nile and in the Sudd, or the *Phragmites* marshes of Mozambique and South Africa all may seem to be key resource areas, but from a herbivore point of view, these areas have little use but for elephants. Where Scholte and Brouwer are of the opinion that it does not matter that Illius and O'Connor focused on primary production, we maintain that it does, and that one should not look at primary production *per se* but at digestible dry-matter production. Here we find another argument to focus on food, not on water.

The major assumption by Illius and O'Connor, viz., that *inside* a key resource area herbivores (a) have a density-dependent relation with their food source; (b) compete for resources because they impact severely enough on the vegetation; and (c) are thus regulated, while *outside* that area, in the surrounding dry lands, herbivores (a) do not have a density-dependent relation with the vegetation; (b) do not compete for resources; and (c) are not regulated there, is not very well addressed by Scholte and Brouwer. Indeed, they suggest that, because there was no strong impact of grazers on the vegetation in West-African floodplains, there was no density dependence in these areas, while there was a strong effect on the vegetation outside these floodplains. The reported intra-specific competition for high-quality forage and adjustment of the length of the stay on resource availability in the key resource area, however, seem to contradict the supposed absence of density dependence.

The conclusions of Scholte and Brouwer appear to be diametrically opposite the assumptions of Illius and O'Connor, but they do not zoom in on this. Similarly, the findings of Scholte et al. (2006) and also Table 10.3 appear to suggest that competition is not stronger in the floodplains than in the surrounding dry lands, again not supporting the major assumptions of Illius and O'Connor. We think that Scholte and Brouwer then should have drawn one of *two* possible conclusions, namely (a) floodplains are local ecological optima with resources that are the key to survival for herbivores and pastoralists depending on them, but they do not support Illius and O'Connor's basic assumptions about what key resource areas are supposed to be, in other words, floodplains are not key resource areas; or (b) Illius and O'Connor's definition of 'key resources' and 'key resource areas' was not adequate. Scholte and Brouwer clearly draw this second conclusion without openly rejecting the first. However, by changing the definition as given by Illius and

O'Connor, they reject the idea that herbivore numbers are regulated in a density-dependent manner by the limited resource in the key resource area where equilibrium conditions exist, while in the largest part of the ecosystem the animal population is uncoupled from the resources, thus easily causing vegetation degradation outside the key resource area. In other words, the resources in the key resource area are not analogous to supplementary feeding, often leading to severe range degradation. This has the important effect that restoring floodplains is perhaps not dangerous for the ecosystem, because if these floodplains were behaving as Illius and O'Connor postulated, then their restoration could lead to severe overgrazing of the surrounding dry lands! The alternative interpretation of Scholte and Brouwer is much less spectacular: perhaps floodplains are just what they are supposed to be, namely a feature in the landscape allowing more animals and more people to live. In that case, restoration is a good idea.

What we clearly miss, though, is good individual-based models coupled to life histories of many marked individuals in these heterogeneous landscapes where we have detailed knowledge of pasture production and individual consumption. Only with these tools can we really choose between the two different interpretations of the landscape.

CHAPTER 11A

STRUCTURING HERBIVORE COMMUNITIES

The role of habitat and diet

SIPKE E. VAN WIEREN AND FRANK VAN LANGEVELDE

Resource Ecology Group, Wageningen University, P.O. Box 47, 6700 AA Wageningen, The Netherlands
E-mail: sip.vanwieren@wur.nl

Abstract. This chapter tries to address the question "Why are there so many species?" with a focus on the diversity of herbivore species. We review several mechanisms of resource specialisation between herbivore species that allow coexistence, ranging from diet specialisation, habitat selection to spatial heterogeneity in resources. We use the ungulate community in Kruger National Park to illustrate approaches in niche differentiation. The habitat overlap of the ungulate species is analysed, continued with the overlap in diet and the spatial heterogeneity in resources. This focus on the constraints on species' exclusive resources is a useful tool for understanding how competitive interactions structure communities and limit species diversity. In explaining community structure of mobile animals, we argue that the existence of exclusive resources governed by spatial heterogeneity plays an important role. Trade-offs between food availability and quality, food availability and predation risk, or food and abiotic conditions (different habitat types) may constrain competitive interactions among mobile animals and allow the existence of exclusive resources. We propose that body mass of the animals considered is crucial here as animals with different body mass use different resources and perceive spatial heterogeneity in resources differently. A functional explanation of the role of body mass in the structuring of communities is still lacking while the study of how much dissimilarity is minimally needed to permit coexistence between strongly overlapping species is still in its infancy. Nevertheless, a theoretical framework is emerging from which testable hypotheses can be generated..

Keywords. assembly rules; body mass; coexistence; competition; diet selection; facilitation; habitat selection

INTRODUCTION

Why are there so many species? For many decades, various theoretical and empirical studies aim at understanding how species that utilise similar resources can coexist. Resource specialisation and competitive interactions between species lie at the foundation of our understanding of the structure and diversity of ecological

H.H.T. Prins and F. van Langevelde (eds.), Resource Ecology: Spatial and Temporal Dynamics of Foraging, 237-262.

communities. These interactions are central in resource ecology. For many sedentary species, such as intertidal organisms and vascular plants, one species may completely eliminate another species when they occur together, i.e., competitive exclusion (Connell 1983; Schoener 1983). Coexistence is then only possible when there is some form of partitioning in resource or habitat, to allow exclusive use of resources. This competitive exclusion is, however, especially rare for mobile animal species (Ritchie 2002). Individuals of mobile species can sample many aspects of their environment and thus have the ability to go to certain locations and avoid others, or to select certain patches or types of resources and ignore others. Based on what cues do animals select locations and is this selection the result of competition? This chapter reviews several mechanisms of resource specialisation between herbivore species that could allow coexistence, ranging from diet specialisation, habitat selection to spatial heterogeneity in resources.

This chapter reviews mechanisms of resource specialization between herbivore species that could allow coexistence

Why are there so many herbivores? In many parts of the world ungulates seem to dominate the green world, especially in the savannas of Africa. African mammalian herbivores are frequently used as a model system for the study of community structure and dynamics (e.g., Prins and Olff 1998; Ritchie and Olff 1999; Olff et al. 2002; Sinclair et al. 2003). A great variety in species can be found there, apparently living peacefully together, sometimes even operating in large mixed herds like the wildebeest, zebra and Thomson's gazelle on the Serengeti plains. Frequently they also seem to utilise the same resource, i.e., grass. The question arises then how all these species can live together, apparently showing no signs of interspecific competition. Or do they? Therefore, the fundamental question at stake here is: do all these species occupy a different niche such that coexistence is possible?

The classical approach to analyse niche differentiation is not without problems. From early on, the niche concept has known two approaches that have been used separately, combined and integrated to various degrees. The first is the niche as the place in an environment that a species occupies (Grinnell 1917). This includes all the necessary conditions for a species' existence. The second is the niche of a species as its functional role within the food cycle and its impact on the environment (Elton 1927). The vagueness of these approaches was greatly improved by the work of Hutchinson, MacArthur and others, who tried to quantify the concept. Hutchinson (1957) defined the niche as the sum of all environmental factors acting on the organism; the niche thus defined is a region of an n-dimensional hypervolume. For each species an explicit number (n) of limiting factors could be determined defining the range of conditions where a species could exist. Hutchinson also introduced the difference between the fundamental niche (all the aspects of the n-dimensional hypervolume in the absence of other species), and the realised niche as the part of the fundamental niche to

The classical approach to analyse niche differentiation is not without problems

which the species is restricted due to interspecific interactions. MacArthur (1958), Levins (1968) and others greatly expanded Hutchinson's approach by emphasising the view that interspecific competition was of great importance in the structuring of natural communities (for review, see Chase and Leibold 2003). The new models were almost always based on the Lotka-Volterra equations. Later on, also these models were criticised for lack of statistical rigour and null models, rigorous testing (Chase and Leibold 2003) and for the strong focus on competitive interactions. In fact, the whole niche concept has been strongly criticised as being irrelevant (Hubbell 2001). Notwithstanding the debate about the relevance of this neutral theory (Chave 2004), a large body of theoretical literature exists nowadays that includes various aspects of species' niches: competitive interactions, effects of predation, and intrinsic and extrinsic spatial and temporal heterogeneity (Kotler and Brown 1999; Ritchie 2002; Chesson 2000a, 2000b). Recently, Chase and Leibold (2003) came with a new definition of the niche as the joint description of the environmental conditions that allow a species to satisfy its minimum requirements so that the birth rate of a local population is equal to or greater than its death rate along with the set of *per capita* effects of that species on these environmental conditions. This definition combines the location and the role of a species and makes rigorous quantitative testing possible.

So far, the quantitative approach, as applied to real communities, is still in its infancy (but see, e.g., Ritchie and Olff 1999; Mendoza et al. 2002). It is still highly theoretical or confined to observational studies or small-scale experiments and it has yet not been applied to ungulate communities. Therefore, we still have to apply a more general approach in which we explore how species can be positioned along a few important resource axes and try to deduce what the likelihood is that species either are clearly separated or show potentially high overlap. First, we expand on the theory about competition among mobile animals. We then use a specific animal community to illustrate approaches in niche differentiation, i.e., the ungulate community in Kruger National Park, South Africa. The habitat overlap of the ungulate species is analysed, continued with the overlap in diet and the spatial heterogeneity in resources.

COMPETITION AMONG MOBILE ANIMALS

The choices for forage, rest, mate, drink, etc., of mobile animals may be constrained by particular physiological and morphological characteristics of the animals, so that differences among species in these characteristics can dictate differences in their choices (Ritchie 2002). Many of these choices are phenotypically or behaviourally plastic or 'adaptive' (Abrams 1988). See Owen-Smith (Chapter 8) for a review on adaptive behaviour. This plasticity allows individuals of a species to avoid competition with other species. Also, constraints of the resource distribution, such as large distances between necessary resources, may negatively affect the choices of animals (Prins and Van Langevelde, Chapter 7; Boone et al., Chapter 9). Extensive field-experimental studies suggest that, among species pairs of mobile animals that seem likely to compete, competition can be detected, such as in terrestrial and aquatic insects (Belovsky 1986; Wissinger et al. 1996), lizards (Petren and Case 1996, 1998), birds (Wiens 1992; Loeb and Hooper 1997), small mammals (Rosenzweig and Abramsky 1997; Morris et al. 2000) and large mammals (Edwards et al. 1996). Competition resulted in competitive exclusion of a species appeared only in very few cases (Ritchie 2002).

Most competition models assume that resources are 'well-mixed' in the environment and that the competing species use all available resources, only at different rates. Tilman (1982) showed that if two species are competing for a single resource, then the species that can persist on the lowest availability of that resource will competitively exclude the other. However, if some resources used by each species do not overlap, then species would not use all available resources (Schoener 1976). Instead, some resources would be exclusively available to individuals of each species. Sufficiently abundant exclusive resources for each species could support at least a certain density of these species regardless of the intensity of competition for shared resources, and thus guarantee coexistence. Because of their ability to move and make choices, mobile species are therefore highly likely to 'find' exclusive resources and thus to coexist (Ritchie 2002).

Most competition models assume that resources are 'well-mixed' in the environment and that the competing species use all available resources

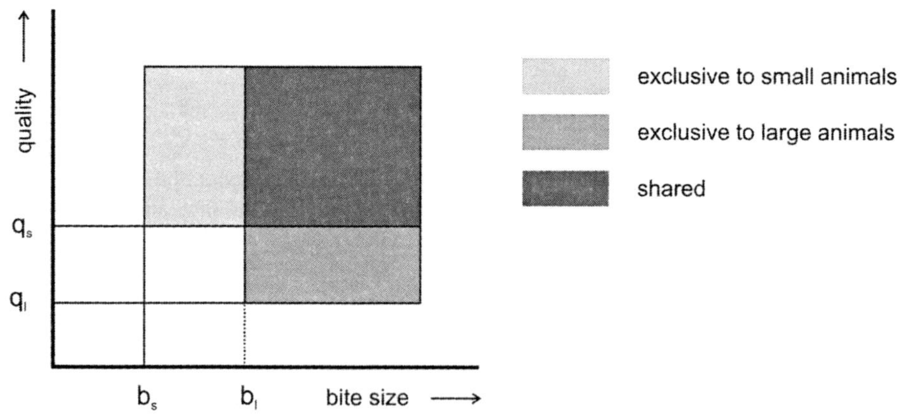

Figure 11.1. *Hypothetical diagram of minimum plant quality (Q_S, Q_L) and bite sizes (B_S, B_L) for small (S) and large (L) herbivores (Belovsky 1986, 1997). The indicated bite sizes are the minimum acceptable item size accepted by a herbivore species. Trade-offs in these minimum thresholds lead to exclusive resources for each species*

Some of the evidence for exclusive resources comes from studies with herbivores. Among herbivores, species of different size may choose resource items of different size and quality because of a trade-off between greater retention and digestion of low-quality food in the digestive tract versus higher metabolic rate, and thus resource requirements, for larger animals (Van Soest 1985; Belovsky 1986, 1997). Different-sized herbivore species, therefore, may partition plant parts (twigs, leaves or parts of leaves) by their relative size and quality (protein, soluble carbohydrate, and secondary chemical content) in a way that generates unique sets of plant parts that are exclusively available to each species (Figure 11.1). Examples for this trade-off generating exclusive resources and competitive coexistence are given by Belovsky (1984, 1986).

Because of their ability to move and make choices, mobile species are therefore highly likely to 'find' exclusive resources and thus to coexist

Segregation in habitat is another mechanism of avoiding competition that can generate exclusive resources and thus coexistence. Habitat selection is often density-dependent (Fretwell 1972). This means that when a population of a species is at sufficiently high densities, individuals may be forced to use less-preferred habitat. Then, they could face competition from other species. Exclusive habitat use among species can arise from trade-offs in their risk of predation, food patch size and quality or different abiotic conditions among different habitats (Kotler and Brown 1999; Olff and Ritchie 1999; Ritchie 2002; Olff et al. 2002).

Exclusive resources imply that trade-offs in species traits do more than just allow them to use resources at different rates as suggested by Tilman (1982). Instead, trade-offs generate access to resources that make coexistence probable (Ritchie 2002). If so, the detailed dynamics and full set of parameters governing competition may be largely irrelevant to understanding competitive outcomes. This implies that perhaps the most important aspect to know about two competing species is what determines their respective sets of exclusive resources (Ritchie 2002). If there are general patterns in these trade-offs, then coexistence models focusing on the constraints on species' exclusive resources may provide a powerful tool for understanding how competitive interactions structure communities and limit species diversity (Ritchie 2002). In Box 11.1, some general principles that structure communities are summarised.

Box 11.1. *Community assembly rules*

An important goal of community ecology is to find out how (local) communities are assembled from (regional) species pools. This goal is based on the idea that communities are characterised by some degree of order as the result of some ordering principle from which an assembly rule can be derived. Keddy and Weiher (1999) distinguish four parts in the procedure of finding assembly rules:
1. Defining and measuring the property of assemblages;
2. Describing patterns in the property;
3. Explicitly stating the rules that govern the expression of the property;
4. Determining the mechanism that causes the pattern.

Species sorting should be based on some measurable property of populations. Various properties thought to play an essential role in the sorting process, have been used; among them are body size, morphological traits like incisor width, and functional group. So far, most work has been done on describing patterns in communities. Within local rodent communities, Dayan and Simberloff (1994) found a much more uniform average ratio of incisor width of succeeding species than expected from communities drawn by chance alone. Similar patterns have been found for body size (Brown and Bowers 1984). Many workers actually use pattern as a synonym for assembly rule (Kelt et al. 1999). Most patterns described are apparently interrelated and this strongly suggests that some common underlying mechanism is operating. The logical candidate for this is interspecific competition because it is strongest among similar species, and it has been demonstrated repeatedly to operate in and to affect local communities (for review see chapters in Keddy and Weiher 1999). It is well recognised that interspecific competition is not the only factor affecting communities, and the effects of historical and geographic processes are not easily to be dismissed (Ricklefs and Schluter 1993). These effects, together with the process of allopatric speciation, might especially affect the composition of regional species pools. Scale effects might also influence community structure, and it is not always easy to define the regional species pool or to make out if the locality of a local community is a representative subset of the region. Increasing the area for the study of local communities might lead to a more 'checkerboard' pattern, permitting more species to be present and leading to communities which are nested (M'Closkey 1978).

Describing and explaining a pattern is not the same as formulating a rule with which communities can be assembled. Community assembly is ultimately driven by the invasion (e.g., immigration, speciation) and extinction of species played out against a complex background of environmental constraints. The environment acts as a filter, eliminating some species and promoting others. Assembly mechanisms are therefore the regulatory agents and processes, which define the suite of plausible system stages or transitions through which a system can proceed (Drake et al. 1999). Drake et al. (1999) define an assembly rule as: an operator, which exists as a function or consequence of some force, dynamic necessity, or context, which provides directionality to a trajectory. The nature of this direction includes movement toward a specific state, or some subset of all possible states. In ecological terms, assembly rules define reachable and unreachable community states, the community being some complete set of species exhibiting limited membership. Within this context, Weiher and Keddy (1999) point out that assembly rules are not so much about recipes for building communities, but, rather, that they are a set of limits that constrain how species can come together to form assemblages. Diamond (1975) is generally cited as the first who tried to formulate a principle with which he could predict bird communities on islands. He matched resource utilisation curves of species to availability of resources (resource production curves), and by subtracting individual resource utility curves from resource production curves till no resource space was left, could make estimates of which combinations of species were either 'forbidden' or 'allowed' on specific islands. This work led to much more research in this field and much discussion among ecologists (see chapters in Weiher and Keddy 1999).

(cont.)

Box 11.1. (cont.)

Despite all this work, surprisingly few assembly rules have been formulated to date. The best-known is the one of Fox (1987, 1989, 1999), who worked on desert rodents and other small-mammal communities. As property he used the functional group (or guild) and he stated a rule by which species belonging to different guilds should enter a developing local community. The rule was: there is a much higher probability that each species entering a (local) community will be drawn from a different functional group until each group is represented, before the cycle repeats (Fox 1999). If some functional group becomes disproportionally represented in a local community, competition lowers the probability that the next species to colonise will belong to that group and raises the probability that it will belong to one of the other groups. A local community is in a 'favoured state' whenever all pairs of functional groups have the same number of species or differ by at most one (if resource availabilities are equal). If the number of species differs by more than one, the local community is in an 'unfavoured' state. The prediction was tested against a null-model of random assembly, and Fox found indeed the rule to apply to the small-mammal communities he studied. The rule has also been found to operate in other communities, e.g., lemurs in Madagascar (Ganzhorn 1997), rodents in Chili (Kelt and Martinez 1991), and shrews in North America (Fox and Brown 1993). Despite this apparent success, Fox's rule has been heavily criticised on various grounds (Stone et al. 2000). The challenges related to both the existence (the rule was probably an 'artefact') and the interpretation of the assembly rule (no evidence that interspecific competition had shaped the local community). In their reactions, Brown et al. (2000) were able to counteract much of the criticism although the debate has most likely not yet fully come to an end. To find the right null-model, for example, is an important consideration before any test can be made. Fox (and many others) tested against communities that were randomly assembled from the regional species pool, but in reality many communities are biased, and then the null-model should reflect this bias and a distribution of the expected number of favoured states should be generated. Even applying randomness is not easy in tests. Empirical data can potentially be compared to multiple random distributions because different random distributions can be expected (e.g., normal, lognormal, broken stick) and different test statistics can be applied (parametrical versus non-parametrical tests, each with unique assumptions).

The fact that, after 30 years of work, still so few real assembly rules have been formulated while the existing ones have not yet been fully accepted either, makes clear that only limited progress has been made in our understanding of communities. Are, then, most (mammal) communities only random assemblages of a regional species pool? Do we need to incorporate many more field data on species and resources before we can make proper analyses? Have history and other stochastic events in many instances blurred regularity in patterns that, without these, would have been revealed? Are ecological communities too complex (Stone et al. 2000)? Surely many communities exhibit some pattern deviating from randomness or some other null-model, yet important properties have been identified and, maybe most important, interspecific competition has been demonstrated to be a key mechanism by which species are structured. This should be sufficient incentive for future ecologists to take up the challenge and pursue the quest for finding fundamental assembly rules governing mammal communities.

OVERLAP IN HABITAT

Although a niche may be composed of many dimensions, the focus in general is on only a few. In ungulate ecology, the most important ones are habitat and diet. If species are completely separated in these aspects, there is no niche overlap and they can coexist. Therefore, we will start with these before adding other dimensions. As a rather arbitrary example, we will focus for our analysis on a community of ungulates living together in a regional setting of ca 19,000 km^2 in South Africa, the Kruger National Park. Thirty species of ungulates can be found here (Table 11.1) which, in principle, have free access to the whole area. One of the conspicuous features of the list in Table 11.1 is the wide range in body sizes that the community comprises: from the 5-kg suni to the 4000-kg African elephant. We cannot specifically analyse this particular system itself in great detail but merely will use this community to demonstrate some general principles which lie at the basis of understanding animal-community structure.

Table 11.1. *Ungulates of Kruger National Park*

Species	Common name	BW (kg)	Species	Common name	BW (kg)
Neotragus moschatus	suni	5	*Damaliscus lunatus*	tsessebe	130
Raphicerus sharpei	Sharpe's grijsbok	11	*Alcelaphus buselaphus*	hartebeest (Licht.)	175
Cephalophus grimmia	common duiker	12	*Kobus ellipsiprymnus*	waterbuck	190
Cephalophus natalensis	red duiker	12	*Hippotragus niger*	sable antelope	210
Oreotragus oreotragus	klipspringer	12	*Tragelaphus strepsiceros*	greater kudu	220
Raphicerus campestris	steenbok	13	*Connochaetes taurinus*	blue wildebeest	230
Ourebia ourebio	oribi	15	*Equus burchelli*	Burchell's zebra	259
Pelea caprealus	vaal rhebok	25	*Hippotragus equinus*	roan antelope	260
Redunca fulvorufula	mountain reedbuck	30	*Taurotragus oryx*	eland	650
Tragelaphus scriptus	bushbuck	45	*Syncerus caffer*	African buffalo	650
Aepyceros melampus	impala	55	*Giraffe camelopardalis*	giraffe	700
Redunca arundinum	common reedbuck	80	*Diceros bicornis*	black rhino	1200
Phacochoerus africanus	warthog	80	*Hippopotamus amphibius*	common hippo	1500
Potamochoerus porcus	bushpig	80	*Ceratotherium simum*	white rhino	1900
Tragelaphus angasi	nyala	90	*Loxodonta africana*	African elephant	4000

BW: body weigh
Licht. stands for Lichtensteins's
From http://www.ecoafrica.com/krugerpark/mammals.htm

In a large area like Kruger various habitats can be distinguished (Table 11.2) and from literature we can derive the habitat preference of the species involved (Table 11.3). The habitats are arranged along a 'cover-axis', going from an open landscape to dense thicket. Table 11.3 makes clear that no species exclusively can make use of its preferred habitat. The minimum number of species in any habitat is 5 while in the open savanna 16 species can be found. Many species can be found in more than one habitat. Nevertheless, with the dimension habitat some differentiation within the community can be made. So it is unlikely that species typical of open landscape (oribi, vaal rhebok, mountain reedbuck, common reedbuck) will ever meet species closely related to habitats with dense cover (common duiker, klipspringer, bushbuck, kudu). So, it is unlikely that interspecific competition between these species occurs.

Table 11.2. *Habitat classification for Kruger National Park ungulates (based on Haltenorth and Diller 1994; Kingdom 1997)*

Code	Habitat	Description
LS	light savanna	open grassland, light savanna, savanna
WS	woodland savanna	woodland savanna, light woodland
Wo	woodland	woodland, dense woodland
Fo	forest	forest
SB	savanna with bush	grass savanna with bushveld, grassland with scattered bush, grassland with thick bush
BS	bush savanna	bush savanna, dense savanna
Bu	bush	bush, bushveld
Th	thicket	thicket, thick bush

With the dimension habitat, some separation can thus be made, but it certainly is not enough as all habitats share a number of species. It also has to be noted that it is not always easy to distinguish between habitats, especially when small, but important, habitat elements are laid out in a fine-grained mosaic with other habitat elements. The small steenbok can be found in rather open woodland savanna, but is nevertheless bound to low bush, which thus needs to be present.

OVERLAP IN DIET

Grass and browse

There is circumstantial evidence for exclusive resources among ungulate-species pairs in field data (Van Wieren 1996b). Food specialisation is likely the most important resource dimension for separating species within communities. While, when considering herbivores as predators, plants are easy to catch, they are much more difficult to process than prey for carnivores. Plants are rich in fibre that requires special adaptations to digest, and this is a slow and not very efficient process. Compared to carnivores, herbivores eat a low-quality diet. Therefore they have to consume large quantities and have to spend many hours a day to meet these requirements. It is thus highly relevant that they search well for the best food (Laca, Chapter 5; Fryxell, Chapter 6; Owen-Smith, Chapter 8).

Food specialisation is likely the most important resource dimension for separating species within communities

The plant world is highly variable in its morphology and chemical composition, but a major distinction in two forage types can be made to which ungulates have adapted themselves through a number of specialisations. There are a few fundamental differences between monocots (grasses and sedges) and dicots (herbaceous and woody plants); they are summarised in Table 11.4.

Grasses are more fibrous than browses, while the cell wall of grasses is thick. The cell wall of grasses is relatively low in lignin, and as lignin is the most important inhibitor for the digestion of the cellulose and the hemicelluloses that make up most of the cell wall (Van Soest 1994), the potential digestibility of the grass cell wall is quite high.

Table 11.3. *Habitat preference of Kruger ungulates (for legend see Table 11.2; based on Haltenorth and Diller 1994; Kingdom 1997)*

	Habitat preference							
	LS	**WS**	**Wo**	**Fo**	**SB**	**BS**	**Bu**	**Th**
suni			●	●				●
Sharpe's grijsbok		●			●	●		
common duiker							●	●
red duiker				●			●	●
klipspringer								●
steenbok		●			●	●	●	
oribi	●							
vaal rhebok	●							
mountain reedbuck	●							
bushbuck						●	●	●
impala	●	●						
common reedbuck	●							
warthog	●	●						
bushpig				●			●	●
nyala			●			●	●	●
tsessebe	●	●			●	●		
hartebeest (Licht.)	●	●	●				●	
waterbuck	●	●			●			
sable antelope			●			●		●
greater kudu							●	●
blue wildebeest	●	●				●		
Burchell's zebra	●	●						
roan antelope		●				●		
eland	●	●	●					
African buffalo	●	●	●	●				●
giraffe	●	●			●	●		
black rhino							●	●
common hippo	●							
white rhino					●	●		
African elephant	●	●	●	●	●	●	●	●

Box 11.2. *Polyphenolics and mammalian herbivores*

Plants can defend themselves against herbivory by means of structural adaptations like hairs and thorns and by certain chemicals that in some way or the other have a deterrent effect on the herbivore that eats them. Many of such secondary compounds are known with widely varying effects. As an example we highlight here the polyphenolic compounds and their effects because they are quantatively the most abundantly produced group of plant allelochemicals and they are widespread throughout the plant kingdom, although mostly found in woody plants like trees and shrubs. They also seem to be one of the earliest groups of allelochemicals (Swain 1978).

Polyphenolics are not a clearly defined group of molecules but they share a high molecular weight, numerous phenolic groups and a lack of other functional groups (Mole 1989). There are two major types of chemically distinct groups, one being the condensed tannins and the other the hydrolysable tannins. These differ in both their component subunits and the type of bonding between these. An important characteristic is that they can form complexes with natural polymers, such as proteins and polysaccharides (starch, cellulose, etc).

The most investigated effect of polyphenolics is their presumed ability to reduce the digestive process in herbivores. This might happen by direct inhibition of digestive enzymes, and by the binding of polyphenols to substances in the diet, thus depriving digestive enzymes of access to them. Robbins et al. (1987) found protein digestion in deer to become reduced (with associated elevated levels of nitrogen in the faeces), while others found that also cell wall digestion decreased with increasing tannin levels because of the inhibiting effect on cellulase functioning (Van Hoven and Furstenburg 1992; McSweeney et al. 2001).

Apart from effects on the digestion process, tannins can also depress voluntary intake (e.g., in goats, Provenza and Malechek 1984), while in snowshoe hares reduced food intake was in proportion to increased levels of polyphenols (Sinclair and Smith 1984). Monogastric herbivores seem to suffer more than polygastric ones, but especially pregastric species seem to be able to metabolise hydrolysable tannins (Mole 1989; McSweeney et al. 2001), while certain species of ruminants are much less affected than others because of the variation in effectiveness of microbial fermentation to deal with a range of dietary components. Grazing species like sheep and cattle seem to be more affected than more browsing species like deer (Barry and Manley 1986; Robbins et al. 1987). Clear evidence is also available that browsers and mixed feeders differ significantly in their ability to cope with tannins (reviewed by Owen-Smith et al. 1993). Production of special proteins that bind with tannins is one way in which animals can gain protection. Salivary proline-rich proteins (which bind strongly to tannins) are being produced in the parotid salivary glands of ruminants, and it has been found that these (per unit body mass) are three times larger in browsers than in grazers, the kudu, however, being a notable exception to this rule (Owen-Smith et al. 1993).

Probably the most widely known effect of tannins is that they cause an astringent sensation in the mouth. The physical basis for this may be that tannins bind to and perhaps precipitate salivary mucoproteins and that they also may bind to taste receptors (Mole 1989). A logical consequence of plants becoming unpalatable is that herbivores try to avoid these plants. Both in feeding trials and in the field, a strong selection against high levels of tannins has been found in numerous cases, e.g., leaf-eating black colobus monkey (Oates et al. 1980), arctic herbivores (Bryant and Kuropat 1980), African buffalo (Field 1976) and deer (Robbins et al. 1987). It is generally thought that herbivores try to limit the intake of tannins by broadening their diet and indeed, many large herbivores are generalist feeders (Freeland and Janzen 1974).

The effects of tannins might even work up to the population level. Freeland et al. (1985) suggested that, because of the sodium-depleting potentials of tannins, these latter might have a population-regulating effect through the limiting of this scarce mineral. Population regulation effects of secondary substances have also been hypothesised to explain microtine cycles (Jensen et al. 1999). During the increasing phase, microtine rodents deplete their resources of preferred foods through heavy grazing, forcing them to switch to plant species that would normally be avoided because of the plant chemicals they contain. Apart from the 'normal' deterrents present, crashes are especially thought to occur because of induced production of toxins as the result of heavy grazing.

(cont.)

Box 11.2. (cont.)

Although generally considered deterrents, tannins may also have positive effects on vertebrates. Complexes of proteins with hydrolysable tannins can hydrolyse in gastric acidity beyond the rumen, releasing protein and amino acids that can then be taken up by the host (Martin 1982; Makkar 2003). Maybe this is the reason that some species actually select for a low level of tannins in their diet (Duncan et al. 1998). In feeding trials, roe deer adjusted their food choice to obtain 4% tannins in their diet (Verheyden-Tixier and Duncan 2000). Tannins have also been found to reduce the detrimental effects of saponins (another secondary metabolite) (Freeland et al. 1985) and to reduce the physiological effects of alkaloids by preventing the absorption of alkaloids in the bloodstream (Freeland and Janzen 1974). Further they are effective anti-oxidants (Hagermann et al. 1998), and they have been found to control intestinal worms (Hodgson et al. 1996).

In contrast, browses have generally less cell wall but these cell walls are high in lignin and have therefore a low to very low digestibility. Only when in a very young stage, browse cell walls can be utilised, otherwise herbivores who want to exploit browse have to draw heavily on the cell contents, of which there is more than in grasses, and which contain much better digestible substances like proteins, sugars and lipids. The two forage classes also differ in their chemical defence structures (see Box 11.2 for different defence mechanisms). Grasses contain considerably fewer secondary compounds than browses. They tend to have a higher silica concentration, which increases tooth wear, but in many browse species, a whole array of chemical substances can be found, ranging from polyphenols (tannins) to toxins like alkaloids. These compounds generally have negative effects on the herbivores through, for example, the lowering of the digestibility of nutrients (tannins), effecting reproduction, appetite and the immune system (see overview in Van Soest 1994; Bailey and Provenza, Chapter 2).

Table 11.4. *Relative differences between chemical and structural characteristics of grasses (monocots) and woody and herbaceous plants (dicots) (Iason and Van Wieren 1999; Van Soest 1994; Shipley 1999)*

Characteristic	Grasses (monocots)	Browses (dicots)
Cell wall	High cell wall content	Low cell wall content
	Thick	Thin
	Low in lignin	High in lignin
	Moderate to high digestibility	Low to moderate digestibility
Defence	Low chemical defence (silica)	High chemical defence (e.g., tannins, terpenes, alkaloids)
Plant architecture	Fine-scaled heterogeneity in nutritional quality within plant	Coarse-scaled heterogeneity in nutritional quality within plant
	New growth added at base	New growth added at tip
	Low growth form	Low to high growth form
Dispersion	Uniform	Dispersed/discrete

Apart from these differences in morphology and chemistry, grasses and browses also differ in architecture, growth form and spatial arrangement. Grasses consist of leaves and stems that differ in quality only over a very small spatial scale. Further, the low growth form of grasses, in which new tillers are added to the base of the plant, creates a rather continuous three-dimensional layer of vegetation with a relatively constant density (Jarman 1974). A grass sward also frequently consists of a mix of different species. These characteristics make that grasses provide a rather homogeneous food source for larger herbivores that face difficulties in selecting the good parts from it, if possible at all (Drescher 2003). Browse on the other hand, tends to contain a more heterogeneous assembly of plant parts of various nutritional quality with buds, leaves and stems. Also the individual browse plants are dispersed in a more discrete fashion and are thus more spread out in space. In browse, the change in quality when bite size increases is also much greater than in grass (Van Wieren 1996a).

Food adaptation in ungulates

Diet overlap can, however, be misleading since species may converge in diet toward the more productive resource type as competition intensifies (Abrams 1990; Ritchie and Tilman 1993) and diet overlap may change with species density. Moreover, it will not be surprising that ungulates have adapted themselves to utilise the various plant resources (Owen-Smith, Chapter 8) and more in particular to the two main forage classes discussed above. It has long been recognised that ungulates differ in the preference they show for certain food categories and that these preferences are related to a number of characteristics which involve both adaptations to the cropping and to the processing of the food base. Cropping and processing abilities have been attributed both to body size and special adaptations irrespective of size. The idea that there are different categories of ungulates, like typical browsers and grazers, which fundamentally differ in their anatomy and digestive system, has both been advocated (Hofmann 1973, 1989; Van Wieren 1996a; Iason and Van Wieren 1999) and challenged (Gordon and Illius 1994; Robbins et al. 1995; Gordon 2003). Fact is that for most characteristics, body size always plays a role because of the enormous range in body sizes that can be found in most ungulate communities (Table 11.1) and because body size affects almost every ecological parameter (Peters 1983). Here we have tried to disentangle the effects of body size and categorical difference (Tables 11.5 and 11.6) but note that there is always an interaction.

Diet overlap can be misleading since species may converge in diet toward the more productive resource type as competition intensifies

Table 11.5. *Some relative differences in characteristics related to the cropping and processing of plants between small and large ungulates (Van Soest 1994; Van Wieren 1996a)*

Characteristic	**Small**	**Large**
Absolute energy requirements	Low	High
Relative energy requirements (per kg $^{0.75}$)	High	Low
Absolute food intake	Low	High
Relative food intake (per kg $^{0.75}$)	High	Low
Gastro-intestinal tract	Small	Large
Muzzle width	Narrow	Broad
Selectivity in feeding	High	Low

In absolute terms, small herbivores need less energy and food than large ones (Table 11.5). Because of allometric scaling, however, small animals require more energy per unit weight, which they can accomplish by eating more or selecting food items of a higher quality. It has been found that relative food intake of smaller ungulates is higher than that of larger ones (Van Wieren unpublished data), but smaller animals are also more selective.

This is facilitated by a smaller muzzle width that allows them to crop smaller food items. High-quality food items are in general both more scarce and smaller than low-quality food items. Although it seems evident that large herbivores are better suited to digest cell walls than smaller ones (Demment and Van Soest 1985), an analysis of more than 200 feeding trials with various ungulate species did not reveal significant differences in the capacity to digest cell walls between small and large animals (Van Wieren 1996a). The various characteristics within a certain size range are, logically, related, and we can state that different sizes produce different adaptive syndromes, which are suited to the exploitation of a certain part of the plant food base. Although size alone has great explanatory power, there are still a number of differences to be found between browsers and grazers (Table 11.6).

Table 11.6. *Relative differences in characteristics related to the cropping and processing of plant material between grazers and browsers, corrected for body-weight effects (Van Wieren 1996a; Iason and Van Wieren 1999; Mendoza et al. 2002; Hofmann 1973; Clauss et al. 2003)*

Characteristic	**Grazers**	**Browsers**
Hypsodonty index	High	Low
Skull length relative to body size	Large	Short
Shape of incisor arcade	Straight	Curved
Muzzle width	Broad	Narrow
Digestive capacity	High	Low
Salivary glands	Small	Large
Liver	Small	Large
Retention time of food	Long	Short
Reticular orifice	Small	Large

Some of these are related to harvesting. Grazers have teeth with higher crowns than browsers, which enables them to utilise the more fibrous and silica-laden grasses (Mendoza et al. 2002). Grazers and browsers also differ in other craniodental features, which make them suitable to exploit either grasses or browses better. Grazers have a relatively large skull, the shape of the incisor arcade is straighter and they have a broader muzzle that makes them fit to exploit the more continuous grass sward, but at the same time reduces their ability to select the smaller, more nutritious parts within the sward (Illius and Gordon 1987; Janis and Ehrhardt 1988). Note that there also are differences related to phylogeny. Suids and equids, in contrast with ruminants, have upper incisors that make them better equipped for tearing off grass leaves.

Browsers, on the other hand, tend to have narrower muzzles and a relatively larger mouth opening that permits sideways stripping of leaves. They frequently have a longer tongue and prehensile lips, which enables them to negotiate thorns and take small bites that are difficult to reach. Browsers also have larger salivary glands and a larger liver (Hofmann 1989; Robbins et al. 1995), both seen as adaptations to deal with the secondary compounds so characteristic of the browse world. Proteins in saliva can bind tannins while many toxic compounds can be detoxified in the liver. Browsers have lower digestive capacity than grazers (Iason and Van Wieren 1999) and are more adapted to utilising cell content than cell wall. Cell content fermentation is rather quick and permits a higher throughput of food through the digestive tract (Clauss et al. 2003), which is facilitated by a large reticular orifice, the opening between the reticulum and omasum.

Diet specialisation within ungulates

The structural differences between grasses and browses have led to various adaptations within ungulates, and this has led to the recognition of a number of dietary strategies (see also Owen-Smith, Chapter 8). A number of classifications based on food preference of ungulates have been proposed (Gagnon and Chew 2000; Hofmann and Stewart 1972; Janis 1995; Langer 1988; Mendoza et al. 2002), but there is no consensus. Here we adopt the classification into six classes of Gagnon and Chew (2000) as it is based on the most extensive survey of the literature (Table 11.7).

Structural differences between grasses and browses have led to various adaptations, which has led to the recognition of dietary strategies

Table 11.7. *Classification of Kruger National Park ungulates based on diet preference (Gagnon and Chew 2000; Van Wieren 1996a)*

Code	Diet
OG	obligate grazers (>90% monocots)
VG	variable grazers (60-90% monocots)
BG	browser-grazer intermediates (30-70% of dicots and monocots, <20% fruits)
GE	generalists (>20% of all food types)
BR	browsers (>70% dicots)
FR	frugivores (70% fruits, little or no monocots)

The various classes of diet specialisation are not randomly distributed across the body size range. Figure 11.2 clearly demonstrates that small animals are more likely to be frugivores and browsers while the very large ones are more of the grazer type. This again points to the important effect that body size has on the structuring of adaptive syndromes. Substantial overlap, however, can be expected in a large portion of the medium size range. We have evidence that browsers are more obligate in their food preference than both pure grazers and animals of the more mixed types (Van Wieren 1996a). This would indicate that pure browsers can be clearly distinguished and separated from all other types. The question how diet preference structures the Kruger ungulate community is illustrated in Table 11.8.

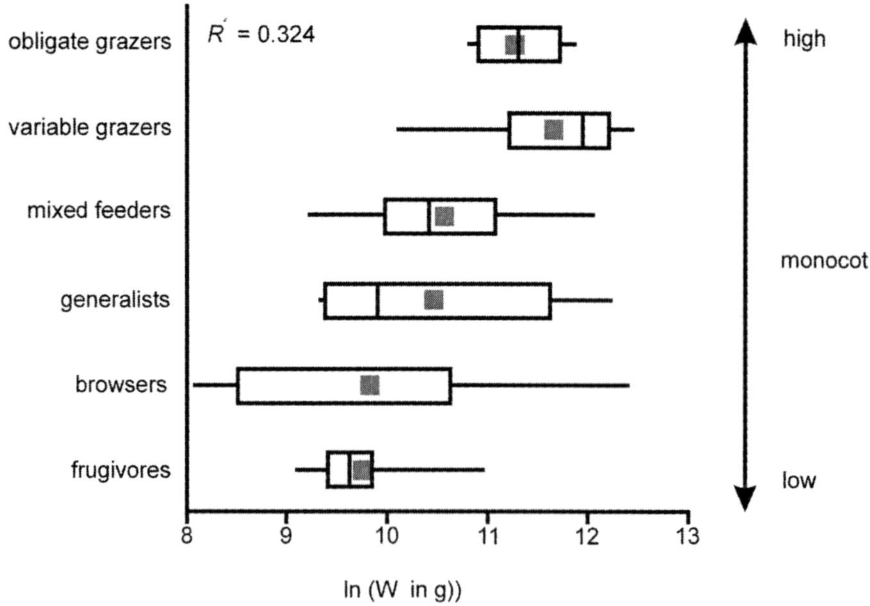

Figure 11.2. *Box plot showing the relationship between body mass (g) and dietary preferences for 76 species of African bovids (Gagnon and Chew 2000)*

Table 11.8. *Classification of Kruger NP ungulates based on dietary preference (for legend see Table 11.7; Gagnon and Chew 2000; Van Wieren 1996a)*

Species	Diet class					
	OG	VG	BG	GE	BR	FR
suni					•	
Sharpe's grijsbok			•			
common duiker					•	
red duiker						•
klipspringer				•		
steenbok			•			
oribi		•				
vaal rhebok			•			
mountain reedbuck	•					
bushbuck					•	
impala			•			
common reedbuck	•					
warthog		•				
bushpig				•		
nyala				•		
tsessebe	•					
hartebeest (Licht.)		•				
waterbuck		•				
sable antelope		•				
greater kudu				•		
blue wildebeest		•				
Burchell's zebra	•					
roan antelope		•				
eland			•			
African buffalo		•				
giraffe					•	
black rhino					•	
common hippo		•				
white rhino	•					
African elephant			•			

Apart from the frugivores, each category contains a number of species. It can further be seen that the 'size rule' is violated in almost each diet category. The small oribi is very much a grazer while the giraffe and the black rhino are obligate browsers. Similar to the classification in habitat, the classification in diet preferences brings about some differentiation but does, in itself, not clearly separate all the species.

HABITAT AND DIET COMBINED

Now we can combine the two most important resource axes into one picture (Table 11.9). The two axes separate more species than either one of them. Quite a few species show no overlap with quite a few other species. According to this analysis,

Table 11.9. *Classification of Kruger National Park ungulates based on both habitat and diet preference. For legend see Tables 11.2 and 11.7. Body weight classes: S = 5-50 kg, M = 51-200 kg, L= >200 kg*

Habitat class	Size class	Diet class OG	VG	BG	GE	BR	FR
LS	S	m-reedbuck	oribi	rhebok			
	M	c-reedbuck tsessebe	warthog hartebeest waterbuck	impala			
	L	zebra	wildebeest buffalo hippo	eland elephant		giraffe	
WS	S			grijsbok steenbok			
	M	tsessebe	warthog hartebeest waterbuck	impala			
	L	zebra	roan wildebeest buffalo	eland elephant		giraffe	
Wo	S					suni	r-duiker
	M		hartebeest		nyala		
	L		sable buffalo	eland elephant			
Fo	S					suni	
	M				bushpig		
	L		buffalo	elephant			
SB	S			steenbok			
	M	tsessebe	waterbuck				
	L	w-rhino		elephant		giraffe	
BS	S			steenbok		bushbuck	
	M	tsessebe			nyala		
	L	w-rhino	sable wildebeest roan	elephant		giraffe	
Bu	S			steenbok		c-duiker bushbuck	r-duiker
	M		hartebeest		bushpig nyala		
	L			elephant	kudu b-rhino		
Th	S				klipspringer	c-duiker suni bushbuck	r-duiker
	M				bushpig nyala		
	L		sable	elephant	kudu		

m = mountain; c = common; w = white; r = red; b = black

however, separation is far from complete. Groups of species can still be found in the various diet–habitat classes. This is not surprising as, in particular, the diet classification is quite general and each class comprises a whole range of plant species. Much more detailed study is needed to try and separate the 'remaining' species while incorporating detailed observations on diet selectivity and a further differentiation in resource axes by including spatial and temporal heterogeneity, and the species' response on these. The effects of size may play an important role here as we will discuss below, but it is furthermore evident that habitat and diet choice most likely are not sufficient to explain the ungulate community completely and that other resource dimensions have to be included too. Nevertheless, this exercise can be used to point to certain combinations of species that likely show large overlap, especially when they are similar in size (e.g., sable antelope – roan antelope – wildebeest – zebra or bushbuck – nyala – common duiker). To explain these combinations poses the greatest challenge to students of ungulate community. In the remainder of this chapter, we will focus on a number of other characteristics of the niche space which need to be incorporated in the analysis and which can help to further separate the species in our ungulate community.

RESOURCE HETEROGENEITY AND BODY SIZE

Especially in the green world resources are not homogeneous entities. Animals can make choices in heterogeneous environments with difference in food types, variation in distribution and density of these food types, differences in food patch size, etc. These possibilities of choice by foraging animals suggest that heterogeneity in the distribution of resources and habitat play a large role in competitive coexistence, because greater heterogeneity implies more available resources (Ritchie 2002).

As indicated in Table 11.4, both grasses and browses are heterogeneous in architecture. Within a sward, patches of different height can be found, and within a plant, different plant parts with frequently varying quality are apparent. This variation in a single resource leads to a further possible niche differentiation within the herbivores exploiting the resource. The factor that best seems to describe this differentiation is body mass. In Table 11.5, a number of vital differences between being either a small or a large animal is given. These differences can be used to explain why small and large herbivores exploit different dimensions of the same resource. Of great importance is the variation in height/biomass of the resource. It is easy to envisage that the giraffe and the steenbok can be separated along a vertical gradient while exploiting the same resource. But separation can also occur when the vegetation is within reach of species of different size (e.g., 0-100 cm). The separation is based on the difference

Heterogeneity in the distribution of resources plays a large role in competitive coexistence, because greater heterogeneity implies more available resources

in the functional response between small and large herbivores when foraging on vegetation of varying height/biomass. This has been best studied in grass swards (Prins and Olff 1998; Drescher 2003).

A schematic representation of the functional response of small and large herbivores across a biomass gradient is given in Figures 11.3 and 11.4. From Figure 11.3, it is clear that in an absolute sense (i.e., per animal), intake rates of larger herbivores are always higher than those of smaller herbivores. However, per unit body mass, intake rates at low plant biomass are higher in small herbivores (Figure 11.4). The reason for this lies in the allometric relationship between body mass and mouth size whereby small species are able to take relatively larger bites at low grass height (Gordon 2003). The asymptote is also reached earlier in small species, to decline thereafter and to become much lower than that of larger species at higher biomass. The reason for this decline may be a relatively larger increase in vigilance behaviour or a relatively greater problem in coping with a decrease in the availability of high-quality food items (Stobbs 1973; Van de Koppel et al. 1997; Drescher 2003). Small herbivores are thus better able to exploit the lower biomass/height ranges of the vegetation and large herbivores the higher.

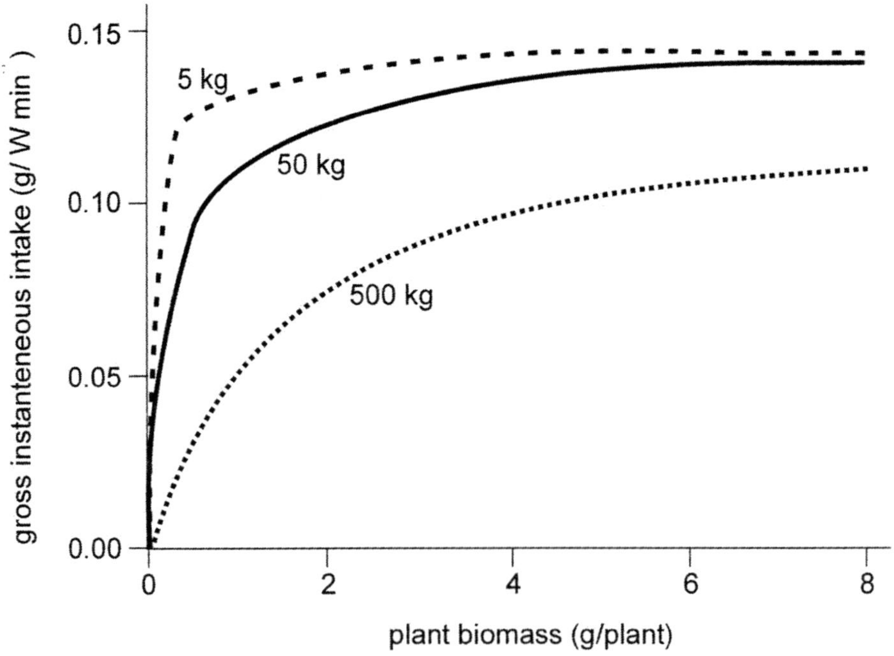

Figure 11.3. Gross instantaneous intake rate as a function of plant biomass at different body mass (Gross et al. 1993, Prins and Olff 1998)

Figure 11.4 can also serve as model for some important interactions between herbivores. Species of similar size are, on the basis of the similarity in their functional response, much more likely to compete than when species differ largely in weight. Further, we can hypothesise that when two species of different size show considerable overlap in diet preference, the smaller species is more likely to suffer from competition from the larger one than *vice versa* because the smaller species is much more overlapped by the larger one than that the smaller one overlaps the larger. On the other hand, it can be envisaged that a large species, by changing the grass biomass/height to a lower level, can facilitate smaller species. Facilitation has been reported between hippo and kob (Verweij et al. 2006), and between other large and small grazers (see Prins and Olff 1998 for more examples). Box 11.3 further discusses facilitation.

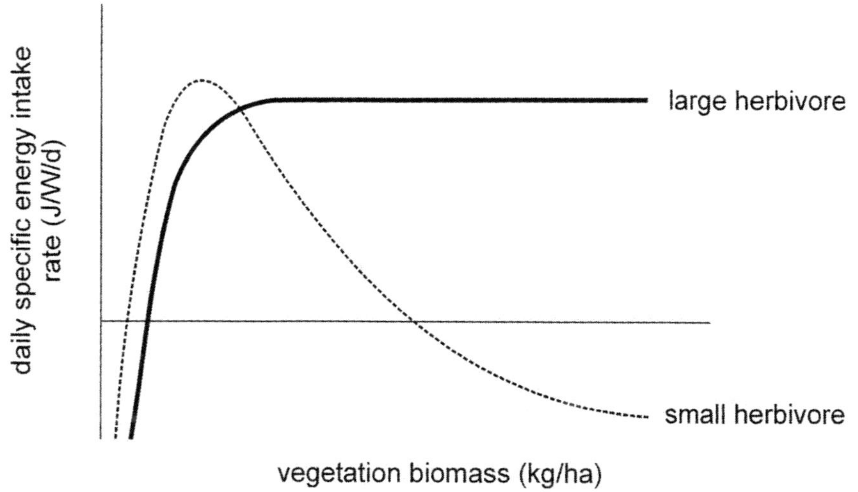

Figure 11.4. *Predicted pattern of functional response of daily specific net energy intake of grazers towards vegetation biomass for a small and a large herbivore, assuming a negative relationship between vegetation biomass and forage quality, and larger problems in the small species in coping with poorer quality forage (Prins and Olff 1998)*

RESOURCE HETEROGENEITY AND SCALE

Heterogeneity in resources exists at different spatial and temporal scales (Skidmore and Ferwerda, Chapter 4). In such heterogeneous environment, the resource distribution is perceived differently by species that differ in body size (Ritchie 1998). In a specific environment, one species, which detects variation at a very small scale of resolution, finds many choices. Another species, which detects variation at a very large scale, may find the environment to be very homogeneous because this species averages across the detailed variation detected by the smaller-scaled species. Size not only determines food requirements, it apparently also determines the scale of resolution at which animals perceive the environment.

Box 11.3. *Facilitation between herbivores*

Facilitation among species is the process whereby one species benefits from the (feeding) activities of another. For herbivores, the interaction is thus strongly mediated through grazing and browsing impacts on the vegetation. Facilitation is generally brought about by the larger species having a positive effect on the smaller ones. Large species have the ability to utilise low-quality food and therefore they can sustain on tall vegetation, while small species require high forage quality, but can tolerate low food levels. Small species may thus benefit from the grazing impacts of the larger species because biomass is reduced and food quality is enhanced. It is well established that grazing stimulates grass regrowth, thereby increasing the nutritional quality of the forage (Vesey-Fitzgerald 1974; McNaughton 1976). Facilitation may act on the population level and, perhaps, even may affect species richness (Prins and Olff 1998; Arsenault and Owen-Smith 2002).

Facilitation may increase access to and quality of resources. On the Island of Rum, Scotland, Gordon (1988) found that areas grazed by Highland cattle during winter had a significantly higher standing crop of green vegetation in the spring, and had a greater proportion of green material, than areas without cattle grazing. Red deer (*Cervus elaphus*) preferred the areas previously grazed by cattle while females produced more calves per female in areas where cattle were reintroduced. Although perhaps not so common, facilitation can also be brought about by small herbivores. On a salt marsh in the Netherlands, it was found that brown hares (*Lepus europaeus*) facilitated brent geese (*Branta bernicla*) by selectively removing the shrub *Atriplex portulacoides*, making the preferred forage grass *Festuca rubra* more accessible (Van der Wal et al. 2000). Because of their size, gregariousness and rather unselective mode of grazing, large grazers are able to create grazing lawns which provide short, high-quality forage over larger areas, attracting smaller grazers (McNaughton 1976, 1986; Fryxell 1991). Especially grazing megaherbivores (>1000 kg) like hippopotamus (*Hippopotamus amphibius*) and white rhinoceros (*Ceratotherium simum*) have been found to create grazing lawns, benefiting short-grass-preferring species like wildebeest, zebra, impala and warthog (Owen Smith 1988).

For facilitative effects on the ecosystem level, affecting community structure and species richness, a keystone species, *sensu* Paine (1969), is needed. This megaherbivore, then, should be able to alter vegetation structure profoundly so as to 'lower the canopy', as the result of which more primary production becomes accessible while variation in available niches should increase. To date, only the elephant has been hypothesised to be able to accomplish this. It is well known that elephants can transform savanna woodlands into either open parkland or shrubland (Laws 1970; Bell 1981; Dublin et al. 1990). The vast increase of elephants in the Tsavo region in Kenya during the 1960s, and consequent opening of dense shrubland, was followed by increases in the abundance of grazers like oryx (*Oryx gazella*) and zebra (*Equus burchelli*), while browsers such as lesser kudu (*Tragelaphus imberbis*) and gerenuk (*Litocranius walleri*) declined (Arsenault and Owen-Smith 2002; Parker 1983). The reverse has also been reported: a decline in some grazing ungulates following the extermination of elephants (Owen-Smith 1989). In a recent analysis of 31 ungulate communities from East and Southern Africa, Fritz et al. (2002) studied the relationship between megaherbivores and mesoherbivores. First, it was established that megaherbivores indeed generally make up the bulk of the ungulate community. Further, the study showed that the various guilds of mesoherbivores reacted differently to the most dominant megaherbivore, the elephant. No relation between elephants and the mesograzers was detected but both the mesobrowsers and the mesomixed feeders were negatively affected by elephants. In this study, the effect on the mesograzers was thus neutral while, like above, the effects on the more browsing species were more competitive. The potential facilitative keystone effect of elephants on the ungulate community at large has thus not unequivocally been substantiated. Furthermore lacking, still, is evidence that megaherbivores can have an effect on the composition and species richness of ungulate communities.

Ritchie and Olff (1999) argue, therefore, that it seems likely that body size, which is often different among coexisting species, influences how much heterogeneity is detected and what choices are available to a species. They provide a mechanism of niche differentiation that could explain coexistence of different-sized animals. This mechanism is based on the assumption that larger-scaled species may only select resource clusters that exceed a certain density, so that small resource clusters are ignored effectively. For non-randomly moving foragers attempting to maximise their encounter with resources, smaller-scaled species should experience a higher average resource density per patch (volume sampled) and greater numbers of acceptable resource patches (Ritchie 1998). Larger species encounter fewer acceptable patches, which contain absolutely more resources per patch (volume sampled), but the resources occur at a lower density per unit volume sampled. If large and small species search the same number of patches per unit time, then the larger species should also sample a greater total volume per unit time. These mathematical outcomes predict that species measure different qualities of a single resource by virtue of their different scales, and may thus differ in the rate of consumption of resources. However, trade-offs in number, resource density and size of resource patches encountered or accepted, as well as search rate, suggest that differences in foraging scale among species provide some potential for coexistence (Ritchie 1998; Ritchie and Olff 1999; Ritchie 2002). These trade-offs generate exclusive resources for species of different size: each species has a unique set of food patches of a particular size and resource concentration.

Body mass of animals could explain exclusive resource use and coexistence

Although this prediction is untested, Ritchie (2002) argued that some field experiments support the assumption that larger-scaled animals select larger resource clusters. In the Negev desert, large gerbils (*Gerbillus* sp.) left patches with more seeds remaining than smaller gerbils (higher giving-up densities, GUDs) (Brown et al. 1994; Garb et al. 2000). This could suggests that larger foragers select only large seed patches and thus perceive the environment in a more coarse-grained manner than smaller rodents. These GUDs directly correspond to a minimum seed consumption rate, and thus could conceivably be used to calculate a minimum patch size.

On a larger scale, Redfern et al. (2006) recently analysed the associations of seven herbivore species (three grazers and four browsers) in the Kruger National Park on three different spatial scales: $5\times5km^2$, $10\times10km^2$ and $15\times15km^2$. It was expected that all herbivore distributions would be similar at large scales because of the influence of abiotic factors, but that at smaller scales the relative importance of guild aggregation versus competition in structuring assemblages would become apparent. For instance, similar-sized species with similar diets are expected to associate less on smaller scales. It was indeed found that similarity was greatest at large scales and that spatial distributions became increasingly disjoint at smaller scales. Larger species had a more even spatial distribution than small species, presumably because of a higher tolerance for low-quality foods.

OTHER NICHE DIMENSIONS

When body mass (through its relationship with resource heterogeneity) is added to the habitat-diet niche space, the community becomes much more structured. Oribi and buffalo will most likely be separated now, like steenbok and elephant, and bushbuck and giraffe, to name a few examples (Table 11.8). Yet, various combinations of species with similar body size remain present in the same habitat-diet group. Apparently more and other niche axes need to be invoked to explain these combinations. Some of these niche axes that can possibly act as niche separators are briefly described below.

Morphological differences

Similar-sized species may differ in mouth structure, which leads to differences in the part of the resource base that can be exploited. Sometimes, these differences are related to phylogeny. So have equids relatively narrower muzzles than ruminants of similar size (Janis and Ehrhardt 1988; MacFadden and Schockey 1997), but differences can also occur within the same taxonomic group. Murray and Brown (1993) compared similar-sized wildebeest, hartebeest and topi, the latter being the same species as the tsessebe from Kruger National Park. From these three species, the wildebeest has the greatest muzzle width, while the muzzle width of topi and hartebeest are similar. Its broad muzzle makes wildebeest better suited for short leafy grasslands while topi, with their relatively pointed jaw, are expected to be better as selective feeders in medium-height upright grasslands (Illius and Gordon 1987). These expected differences were also found in field observations (Bell 1970; Duncan 1975; Murray and Brown 1993).

Similar-sized species may differ in mouth structure, which leads to differences in the part of the resource base that can be exploited

Although the muzzle width of hartebeest is similar to that of topi, their foraging behaviour is nevertheless different. Hartebeest were always less successful in a series of feeding trials, selecting less leaf, securing smaller bites at a slower bite rate (Murray and Brown 1993). Also in the field, hartebeest and topi differ in foraging behaviour, especially in the dry season. Thus, even while hardly distinguishable morphologically, topi and hartebeest are specialised on a different growth stage of the grass sward (Stanley Price 1974), indicating that spatial and vertical arrangements in and between grass swards are thus additional dimensions of niche space.

More differences and Gause's axiom

When species, even after the most important resource axes have been applied, still show considerable overlap, then the analysis becomes fuzzier. To circumvent the Gausian paradigm that two completely overlapping species cannot coexist, ever more niche dimensions need to be invoked because, by definition, they must be there. And, indeed, the more we study and compare species, we will always find plenty of differences because no two species are alike! Species may differ in water dependence (Sahid 2003), in efficiency with which they utilise acquired resources (e.g., similar-sized ruminants versus non ruminants, Van Wieren 1996a), in efficiency with which they can cope with predation (Sinclair et al. 2003.), in metabolic rate (e.g., wildebeest, eland and hartebeest, Owen-Smith 1985), in heat tolerance (oryx and wildebeest: Sahid 2003), and a host of other characteristics. No doubt, all these differences are real but the problem is that we may end up in the tautological backwater of the Gausian axiom with a potentially never-ending search for the niche axis along which two species might differ. Although this problem has been considered real (Hubbell 2001; Chase and Leibold 2003), solutions have been sought and found (for a review see Chase and Hubbell 2003).

Problem is to end up in the tautological backwater of the Gaussian axiom with a never-ending search for the niche axis along which two species might differ

SYNTHESIS

In this chapter, many factors that possibly shape the (co)existence of ungulate communities have passed by. Although we can not yet predict the composition and structure of a community in a given biome, we can hypothesise about the principal processes involved and put them in a coherent framework. From this scheme, testable hypotheses can be generated.

Despite difficulties, our starting points are the niche concept and the strive between species for the acquisition of resources. Our principal hypothesis is that species can only exist if they have exclusive access to resources that cannot be used by others. The playing field for competition is then confined to resources shared with other species, and coexistence and equilibrium densities of populations become a function of the amount of shared resources and the exclusive resources of other species (Ritchie 2002).

The second crucial factor affecting community richness and structure is related to the distribution of resources in space: spatial heterogeneity. Resources are not homogeneously distributed in space but density and extent of density vary, leading to a patchy landscape with patches of varying size and resource density. This patchiness can be described by fractal geometry (Ritchie 1998, 2002). Especially for mobile animals like ungulates this patchiness is important because they can move through the landscape and are thus able to make choices. By making choices they may try to evade competition with others and, even more important, the scale of resolution with which animals perceive the landscape depends on their size, leading to differential use of a patchy landscape by small and large species (Ritchie 2002). Adding heterogeneity thus both underlines the fundamental role of body size and also can explain the presence of the large body weight ranges so characteristic of the richer ungulate communities. It can also be deduced that heterogeneity and ungulate species richness are positively related. Most of this is, however, still hypothetical, and although the structure of a number of communities could be explained by this theory (Ritchie 2002), for ungulate communities the work still has to be done. A complicating factor here is that the resource (grass, browse) is not so discrete (unlike seeds) and that foraging can change the amount and quality (density) of the resource, possibly affecting the amount of exclusive resources for species.

Therefore, much still needs to be done, and we suggest that in order to make real progress in this field, future research should focus on the following topics. First, define the fundamental niche (adaptive syndrome) of species regarding resource and habitat requirements, based on predominantly physiological and morphological characteristics. A start has been made by Chase and Leibold (2003). Second, describe the heterogeneity of the landscape in terms of patchiness and patch resource density using fractal geometry, and, third, estimate the minimum threshold patch size and the minimum resource density for each species. If the above programme can be executed, then the road is open to analyse fundamental relationships and to test crucial hypotheses which are now coming up in the mind of the reader. Some hypotheses for future research are formulated in Box 11.4.

Box 11.4. *Testable hypotheses for future research*

Hypothesis 1. A species can only exist if it has exclusive access to resources that cannot be used by others.
Hypothesis 2. Increasing heterogeneity leads to a higher species packing, greater species richness, and greater size range of ungulate communities.

CHAPTER 11B

COMMENTS ON "STRUCTURING HERBIVORE COMMUNITIES: THE ROLE OF HABITAT AND DIET"

HERBERT H.T. PRINS, AREND M.H. BRUNSTING AND WILLEM F. DE BOER

Resource Ecology Group, Wageningen University, P.O. Box 47, 6700 AA Wageningen, The Netherlands
E-mail: herbert.prins@wur.nl

The question Van Wieren and Van Langevelde (Chapter 11) are trying to find an answer to, namely "Why are there so many species?", and especially "Why are there so many herbivore species at some location?" is an intriguing one, but not a simple one. To solve this question, one must first look into the exact articulation of the question. The word "Why" is particularly knotty. The question could be reformulated as "What causes the existence of so many species?", but also as "What is the function of so many species?" or even "How did so many species evolve?". At first sight, Van Wieren and Van Langevelde deal with the first question, about the cause. However, a closer look at the text reveals that they try to find an answer to another question than the one they pose, namely, "What allows so many different species to co-exist?". This is a pity, because if they had tried to find an answer to the question about causality, then they would have taken, hopefully, an evolutionary and dynamic approach. Now their approach is static, and focuses on the conditionality instead of the causality.

In their search for conditionalities of co-existence, they justifiably concentrate on herbivores, and especially African herbivores, because of the superb species richness of this group. They then tackle their question armed with niche theory. The 'niche' concept is fraud with difficulties, and they rely heavily on Chase and Leibold (2003). It was good to see that they do refer back to the early meanings of the word 'niche' by referring to Grinnell (1917) and Elton (1927); even today, a dictionary gives both sets of meanings ('place, position, slot, alcove, nook' reflecting the Grinnellian point of view versus 'function, role', which is the Eltonian one). Van Wieren and Van Langevelde then follow Chase and Leibold's (2003) new definition of 'niche' as "the joint description of the environmental conditions that allow a

H.H.T. Prins and F. van Langevelde (eds.), Resource Ecology: Spatial and Temporal Dynamics of Foraging, 263-266.

species to satisfy its minimum requirements so that the birth rate of a local population is equal to or greater than its death rate along with the set of per capita effects of that species on these environmental conditions" (the latter part of the definition is especially of importance for herbivores that can modify their resources). But, do they do so?

Their apportioning of the component species from an exemplar assemblage, that of Kruger National Park, over a set of niches is very Grinellian! Every species is allocated a place within the system based on its location (and not on its function), which tells little to nothing about the above-mentioned "minimum requirements so that the birth rate of a local population is equal to or greater than its death rate". The same is the case for assigning the species to diet classes. A combination of the two may be a first step leading to a Hutchinsonian *n*-dimensional hyperspace (Hutchinson 1957), but it is not a step towards solving the question of species richness. Indeed, one must seriously question whether diet class, just as body mass, is a proper niche parameter! Diet selection is the resultant from the interface between evolutionary derived predisposition and the local flora (possibly modified by the impact of other herbivores), while body mass is the result of evolution alone and a characteristic of the animal, not of its niche. There are in our view more problems with their analysis. We think that a niche dimension should be defined independently of the species making use of it. So this is another reason why diet is a dicey niche dimension, because it is not independently defined. They should have stuck to quantity and quality parameters, like they did so well at the beginning of the chapter, and on which especially Van Wieren is a rare specialist. Van Wieren and Van Langevelde use another niche dimension, namely 'habitat', to separate the different species of the Kruger assemblage. For that they chose to use the habitat classification of Table 11.2. In a strict Hutchinsonian sense, their eight categories do not lie on one dimension though, but they lie on 4 dimensions, namely grass cover, shrub cover, tree cover and woody species height (e.g., Loth and Prins 1986). Habitat thus does not form a niche dimension. However, strictly speaking, habitat is not a resource either, and a habitat category can thus not be an exclusive resource. Moreover, the habitat and diet classes used in their analysis are subjective, as one could also argue that both classes only comprise two major distinctions, namely forest versus grassland, or browse versus grass.

It thus appears as if Van Wieren and Van Langevelde (Chapter 11) have chosen to ignore the challenge they set themselves. Their chapter is neither addressing the question about *why* there are so many species, nor have they met the challenge put forward by the new definition of 'niche' by Chase and Leibold (2003). Their approach seems *ad hoc*, trying to find dimensions that separate the species and can explain the co-existence. This concept lacks rigorous scientific testing and null hypotheses. From their descriptive approach no conclusions can therefore be drawn, at best hypotheses could be formulated that could direct future research.

But even if habitat the way they use it could be considered a niche axis along which species have to find a place, how then does one explain the next issue of concern? If one studies Table 11.3 then the following emerges. On average a species makes use of more than one habitat (namely, of 2.7); elephant and buffalo are very catholic and use 8 and 5 habitats, respectively, and if we leave these two species out, the average species utilizes 2.5 habitats. This one would not have expected: one would have anticipated that species would be mutually exclusive due to competitive exclusion (which can only be studied of course by looking at the dynamics of the system). A finer subdivision of the 'niche space', as done in Table 11.9, does not help. If one leaves body mass out (because it is not a niche dimension), then the average number of species in one combinational category of 'habitat' and 'diet class' is 2.1 species (with a maximum of 7 species in the combination of 'variable grazers' and 'light savanna'). So, 7 African herbivore species in one locality can share a 'niche'. Finally, if one wants to be very generous, and if one would accept that body mass could be equated with a niche dimension, even then in Table 11.9 the maximum number of species sharing a combinational category of 'habitat', 'diet class' and 'body mass class' is 1.4 species, but with still maximally 3 species in a 'unique' combination (i.e., small-sized browsers in thickets: common duiker, suni and bushbuck, medium-sized variable grazers in a light savanna: warthog, hartebeest and waterbuck, the same combination in a woodland savanna, or large-sized variable grazers in such a savanna: wildebeest, buffalo and hippo, and finally, large-sized variable grazers in a bush savanna: sable, wildebeest and roan antelope). The hypothesis of Van Wieren and Van Langevelde that "a species can only exist if it has exclusive access to resources that cannot be used by others" is thus not supported by their own ordering and grouping of species, and more importantly, the justification behind the hypothesis is lacking, as is the hypothesis testing. The resulting categories simply do not equate to niches, and their categorization does not lead to the identification of resource dimensions.

When Van Wieren and Van Langevelde write that "When body mass … is being added to the habitat–diet niche space, the community becomes much more structured", then they are, in a way right: indeed, every individual species gets an increasingly larger chance to become uniquely circumscribed if there are more unique combinations of descriptive categorical axes. However, we do not think that they have approached the answer to their question any closer. The basis for our assertion is that they have not chosen a dynamic or evolutionary tack. A dynamic course would have been necessary to meet Chase and Leibold's (2003) new definition, and an evolutionary approach would be indicated by the fact that body mass (and to a lesser extent habitat choice or food adaptations) are the outcomes of a competition between species over time.

The allotment of species to a Hutchinsonian-type multi-dimensional hyperspace is reminiscent of mediaeval Italian noble families living in their tower-like townhouses. Every family had its own townhouse, but did this description explain why there were so many towers? Or why in one town there were 72 of these towers (in San Gimignano), and in another 170 (in Lucca, inclusive of church towers)? Of course not! A description of the number and locations of these towers gives a description of the structure of such an Italian town, but this does not yield an explanation. The enlightenment comes from the dynamics, the evolution: which family was able to protect its locality, what tower was torn down by a competing family, and between which neighbours could one settle? Competition and facilitation (Prins and Olff 1998) are key concepts here, and that is why Chase and Leibold (2003) focus on birth rate and death rates of local populations.

The hypotheses and conclusions put forward in their chapter need to be reconsidered. It is not clear whether the hypothesis that species can only co-exist if they have exclusive resource access is merely a reformulation of Gause's axiom. For instance, spatial or temporal heterogeneity in resource abundance can also promote species co-existence of two species that exploit a single resource (Ritchie and Olff 1999; Chesson 2000a; 2000b). The hypothesis that heterogeneity increases the size range of species needs further thoughts: what is the basis for this hypothesis? How would we be able to test these hypotheses? Another problem is that it is still unclear how important body size is in determining niche dimensions. It is clear that our quest for the mechanisms that determine species co-existence is a long one, and can only be completed by way of careful formulation and testing of hypotheses.

We do not fully support the research agenda which Van Wieren and Van Langevelde propose at the end of their chapter, because it harkens back to the Grinnellian and Hutchinsonian definitions of niche. What we think is needed, over and above (or perhaps even in place of) what they propose, is controlled experiments in which they bring Chase and Leibold's (2003) new definition of niche to the test. What is thus needed is that for different herbivores of the same or different body mass and with the same or a different 'adaptive syndromes' (Van Wieren and Van Langevelde, under different sets of environmental conditions, the minimum requirements of these herbivores under which the birth rate is equal to or greater than their death rate are determined. Then at a next stage it has to be determined what the sets of *per capita* effects of these chosen species are on these experimental environmental conditions. Penultimately, this has to be coupled with a phylogenetic study to understand the evolution of niche use better, and also to understand the body mass development over geological time for the different species. Finally, this research agenda calls for a careful study of phenotypic plasticity of body mass (and associated body measurements such as incisor arc width) and character displacement of species living with different competitors in different assemblages over the whole geographic range of the component species. What we propose is thus a research agenda that may ultimately answer one of the holy grails of modern science, namely "What causes the (co-)existence of so many species?".

CHAPTER 12

PROSPECTS FOR FURTHER DEVELOPMENT OF RESOURCE ECOLOGY

HERBERT H.T. PRINS AND FRANK VAN LANGEVELDE

Resource Ecology Group, Wageningen University, P.O. Box 47,
6700 AA Wageningen, The Netherlands
E-mail: herbert.prins@wur.nl

This book is about 'resource ecology', which we defined in our introductory chapter as "the ecology of trophic interactions between consumers and their resources". We have chosen to focus on a particular class of consumers, namely, large mammalian placental herbivores. All chapters in this book deal actually with ungulates (in the broad sense, thus including elephant), whether free-ranging or domestic, but we are certain that every chapter is of much use to understand other classes of herbivores, such as marsupials, herbivorous birds or even herbivorous reptiles. In the comment on Chapter 4, the definition of 'resource' is given as "usable energy or any biotic or abiotic substance directly exploited by an organism, which includes food, nutrients, water, atmospheric gas compounds, as well as light, and the use of which can lead to the (temporary) exhaustion of that resource". The essence of the concept of 'resource' is that organisms can compete for a resource and that it can be limiting the growth of individual organisms or of populations. In herbivores, the resource that is most interesting from a conceptual point of view, is herbage, because the feedback relations that exist between consumers and this type of resource (see for instance Van de Koppel et al. 2002; Rietkerk et al. 2002a; Van Langevelde et al. 2003). This of course does not deny the fact that other resources, such as water, or environmental conditions, such as temperature, can be very important factors to understand the distribution of herbivores (Bailey and Provenza, Chapter 2; Stein and Georgiadis, Chapter 3). From the consumer's perspective, acquiring sufficient resources, such as energy, nutrients and water, are conditions for life and reproduction. In resource ecology, foraging is the central process because it leads to growth, survival and reproduction of the animal. This book deals with foraging, and it ignores predation or disease and highlights only a restricted set of fitness parameters of the consumer.

H.H.T. Prins and F. van Langevelde (eds.), Resource Ecology: Spatial and Temporal Dynamics of Foraging, 267-271.

In nature, resources are always heterogeneously distributed and exposed to changing conditions (Skidmore and Ferwerda, Chapter 4). Seeking out food by animals is thus inevitably uneven in space and over time. For many years, foraging theory concentrates on issues how animals search and forage and what they should do so as to maximise their fitness, based on problems (i) how different possible behaviours affect fitness and (ii) what the decision variables are to maximise fitness. Foraging theory has made substantial achievements explicating observations of foraging behaviour. As foraging behaviour is largely determined by the spatial distribution and variableness in time of the resources, many studies have recently been done to investigate movements and spatial decisions in foraging at various spatial scales and under variability in time. Accordingly, resource ecology forms a bridge between the well-developed foraging theory and the emerging field of spatial ecology.

New technologies allow a much better description of both resource distribution and animals across the landscape

From the contributions in this book a number of issues arise. The *first* is that new technologies, including new algorithms, allow a much better description of both resource distribution and of animals across the landscape. Bailey and Provenza (Chapter 2) show how GPS technology allows tracking of individual animals over a landscape. Different types of transponders are now so good that even information about physiological states of free-ranging individuals can be followed real-time. The accuracy is now so great that this can be done with a precision of metres while the time frames are down to minutes and even seconds. Stein and Georgiadis (Chapter 3) highlight new statistical techniques to describe aggregation patterns of animals. During the last few years, the development of new and better/faster algorithms even outstrips the speed of development of electronic hardware. The combination of new algorithms with faster and more precise hardware allows for the first time data collection and data handling surpassing theoretical insights. Skidmore and Ferwerda (Chapter 4) show how hyperspectral remote sensing allows an incredibly precise spatial description of the amounts of food, and even the quality of that food or the different plant species. We as scientists can now make accurate maps with every individual plant or clump of vegetation with a precision of centimetres and an extent of tens of kilometres within a brief span of time. It further develops theory by formulating new hypotheses to be tested. As the comment on Chapter 4 brings to the fore, it is now the task of students of animal behaviour to harness this new way of looking at resources or animals, and to use this plethora of data for testing ecological theories and for yielding a better understanding of resource ecology. Data now can become an embarrassment of riches in ecology.

Linked to this is the *second* conclusion to be drawn from the first three chapters: we can now describe both the distribution of resources and animals in such a more detailed and repeatable way that simple optimal foraging theory is insufficient to deal with the richness of the spatial data. Prior to the explosion of detailed knowledge of resource distribution over time and in space, optimal foraging theory provided a very useful 'harness' for the analysis of foraging between a few different patches containing food, but now it only does so in simple experimental set-ups. In that context it can still provide very useful insights, just as basic insights in economy are still very useful in understanding some behaviour of consumers. Indeed, optimal foraging theory is not dead or outdated. In Chapters 2 and 6, the authors show how new insights can be gained to understand the giving-up time for patches that are visited by herbivores. It is important perhaps here to give a definition of what a 'patch' is, since it is such a central concept in this book and in resource ecology in general. Patches are defined as regions that are more or less homogeneous with respect to a measured variable. There are four general approaches to defining patches, namely, simple aggregation of like-valued regions, moving- or split-window methods, global zonation and finally spatially constrained clustering (see, e.g., Pielou 1984; Legendre and Fortin 1989; Fortin 1994; Fortin and Drapeau 1995). Understanding giving-up times is an essential part of the development of optimal foraging theory. Bailey and Provenza (Chapter 2) point out that giving-up rules are not very robust when food in patches is plentiful. This clearly needs more attention in future. Yet, spatial ecology now can become a mature science since it is realised that scale issues become of overriding importance in ecology (see Rietkerk et al. 2002b). Indeed, numerous studies demonstrate the significance of taking into account the spatial structure of resources, including scale, both on the population dynamics of individual species as on assemblage structure of consumers. The link between foraging theory and spatial ecology sets resource ecology in a new context from which new theory can emerge.

Simple optimal-foraging theory is insufficient to deal with the richness of the spatial data

The *third* conclusion we can draw is brought to the fore in Chapter 5 by Laca, namely, that large herbivores, like all other consumers, interact with their resources through a series of nested processes such as ingestion, searching, digestion and resting, which define relevant scales (see also Chapter 2, where definitions are given) for research but also for understanding what herbivores do. Classical optimal-foraging theory does not address the issue of scale. In the comment on Laca's Chapter 5, the potential consequence of size and mobility on the perception of heterogeneity, diet choice and patch choice in large herbivores is discussed. In that comment reference is made to relevant experiments (Drescher 2003). The point is that because resource distribution has to be understood at a nested series of scales, new theory is needed to cope with that. Murwira's (2003) work on using wavelets to describe resource heterogeneity, referred to in Chapter 4, is a possible way to use new mathematical techniques to understand the resource heterogeneity better. Not only spatial scale is relevant to understand animal foraging, Owen-Smith (Chapter 8) clearly demonstrates that temporal change in food quality and availability largely explains foraging. Different time scales may simultaneously interact: resource depletion and regrowth, day–night variation, even up to seasonal variation (see also Boone et al., Chapter 9).

Animals interact with their resources through a series of nested processes, which define relevant scales for research but also for understanding what herbivores do

The *fourth* generalization is that we begin to understand upscaling of foraging decisions much better than a few years ago. Several chapters (for instance, Chapters 5, 6, 7, 8 and 9) deal with this. In Chapter 9, Boone et al. point out that even the functional-response curve describing the relation between intake and biomass on offer is scale-dependent.

We begin to understand upscaling of foraging decision much better than a few years ago

In an elegant application of the SAVANNA-model, they show that subdivision of land leads to a lower total herbivore biomass offtake. Similar results were found by Kramer et al. (2006) when they investigated the interaction between resources and a set of temperate herbivores using the FORSPACE-model (see also Groot-Bruinderink et al. 1999). Linked to these issues of nestedness and upscaling is the *fifth* important theme emerging from this book, which is non-linearity. Chapter 5 addresses this issue, but it is really emphasized in Chapter 9. Perhaps we can safely draw the conclusion that neither resource utilization nor feedback between herbivores and resources can be linearly upscaled. Chapters 7, 9 and 10 deal with animal movements. Mobility is the important differentiating factor between plants and sedentary organisms on the one hand and most animals on the other. These chapters lead to the *sixth* general conclusion, namely that mobility can counteract resource heterogeneity. These chapters offer the building blocks for new theory. This novel theory has not crystallised yet but we are sure that it will entail the following hypothesis, namely "the relative abundance of large animals (in contrast to small ones) in assemblages increases with spatial variation and sudden fluctuations in resource availability that result in longer time periods between foraging events". The rationale behind this hypothesis is that larger animals are better in dealing with increased spatial variation in resource availability (Prins and Van Langevelde, Chapter 7; Morse et al. 1985; Ritchie and Olff 1999) and in buffering drops in resource availability of longer duration than are smaller ones (Owen-Smith, Chapter 8; Biddanda et al. 2001; Li 2002; Li et al. 2004), as their metabolic energy use per unit of mass and per unit of time is lower than for smaller animals. Increased spatial variation may lead to larger distances between food patches, resulting in decreased food intake (Laca, Chapter 5; Fryxell, Chapter 6; Prins and Van Langevelde, Chapter 7). Food intake also decreases with more fluctuations in resource availability (Owen-Smith, Chapter 8).

Animal's mobility can counteract resource heterogeneity

Both increased spatial variation and more fluctuations lead to longer time periods between foraging events, and could even lead to starvation. Larger animals can move larger distances between spatially distributed resources (have larger home ranges, Haskell et al. 2002; Jetz et al. 2004, and higher movement speed, Prins and Van Langevelde, Chapter 7; Jetz et al. 2004), and buffer sudden drops in resources over a longer time than smaller species (Owen-Smith, Chapter 8; Dunbrack and Ramsay 1993) as they have more fat reserves per unit mass, hence higher starvation resistance. Although there are many strategies that animals can adopt to cushion the stress of seasonality (e.g., migration and hibernation), these strategies fail when resource availability becomes increasingly irregular.

By applying the allometric scaling laws for mobility and starvation resistance from Chapter 7 of Prins and Van Langevelde, one could model food intake as a function of spatial variation and variation over time in resource availability. By including this intake model into population models one could then make predictions about abundances as function of body mass and spatial and temporal variation in resource availability (Prins and Van Langevelde, Chapter 7; Van Langevelde et al. submitted), which will perhaps be the breakthrough needed. Science is in progress, and if we were certain then it would have been published already.

Perhaps the two most important elements in the present volume are the commentaries and the hypotheses. All contributors to the present volume sincerely hope that the commentaries will stimulate discussions or provoke new insights. The commentaries were not written to please the authors but they were put in writing so as to put on view disparities in opinion. Indeed, in science the age-old method of dialectics with its formulation of thesis and antithesis is still valid, and we hope that readers will further contribute to this dialectic discourse so as to arrive at new insights. The second important element is comprised of the sets of hypotheses. We are acutely aware that the purpose of mathematical theory is to deal with 'all possible worlds' and the purpose of experiments and fieldwork is to deal with the real world. We believe with Wilson and Bossert (1971) that to measure the parameters, to search for new parameters and to improve theory is ultimately the most effective way of viewing the real world. Well-formulated hypotheses can be a sure way of organising one's research, and can help searching for new parameters and measuring them well. All in all, the authors have formulated some 30-odd hypotheses to further develop theory on resource ecology. Some of them can be tested quite straightforward from observational data. Other hypotheses need careful experimentation. We believe, however, that these hypotheses and propositions have been formulated in such a way that they can and must be tested.

LIST OF REFERENCES

Abrams, P.A., 1988. How should resources be counted? *Theoretical Population Biology*, 33 (2), 226-242.

Abrams, P.A., 1990. The effects of adaptive behavior on the type-2 functional response. *Ecology*, 71 (3), 877-885.

Abrams, P.A., 1995. Implications of dynamically variable traits for identifying, classifying, and measuring direct and indirect effects in ecological communities. *American Naturalist*, 146 (1), 112-134.

Abrams, P.A., 2000. The impact of habitat selection on the spatial heterogeneity of resources in varying environments. *Ecology*, 81 (10), 2902-2913.

Aerts, B.A., Esselink, P. and Helder, G.J.F., 1996. Habitat selection and diet composition of greylag geese *Anser anser* and barnacle geese *Branta leucopsis* during fall and spring staging in relation to management in the tidal marshes of the Dollard. *Zeitschrift für Ökologie und Naturschutz*, 5 (2), 65-75.

Albon, S.D. and Langvatn, R., 1992. Plant phenology and the benefits of migration in a temperate ungulate. *Oikos*, 65 (3), 502-513.

Allen, A.W., Cook, J.G. and Armbruster, M.J., 1984. *Habitat suitability index models: pronghorn*. US Fish and Wildlife Service no. FWS/OBS-82/10.65. [http://el.erdc.usace.army.mil/emrrp/emris/EMRIS_PDF/Pronghorn.pdf]

Alonso, D., Bartumeus, F. and Catalan, J., 2002. Mutual interference between predators can give rise to Turing spatial patterns. *Ecology*, 83 (1), 28-34.

Anderson, A., 1984. The extinction of moa in southern New Zealand. *In:* Martin, P.S. and Klein, R.G. eds. *Quaternary extinctions: a prehistoric revolution* University of Arizona Press, Tucson, 728-740.

Anderson, J.A., 1995. *An introduction to neural networks*. MIT Press, Cambridge.

Andresen, H., Bakker, J.P., Brongers, M., et al., 1990. Long-term changes of salt marsh communities by cattle grazing. *Vegetatio*, 89 (2), 137-148.

ARC, 1980. *The nutrient requirements of ruminant livestock: technical review by an Agricultural Research Council working party*. Agricultural Research Council, Commonwealth Agricultural Bureaux, Slough.

Arditi, R. and Dacorogna, B., 1988. Optimal foraging on arbitrary food distributions and the definition of habitat patches. *American Naturalist*, 131 (6), 837-846.

Arnold, G.W., 1985. Regulation of forage intake. *In:* Hudson, R.J. and White, R.G. eds. *Bioenergetics of wild herbivores*. CRC Press, Boca Raton, 81-101.

Arnold, W., Ruf, T., Reimoser, S., et al., 2004. Nocturnal hypometabolism as an overwintering strategy of red deer (*Cervus elaphus*). *American Journal of Physiology*, 286 (1), R174-R181.

Arsenault, R. and Owen-Smith, N., 2002. Facilitation versus competition in grazing herbivore assemblages. *Oikos*, 97 (3), 313-318.

Asner, G.P., 1998. Biophysical and biochemical sources of variability in canopy reflectance. *Remote Sensing of Environment*, 64 (3), 234-253.

Astrom, M., Lunberg, P. and Danell, K., 1990. Partial prey consumption by browsers: trees as patches. *Journal of Animal Ecology*, 59 (1), 287-300.

Atkinson, M.D., Jervis, A.P. and Sangha, R.S., 1997. Discrimination between *Betula pendula*, *Betula pubescens*, and their hybrids using near-infrared reflectance spectroscopy. *Canadian Journal of Forest Research*, 27 (11), 1896-1900.

Atwood, S.B., Provenza, F.D., Wiedmeier, R.D., et al., 2001a. Changes in preferences of gestating heifers fed untreated or ammoniated straw in different flavours. *Journal of Animal Science*, 79 (12), 3027-3033.

Atwood, S.B., Provenza, F.D., Wiedmeier, R.D., et al., 2001b. Influence of free-choice vs. mixed-ration diets on food intake and performance of fattening calves. *Journal of Animal Science*, 79 (12), 3034-3040.

Augustin, N.H., Mugglestone, M.A. and Buckland, S.T., 1996. An autologistic model for the spatial distribution of wildlife. *Journal of Applied Ecology*, 33 (2), 339-347.

Bailey, D.W., 1995. Daily selection of feeding areas by cattle in homogeneous and heterogeneous environments. *Applied Animal Behaviour Science,* 45 (3/4), 183-199.

Bailey, D.W., 2004. Management strategies for optimal grazing distribution and use of arid rangelands. *Journal of Animal Science,* 82 (E-Suppl.), E147-E153.

Bailey, D.W., Gross, J.E., Laca, E.A., et al., 1996. Mechanisms that result in large herbivore grazing distribution patterns. *Journal of Range Management,* 49 (5), 386-400.

Bailey, D.W., Howery, L.D. and Boss, D.L., 2000. Effects of social facilitation for locating feeding sites by cattle in an eight-arm radial maze. *Applied Animal Behaviour Science,* 68 (2), 93-105.

Bailey, D.W., Keil, M.R. and Rittenhouse, L.R., 2004. Research observation: daily movement patterns of hill climbing and bottom dwelling cows. *Journal of Range Management,* 57 (1), 20-28.

Bailey, D.W., Kress, D.D., Anderson, D.C., et al., 2001a. Evaluation of F1 crosses from Angus, Charolais, Salers, Piedmontese, Tarentaise and Hereford sires. V. Grazing distribution patterns. *Proceedings of the Western Section American Society of Animal Science,* 52, 110-113.

Bailey, D.W., Kress, D.D., Anderson, D.C., et al., 2001b. Relationship between terrain use and performance of beef cows grazing foothill rangeland. *Journal of Animal Science,* 79 (7), 1883-1891.

Bailey, D.W., Rittenhouse, L.R., Hart, R.H., et al., 1989a. Characteristics of spatial memory in cattle. *Applied Animal Behaviour Science,* 23 (4), 331-340.

Bailey, D.W., Rittenhouse, L.R., Hart, R.H., et al., 1989b. Association of relative food availabilities and locations by cattle. *Journal of Range Management,* 42 (6), 480-482.

Bailey, D.W. and Sims, P.L., 1998. Association of food quality and locations by cattle. *Journal of Range Management,* 51 (1), 2-8.

Bailey, D.W. and VanWagoner, H., 2004. Can we improve livestock distribution with selection? *In: Abstracts of the International Society for Range Management Annual Meeting.* Salt Lake City.

Bailey, D.W., Walker, J.W. and Rittenhouse, L.R., 1990. Sequential analysis of cattle location: day-to-day movement patterns. *Applied Animal Behaviour Science,* 25 (1/2), 137-148.

Bailey, D.W. and Welling, G.R., 1999. Modification of cattle grazing distribution with dehydrated molasses supplement. *Journal of Range Management,* 52 (6), 575-582.

Bailey, D.W., Welling, G.R. and Miller, E.T., 2001c. Cattle use of foothills rangeland near dehydrated molasses supplement. *Journal of Range Management,* 54 (4), 338-347.

Bailey, R.G., 1996. *Ecosystem geography.* Springer, New York.

Baker, D.L. and Hobbs, N.T., 1987. Strategies of digestion: digestive efficiency and retention time of forage diets in montane ungulates. *Canadian Journal of Zoology,* 65 (8), 1978-1984.

Bakker, J.P., 1989. *Nature management by grazing and cutting: on the ecological significance of grazing and cutting regimes applied to restore former species-rich grassland communities in the Netherlands.* Kluwer Academic Publishers, Dordrecht. Geobotany no. 14.

Ball, J.P., Danell, K. and Sunesson, P., 2000. Response of a herbivore community to increased food quality and quantity: an experiment with nitrogen fertilizer in a boreal forest. *Journal of Applied Ecology,* 37 (2), 247-255.

Baret, F., Jacquemoud, S., Guyot, G., et al., 1992. Modeled analysis of the biophysical nature of spectral shifts and comparison with information content of broad bands. *Remote Sensing of Environment,* 41 (2/3), 133-142.

Barry, T.N., Manley, T.R. and Duncan, S.J., 1986. The role of condensed tannins in the nutritional value of *Lotus pedunculatus* for sheep. *British Journal of Nutrition,* 55 (1), 123-137.

Bashaw, M.J., Tarou, L.R., Maki, T.S., et al., 2001. A survey assessment of variables related to stereotypy in captive giraffe and okapi. *Applied Animal Behaviour Science,* 73 (3), 235-247.

Bassett, T. and Koli Bi, Z., 1999. Fulbe livestock raising and environmental change in northern Côte d'Ivoire. *In:* Azarya, V., Breedveld, A., De Bruijn, M., et al. eds. *Pastoralists under pressure? Fulbe societies confronting change in West Africa.* Brill Academic Publishers, Leiden, 139-159.

Baumont, R., Prache, S., Meuret, M., et al., 2000. How forage characteristics influence behaviour and intake in small ruminants: a review. *Livestock Production Science,* 64 (1), 15-28.

Bayer, W. and Waters-Bayer, A., 1994. Forage alternatives from range and field: pastoral forage management and improvement in the African drylands. *In:* Scoones, I. ed. *Living with uncertainty: new directions in pastoral development in Africa.* Intermediate Technology Publications, London, 58-78.

Beekman, J.H. and Prins, H.H.T., 1989. Feeding strategies of sedentary large herbivores in East Africa with emphasis on the African buffalo, *Syncerus caffer. African Journal of Ecology,* 27, 129-147.

Behnke, R.H. and Scoones, I.E., 1993. Rethinking range ecology: implications for rangeland management in Africa. *In:* Behnke, R.H. and Scoones, I.E. eds. *Range ecology at disequilibrium: new models of natural variability and pastoral adaptation in Africa savannas.* Overseas Development Institute, London, 1-30.

Behnke, R.H., Scoones, I.E. and Kerven, C., 1993. *Range ecology at disequilibrium: new models of natural variability and pastoral adaptation in African savannas.* International Livestock Center for Africa ILCA, Addis Ababa.

Bekure, S., De Leeuw, P.N., Grandin, B.E., et al. (eds.), 1991. *Maasai herding: an analysis of the livestock production system of Maasai pastoralists in eastern Kajiado District, Kenya.* International Livestock Center for Africa ILCA, Addis Ababa. ILCA Systems Study no. 4.

Bell, R.H.V., 1970. The use of herbaceous layers by grazing ungulates in the Serengeti. *In:* Watson, A. ed. *Animal populations in relation to their food resources: a symposium of the British Ecological Society, Aberdeen, 24-28 March 1969.* Oxford, Blackwell, 111-124.

Belovsky, G.E., 1978. Diet optimization in a generalist herbivore: the moose. *Theoretical Population Biology,* 14 (1), 105-134.

Belovsky, G.E., 1984. Herbivore optimal foraging: a comparative test of three models. *American Naturalist,* 124 (1), 97-115.

Belovsky, G.E., 1986. Optimal foraging and community structure: implications for a guild of generalist grassland herbivores. *Oecologia,* 70 (1), 35-52.

Belovsky, G.E. and Schmitz, O.J., 1994. Plant defenses and optimal foraging by mammalian herbivores. *Journal of Mammalogy,* 75 (4), 816-832.

Belovsky, G.E. and Slade, J.B., 1986. Time budgets of grassland herbivores: body size similarities. *Oecologia,* 70 (1), 53-62.

Bergman, C.M., Fryxell, J.M. and Gates, C.C., 2000. The effect of tissue complexity and sward height on the functional response of wood bison. *Functional Ecology,* 14 (1), 61-69.

Bergman, C.M., Fryxell, J.M., Gates, C.C., et al., 2001. Ungulate foraging strategies: energy maximizing or time minimizing? *Journal of Animal Ecology,* 70 (2), 289-300.

Bernus, E., 1979. Le contrôle du milieu naturel et du troupeau par les éleveurs touaregs sahéliens. *In:* L'Equipe écologie et anthropologie des sociétés pastorales ed. *Pastoral production and society=Production pastorale et société: proceedings of the international meeting on nomadic pastoralism, Paris 1-3 Dec. 1976.* Cambridge University Press, Cambridge, 67-74.

Berteaux, D., Crete, M., Huot, J., et al., 1998. Food choice by white-tailed deer in relation to protein and energy content of the diet: a field experiment. *Oecologia,* 115 (1/2), 84-92.

Biddanda, B., Ogdahl, M. and Cotner, J., 2001. Dominance of bacterial metabolism in oligotrophic relative to eutrophic waters. *Limnology and Oceanography,* 46 (3), 730-739.

Biondini, M.E., Steuter, A.A. and Hamilton, R.G., 1999. Bison use of fire-managed remnant prairies. *Journal of Range Management,* 52 (5), 454-461.

Black, J.L. and Kenney, P.A., 1984. Factors affecting diet selection by sheep. 2. Height and density of pasture. *Australian Journal of Agricultural Research,* 35 (4), 565-578.

Blackman, F.F., 1905. Optima and limiting factors. *Annals of Botany,* 19, 281-295.

Boice, R., 1972. Some behavioural tests of domestication in Norway rats. *Behaviour,* 42 (3/4), 198-231.

Bokdam, J. and Wallis De Vries, M.F., 1992. Forage quality as a limiting factor for cattle grazing in isolated Dutch nature reserves. *Conservation Biology,* 6 (3), 399-408.

Boomker, E.A., 1987. *Fermentation and digestion in the kudu.* PhD Thesis, University of Pretoria, Pretoria.

Boone, R.B., 2000. *Integrated management and assessment system balancing food security, conservation, and ecosystem integrity in East Africa: training manual.* Global Livestock Collaborative Research Support Program, University of California, Davis. [http://www.nrel.colostate.edu/projects/imas/prods/rboone/Workshop.html]

Boone, R.B., BurnSilver, S.B., Thornton, P.K., et al., 2005. Quantifying declines in livestock due to land subdivision. *Rangeland Ecology and Management,* 58 (5), 523-532.

Boone, R.B., Coughenour, M.B., Galvin, K.A., et al., 2002. Addressing management questions for Ngorongoro Conservation Area, Tanzania, using the Savanna modelling system. *African Journal of Ecology,* 40 (2), 138-150.

Boone, R.B., Galvin, K.A., Smith, N.M., et al., 2000. Generalizing El Niño effects upon Maasai livestock using hierarchical clusters of vegetation patterns. *Photogrammetric Engineering and Remote Sensing,* 66 (6), 737-744.

Boone, R.B. and Hobbs, N.T., 2004. Lines around fragments: effects of fencing on large herbivores. *African Journal of Range and Forage Science,* 21 (3), 147-158.

Borchers, D.L., Buckland, S.T. and Zucchini, W., 2002. *Estimating animal abundance: closed populations.* Springer, London.

Bos, D., 2002. *Grazing in coastal grasslands: Brent Geese and facilitation by herbivory.* PhD Thesis, Rijksuniversiteit Groningen, Groningen.

Boughey, A.S., 1973. *Ecology of populations.* 2nd edn. MacMillan, New York.

Bowers, M.A. and Brown, J.H., 1982. Structure in a desert rodent community: use of space around *Dipodomys spectabilis* mounds. *Oecologia,* 92 (2), 242-249.

Box, E.O., Holben, E.N. and Kalb, V., 1989. Accuracy of the AVHRR vegetation index as a predictor of biomass, primary productivity and net CO_2 flux. *Vegetatio,* 80 (2), 71-89.

Bradbury, J.W., Vehrencamp, S.L., Clifton, K.E., et al., 1996. The relationship between bite rate and local forage abundance in wild Thomson's gazelles. *Ecology,* 77 (7), 2237-2255.

Braithwaite, L.W., Dudzinski, M.L. and Turner, J., 1983. Studies on the arboreal marsupial fauna of eucalypt forests being harvested for wood pulp at Eden, N.S.W. II. Relationship between the fauna density, richness and diversity, and measured variables of the habitat. *Australian Wildlife Research,* 10 (2), 231-247.

Breman, H. and De Ridder, N., 1991. *Manuel sur les pâturages des pays sahéliens.* CTA, Wageningen.

Briske, D.D., Fuhlendorf, S.D. and Smeins, F.E., 2003. Vegetation dynamics on rangelands: a critique of the current paradigms. *Journal of Applied Ecology,* 40 (4), 601-614.

Brock, B.L. and Owensby, C.E., 2000. Predictive models for grazing distribution: a GIS approach. *Journal of Range Management,* 53 (1), 39-46.

Brown, J.H., Fox, B.J. and Kelt, D.A., 2000. Assembly rules: desert rodent communities are structured at scales from local to continental. *American Naturalist,* 156 (3), 314-321.

Brown, J.S., 1999. Vigilance, patch use and habitat selection: foraging under predation risk. *Evolutionary Ecology Research,* 1 (1), 49-71.

Brown, J.S., Kotler, B.P. and Mitchell, W.A., 1994. Foraging theory, patch use, and the structure of a Negev Desert granivore community. *Ecology,* 75 (8), 2286-2300.

Brown, J.S., Kotler, B.P., Smith, R.J., et al., 1988. The effects of owl predation on the foraging behavior of heteromyid rodents. *Oecologia,* 76 (3), 408-415.

Bruce, A. and Hong-Ye, G., 1996. *Applied wavelet analysis with S-plus.* Springer, New York.

Bryant, J.P. and Kuropat, P.J., 1980. Selection of winter forage by subarctic browsing vertebrates: the role of plant chemistry. *Annual Review of Ecology and Systematics,* 11, 261-285.

Bryant, J.P., Provenza, F.D., Pastor, J., et al., 1991. Interactions between woody plants and browsing mammals mediated by secondary metabolites. *Annual Review of Ecology and Systematics,* 22, 431-446.

Buckland, S.T. and Elston, D.A., 1993. Empirical models for the spatial distribution of wildlife. *Journal of Applied Ecology,* 30 (3), 478-495.

Bunnell, F.L. and Gillingham, M.P., 1985. Foraging behaviour: dynamics of dining out. *In:* Hudson, R.J. and White, R.G. eds. *Bioenergetics of wild herbivores.* CRC Press, Boca Raton, 53-79.

Burditt, L., Buchanan, D. and Fitch, J., 1995. *Breeds of livestock: Cattle [website].* Oklahoma State University, Stillwater. [http://www.ansi.okstate.edu/breeds/cattle]

Burnham, K.P. and Anderson, D.R., 2002. *Model selection and multimodel inference: a practical information-theoretic approach.* Springer, New York.

BurnSilver, S., Boone, R.B. and Galvin, K.A., 2004. Linking pastoralists to a heterogeneous landscape: the case of four Maasai group ranches in Kajiado District, Kenya. *In:* Fox, J., Mishra, V., Rindfuss, R., et al. eds. *Linking household and remotely sensed data: methodological and practical problems.* Kluwer Academic Publishing, Boston, 173-199.

Burritt, E.A. and Provenza, F.D., 2000. Role of toxins in intake of varied diets by sheep. *Journal of Chemical Ecology,* 26 (8), 1991-2005.

Buttolph, L.P. and Coppock, D.L., 2001. Project Alpaca. *Rangelands,* 23 (2), 10-13.

Campling, R.C., 1970. Physical regulation of voluntary intake. *In:* Phillipson, A.J. ed. *Physiology of digestion and metabolism in the ruminant.* Oriel Press, Newcastle, 226-234.

Caughley, G., 1976. Plant-herbivore systems. *In:* May, R.M. ed. *Theoretical ecology.* Blackwell, Oxford, 94-113.

Caughley, G., 1977. Sampling in aerial survey. *Journal of Wildlife Management,* 41 (4), 605-615.

Chan-McLeod, A.C.A., White, R.G. and Holleman, D.F., 1994. Effects of protein and energy intake, body condition, and season on nutrient partitioning and milk production in caribou and reindeer. *Canadian Journal of Zoology,* 72 (5), 938-947.

Chang, C., 1993. Pastoral transhumance in the southern Balkans as a social ideology: ethnoarcheological research in northern Greece. *American Anthropologist,* 95 (3), 687-703.

Charnov, E.L., 1976. Optimal foraging, the marginal value theorem. *Theoretical Population Biology,* 9 (2), 129-136.

Chase, J.M. and Leibold, M.A., 2003. *Ecological niches: linking classical and contemporary approaches.* University of Chicago Press, Chicago.

Chatty, D. and Colchester, M. (eds.), 2002. *Conservation and mobile indigenous peoples: displacement, forced settlement and sustainable development.* Berghahn, New York. Studies in Forced Migration, no. 10.

Chesson, P., 2000a. General theory of competitive coexistence in spatially-varying environments. *Theoretical Population Biology,* 58 (3), 211-237.

Chesson, P., 2000b. Mechanisms of maintenance of species diversity. *Annual Review of Ecology and Systematics,* 31, 343-366.

Cibils, A.F., Howery, L.D. and Ruyle, G.B., 2004. Diet and habitat selection by cattle: the relationship between skin- and gut-defense systems. *Applied Animal Behaviour Science,* 88 (3/4), 187-208.

Clanet, J.C., 1996. L'évolution récente des mouvements pastorales dans le Bassin Conventionel du Lac Tchad. *In: CIRAD/CTA Livestock atlas of the Lake Chad basin.* CTA, Wageningen, 97-103.

Clark, C.W. and Mangel, M., 2000. *Dynamic state variable models in ecology: methods and applications.* Oxford University Press, New York.

Clark, D.A. and Harris, P.S., 1985. Composition of the diet of sheep grazing swards of differing white clover content and spatial distribution. *New Zealand Journal of Agricultural Research,* 28, 233-240.

Clarke, J.L., Welch, D. and Gordon, I.J., 1995. The influence of vegetation pattern on the grazing of heather moorland by red deer and sheep. I. The location of animals on grass/heather mosaics. *Journal of Applied Ecology,* 32 (1), 166-176.

Clauss, M., Frey, R., Kiefer, B., et al., 2003a. The maximum attainable body size of herbivorous mammals: morphophysiological constraints on foregut, and adaptations of hindgut fermenters. *Oecologia,* 136 (1), 14-27.

Clauss, M., Lason, K., Gehrke, J., et al., 2003b. Captive roe deer (*Capreolus capreolus*) select for low amounts of tannic acid but not quebracho: fluctuation of preferences and potential benefits. *Comparative Biochemistry and Physiology. B. Biochemistry and Molecular Biology,* 136 (2), 369-382.

Clevers, J.G.P.W., 1988. The derivation of a simplified reflectance model for the estimation of leaf area index. *Remote Sensing of Environment,* 25 (1), 53-69.

Clevers, J.G.P.W., 1989. The application of a weighted infrared-red vegetation index for estimating leaf area index by correcting for soil moisture. *Remote Sensing of Environment,* 29 (1), 25-37.

Clevers, J.G.P.W. and Jongschaap, R., 2001. Imaging spectrometry for agricultural applications. *In:* Van der Meer, F.D. and De Jong, S.M. eds. *Imaging spectrometry: basic principles and prospective applications.* Kluwer Academic Publishers, Dordrecht, 157-199. Remote Sensing and Digital Image Processing no. 4.

Clevers, J.P.G.W., 1994. Imaging spectroscopy in agriculture: plant vitality and yield indicators. *In:* Hill, J. and Megier, J. eds. *Imaging spectrometry: a tool for environmental observations.* Kluwer Academic Publishers, Dordrecht, 193-219.

Coe, M.J., 1983. Large herbivores and food quality. *In:* Lee, J.A., McNeill, J. and Rorison, I.H. eds. *Nitrogen as an ecological factor: the 22nd symposium of the British Ecological Society, Oxford, 1981.* Blackwell, Oxford, 345-368.

Coe, M.J., Cumming, D.H. and Phillipson, J., 1976. Biomass and production of large African herbivores in relation to rainfall and primary production. *Oecologia,* 22 (4), 341-354.

Connell, J.H., 1983. On the prevalence and relative importance of interspecific competition: evidence from field experiments. *American Naturalist,* 122 (5), 661-696.

Cooper, S.D.B., Kyriazakis, I., Anderson, D.H., et al., 1993. The effect of physiological state (late pregnancy) on the diet selection of ewes. *Animal Production,* 56, 469A.

Cooper, S.D.B., Kyriazakis, I., Nolan, J.V., et al., 1995. Diet selection in sheep: the role of the rumen environment in the selection of a diet from two feeds that differ in their energy density. *British Journal of Nutrition,* 74 (1), 39-54.

Cooper, S.M. and Owen-Smith, N., 1986. Effects of plant spinescence on large mammalian herbivores. *Oecologia,* 68 (3), 446-455.

Cooper, S.M., Owen-Smith, N. and Bryant, J.P., 1988. Foliage acceptability to browsing ruminants in relation to seasonal changes in the leaf chemistry of woody plants in a South African savanna. *Oecologia,* 75 (3), 336-342.

Coppedge, B.R. and Shaw, J.H., 1998. Bison grazing patterns on seasonally burned tallgrass prairie. *Journal of Range Management,* 51 (3), 258-264.

Coppock, D.L., Ellis, J.E., Detling, J.K., et al., 1983. Plant-herbivore interactions in a North American mixed-grass prairie. II. Responses of bison to modification of vegetation by prairie dogs. *Oecologia,* 56 (1), 10-15.

Corsi, F., De Leeuw, J. and Skidmore, A.K., 2000. Modelling species distribution with GIS. *In:* Boitani, L. and Fuller, T.K. eds. *Research techniques in animal ecology controversies and consequences.* Columbia University Press, New York, 389-434.

Cosner, C., Deangelis, D.L., Ault, J.S., et al., 1999. Effects of spatial grouping on the functional response of predators. *Theoretical Population Biology,* 56 (1), 65-75.

Coughenour, M.B., 1991. Spatial components of plant-herbivore interactions in pastoral, ranching, and native ungulate ecosystems. *Journal of Range Management,* 44 (6), 530-542.

Cowling, R.M., 2000. Challenges to the 'new' rangeland science. *Trends in Ecology and Evolution,* 15 (8), 303-304.

Crawley, M.J., 1983. *Herbivory : the dynamics of animal-plant interactions.* Blackwell, Oxford. Studies in Ecology no. 10.

Creel, S., 2001. Four factors modifying the effect of competition on Carnivore population dynamics as illustrated by African wild dogs. *Conservation Biology,* 15 (1), 271-274.

Critchley, H.D. and Rolls, E.T., 1996. Hunger and satiety modify the responses of olfactory and visual neurons in the primate orbitofrontal cortex. *Journal of Neurophysiology,* 75 (4), 1673-1686.

Curran, P.J., 1988. The semivariogram in remote sensing: an introduction. *Remote Sensing of Environment,* 24 (3), 493-507.

Dahl, G. and Hjort, A., 1976. *Having herds: pastoral herd growth and household conomy.* University of Stockholm, Stockholm. Stockholm Studies in Social Anthropology no. 2.

Dale, M.R.T., 1999. *Spatial pattern analysis in plant ecology.* Cambridge University Press, Cambridge.

Davies, R.A. and Skinner, J.D., 1986. Temporal activity patterns of springbok and merino sheep during a karoo drought. *Transactions of the Royal Society of South Africa,* 46, 133-147.

Dayan, T. and Simberloff, D., 1994. Character displacement, sexual dimorphism, and morphological variation among British and Irish mustelids. *Ecology,* 75 (4), 1063-1073.

De Bie, S., 1991. *Wildlife resources of the West African savanna.* PhD Thesis, Wageningen University, Wageningen.

De Boer, W.F. and Prins, H.H.T., 1990. Large herbivores that strive mightily but eat and drink as friends. *Oecologia,* 82 (2), 264-274.

De Haan, C., Steinfeld, H. and Blackburn, H., 1997. *Livestock and the environment: finding a balance.* Commission of the European Communities, Brussels.

De Leeuw, J., Prins, H.H.T., Njuguna, E.C., et al., 1998. *Interpretation of DRSRS: animal counts (1977-1997) in the rangeland districts of Kenya.* ITC, Enschede.

DeAngelis, D.L., 1992. *Dynamics of nutrient cycling and food webs.* Chapman and Hall, London. Population and Community Biology Series no. 9.

Delany, M. and Happold, D.C.D., 1979. *Ecology of African mammals.* Longman, London.

Demment, M.W. and Van Soest, P.J., 1985. A nutritional explanation for body-size patterns of ruminant and nonruminant herbivores. *American Naturalist,* 125 (5), 641-672.

Denny, P., 1993. Wetlands of Africa: introduction. *In:* Whigham, D., Dykyjova, D. and Hejny, S. eds. *Wetlands of the world: inventory, ecology and management.* Kluwer, Dordrecht, 1-31. Handbook of Vegetation Science no. 15/2.

Desforges, M.F. and Wood Gush, D.G.M., 1976. A behavioural comparison of domestic and mallard ducks: habituation and flight reactions. *Animal Behaviour,* 23, 692-697.

Diamond, J.M., 1975. Assembly of species communities. *In:* Cody, M.L. and Diamond, J.M. eds. *Ecology and evolution of communities.* Harvard University Press, Cambridge, 342-444.

Diggle, P.J., 2003. *Statistical analysis of spatial point patterns.* 2nd edn. Academic Press, London.

Diggle, P.J. and Chetwynd, A.G., 1991. Second-order analysis of spatial clustering for inhomogeneous populations. *Biometrics,* 47 (3), 1155-1163.

Distel, R.A., Griggs, T.C., Laca, E.A., et al., 1991. Cattle prefer patches that yield greater instantaneous intake rate in patchy ryegrass swards. *In: Proceedings of the second grazing livestock nutrition conference.* Oklahoma State University, 171. MP / Oklahoma State University, Agricultural Experiment Station, Division of Agriculture no. MP-133.

Distel, R.A., Laca, E.A., Griggs, T.C., et al., 1995. Patch selection by cattle: maximization of intake rate in horizontally heterogeneous pastures. *Applied Animal Behaviour Science,* 45 (1/2), 11-21.

Drake, J.A., Zimmermann, C.R., Purucker, T., et al., 1999. On the nature of the assembly trajectory. *In:* Weiher, E. and Keddy, P. eds. *Ecological assembly rules: perspectives, advances, retreats.* Cambridge University Press, Cambridge, 233-250.

Drent, R.H. and Prins, H.H.T., 1987. The herbivore as prisoner of its food supply. *In:* Van Andel, J., Bakker, J. and Snaydon, R.W. eds. *Disturbance in grasslands: species and population responses.* Dr. W. Junk Publishing Company, Dordrecht, 133-149.

Drescher, M., 2003. *Grasping complex matter: large herbivore foraging in patches of heterogeneous resources*. PhD Thesis, Wageningen University.

Drescher, M., Heitkönig, I.M.A., Raats, J.G., et al., 2006a. The role of grass stems as structural foraging deterrents and their effects on the foraging behaviour of cattle. *Applied Animal Behaviour Science,* 101 (1/2), 10-26.

Drescher, M., Heitkönig, I.M.A., Van den Brink, P.J., et al., 2006b. Effects of sward structure on herbivore foraging behaviour in a South African savanna: an investigation of the forage maturation hypothesis. *Austral Ecology,* 31 (1), 76-87.

Du Toit, J.G., 1996. Water requirements. *In:* Bothma, J.d.P. ed. *Game ranch management.* 3rd edn. Van Schaik JL, Pretoria, 95-98.

Dublin, H., 1995. Vegetation dynamics in the Serengeti-Mara ecosystem: the role of elephants, fire, and other factors. *In:* Sinclair, A.R.E. and Arcese, P. eds. *Serengeti 2: dynamics, management, and conservation of an ecosystem [based on papers at a workshop held in December 1991 at the Serengeti Research Institute, Tanzania*. University of Chicago Press, Chicago, 71-90.

Dublin, H.T., Sinclair, A.R.E. and McGlade, J., 1990. Elephants and fire as causes of multiple stable states in the Serengeti-Mara woodlands. *Journal of Animal Ecology,* 59 (3), 1147-1164.

Dumont, B. and Boissy, A., 2000. Grazing behaviour of sheep in a situation of conflict between feeding and social motivations. *Behavioural Processes,* 49 (3), 131-138.

Dumont, B., Maillard, J.F. and Petit, M., 2000. The effect of the spatial distribution of plant species within the sward on the searching success of sheep when grazing. *Grass and Forage Science,* 55 (2), 138-145.

Dumont, B. and Petit, M., 1998. Spatial memory of sheep at pasture. *Applied Animal Behaviour Science,* 60 (1), 43-53.

Dunbrack, R.L. and Ramsay, A., 1993. The allometry of mammalian adaptations to seasonal environments: a critique of the fasting endurance hypothesis. *Oikos,* 66 (2), 336-342.

Duncan, A.J. and Gordon, I.J., 1999. Habitat selection according to the ability of animals to eat, digest and detoxify foods. *Proceedings of the Nutrition Society,* 58 (4), 799-805.

Duncan, A.J., Hartley, S.E. and Iason, G.R., 1998. The effect of previous browsing damage on the morphology and chemical composition of Sitka spruce (*Picea sitchensis*) saplings and on their subsequent susceptibility to browsing by red deer (*Cervus elaphus*). *Forest Ecology and Management,* 103 (1), 57-67.

Duncan, A.J., Mayes, R.W., Young, S.A., et al., 2001. Choice of foraging patches by hill sheep given different opportunities to seek shelter and food. *Animal Science,* 73 (3), 563-570.

Duncan, A.J., Reid, S.A., Thoss, V., et al., 2005. Browse selection in response to simulated seasonal changes in diet quality through postingestive effects. *Journal of Chemical Ecology,* 31 (4), 729-744.

Duncan, P., 1975. *Topi and their food supply*. PhD Thesis, University of Nairobi, Nairobi.

Duncan, P., 1983. Determinants of the use of habitat by horses in a Mediterranean wetland. *Journal of Animal Ecology,* 52, 93-109.

Duncan, P., 1985. Time budgets of Camargue horses. III. Environmental influences. *Behaviour,* 92, 188-208.

Durant, S.M., 1998. Competition refuges and coexistence: an example from Serengeti carnivores. *Journal of Animal Ecology,* 67 (3), 370-386.

Dury, S., Turner, B., Foley, B., et al., 2001. Use of high spectral resolution remote sensing to determine leaf palatability of eucalypt trees for folivorous marsupials. *International Journal of Applied Earth Observation and Geoinformation,* 3 (4), 328-336.

Dyson-Hudson, N., 1980. Strategies of resource exploitation among East African savanna pastoralists. *In:* Harris, D.R. ed. *Human ecology in savanna environments*. Academic Press, London, 171-184.

Dyson-Hudson, R. and Dyson-Hudson, N., 1980. Nomadic pastoralism (Africa). *Annual review of anthropology,* 9, 15-61.

Early, D.M. and Provenza, F.D., 1998. Food flavour and nutritional characteristics alter dynamics of food preference in lambs. *Journal of Animal Science,* 76 (3), 728-734.

East, R., 1984. Rainfall, soil nutrient status and biomass of large African savanna mammals. *African Journal of Ecology,* 22 (4), 245-270.

Edwards, G.R., Newman, J.A., Parsons, A.J., et al., 1996. The use of spatial memory by grazing animals to locate food patches in spatially heterogeneous environments: an example with sheep. *Applied Animal Behaviour Science,* 50 (2), 147-160.

Egan, A.R., 1980. Host animal-rumen relationships. *Proceedings of the Nutrition Society,* 39 (1), 79-87.

Eklundh, L., 1998. Estimating relations between AVHRR NDVI and rainfall in East Africa at 10-day and monthly time scales. *International Journal of Remote Sensing,* 19 (3), 563-568.

Ellenbroek, G.A., 1987. *Ecology and productivity of an African wetland system: the Kafue Flats, Zambia*. Junk, Dordrecht. Geobotany no. 9.

Ellis, J.E., 1995. Climate variability and complex ecosystem dynamics: implications for pastoral development. *In:* Scoones, I. ed. *Living with uncertainty: new directions in pastoral development in Africa*. Intermediate Technology Publications, London, 37-46.

Ellis, J.E. and Coughenour, M.B., 1998. The SAVANNA integrated modelling system: an integrated remote sensing, GIS and spatial simulation modelling approach. *In:* Squires, V.R. and Sidahmed, A.E. eds. *Drylands: sustainable use of rangelands into the twenty-first century*. IFAD, Rome, 97-106.

Ellis, J.E. and Swift, D.M., 1988. Stability of African pastoral ecosystems: alternate paradigms and implications for development. *Journal of Range Management,* 41 (6), 450-459.

Elton, C., 1927. *Animal ecology*. Sidgwick and Jackson, London.

Elvidge, C.D., 1990. Visible and near infrared reflectance characteristics of dry plant materials. *International Journal of Remote Sensing,* 11 (10), 1775-1795.

Elvidge, C.D., Zhikang, C. and Groeneveld, D.P., 1993. Detection of trace quantities of green vegetation in 1990 AVIRIS data. *Remote Sensing of Environment,* 44 (2/3), 271-279.

Engel, C., 2002. *Wild health: how animals keep themselves well and what we can learn from them*. Houghton Mifflin Co., New York.

Enkh-Amgalan, A., 2002. Change and sustainability of pastoral land use systems in Mongolia. *In:* Chuluun, T. and Ojima, D. eds. *Fundamental issues affecting sustainability of the Mongolian steppe*. International Institute for Study of Nomadic Civilizations, Ulaanbaatar, 228-231.

Enkhtuvshin, B. and Tumujav, M., 2002. World nomads and Mongolian traditional pasture use. *In:* Chuluun, T. and Ojima, D. eds. *Fundamental issues affecting sustainability of the Mongolian steppe*. International Institute for Study of Nomadic Civilizations, Ulaanbaatar, 19-25.

Estes, J.E. and Mooneyhan, D.W., 1994. Of maps and myths. *Photogrammetric Engineering and Remote Sensing,* 60 (5), 517-524.

Etzenhouser, M.J., Owens, M.K., Spalinger, D.E., et al., 1998. Foraging behaviour of browsing ruminants in a heterogeneous landscape. *Landscape Ecology,* 13 (1), 55-64.

Evans-Pritchard, E.E., 1940. *The nuer: a description of the modes of livelihood and political institutions of a Nilotic people*. Oxford University Press, Oxford.

Evans, M., Green, B. and Newgrain, K., 2003. The field energetics and water fluxes of free-living wombats (Marsupialia: Vombatidae). *Oecologia,* 137 (2), 171-180.

Fancy, S.G. and White, R.G., 1985. Incremental cost of activity. *In:* Hudson, R.J. and White, R.G. eds. *Bioenergetics of wild herbivores*. CRC Press, Boca Raton, 143-159.

FAO, 2001. *Pastoralism in the new millenium*. FAO, Rome. FAO Animal Production and Health Paper no. 150.

Farnsworth, K.D. and Beecham, J.A., 1999. How do grazers achieve their distribution? A continuum of models from random diffusion to the ideal free distribution using biased random walks. *American Naturalist,* 153 (5), 509-526.

Farnsworth, K.D. and Illius, A.W., 1996. Large grazers back in the fold: generalizing the prey model to incorporate mammalian herbivores. *Functional Ecology,* 10 (5), 678-680.

Ferguson, S.H., Bergerud, A.T. and Ferguson, R., 1988. Predation risk and habitat selection in the persistence of a remnant caribou population. *Oecologia,* 76 (2), 236-245.

Fernandez-Gimenez, M.E. and Allen-Diaz, B., 1999. Testing a non-equilibrium model of rangeland vegetation dynamics in Mongolia. *Journal of Applied Ecology,* 36 (6), 871-885.

Ferwerda, J.G., 2005. *Charting the quality of forage: measuring and mapping the variation of chemical components in foliage with hyperspectral remote sensing*. PhD Thesis, Wageningen University, Wageningen.

Ferwerda, J.G., Skidmore, A.K., Van Wieren, S.E., et al., 2002. Field spectrometry of leaves for nutrient assesment. *In:* Navalgund, R.R., Nayak, S.R., Sudarshana, R., et al. eds. *ISPRS 2002 : proceedings of ISPRS commission VII international symposium : resource and environmental monitoring, December 3-6 2002, Hyderabad, India*. National Remote Sensing Agency, Hyderabad.

Feyaerts, F. and Van Gool, L., 2001. Multi-spectral vision system for weed detection. *Pattern Recognition Letters,* 22 (6/7), 667-674.

Field, C. and Mooney, H.A., 1986. The photosynthesis-nitrogen relationship in wild plants. *In:* Givnish, T.J. ed. *On the economy of plant form and function: proceedings of the sixth Maria Moors Cabot symposium, "Evolutionary constraints on primary productivity: adaptive patterns of energy capture in plants," Harvard Forest, August 1983*. Cambridge University Press, Cambridge, 25-55.

Field, C.R., 1976. Palatability factors and nutrive values of the food of buffaloes (*Syncerus caffer*) in Uganda. *East African Wildlife Journal,* 14, 181-201.

Fisher, D.S., 1996. Modeling ruminant feed intake with protein, chemostatic, and distention feedbacks. *Journal of Animal Science,* 74 (12), 3076-3081.

Focardi, S. and Marcellini, P., 1995. A mathematical framework for optimal foraging of herbivores. *Journal of Mathematical Biology,* 33 (4), 365-387.

Focardi, S., Marcellini, P. and Montanaro, P., 1996. Do ungulates exhibit a food density threshold? A field study of optimal foraging and movement patterns. *Journal of Animal Ecology,* 65 (5), 606-620.

Gordon, I.J., 2003. Browsing and grazing ruminants: are they different beasts? *Forest Ecology and Management,* 181 (1/2), 13-21.

Gordon, I.J., Duncan, P., Grillas, P., et al., 1990. The use of domestic herbivores in the conservation of the biological richness of European wetlands. *Bulletin d'Ecologie,* 21 (3), 49-60.

Gordon, I.J. and Illius, A.W., 1988. Incisor arcade structure and diet selection in ruminants. *Functional Ecology,* 2, 15-22.

Gordon, I.J. and Illius, A.W., 1994. The functional significance of the browser-grazer dichotomy in African ruminants. *Oecologia,* 98 (2), 167-175.

Gordon, I.J. and Illius, A.W., 1996. The nutritional ecology of African ruminants: a reinterpretation. *Journal of Animal Ecology,* 65 (1), 18-28.

Gordon, I.J., Illius, A.W. and Milne, J.D., 1996. Sources of variation in the foraging efficiency of grazing ruminants. *Functional Ecology,* 10 (2), 219-226.

Grace, J.B., 1995. On the measurement of plant competition intensity. *Ecology,* 76 (1), 305-308.

Graham, R.W. and Lundelius, E.L., 1984. Coevolutionary disequilibrium and Pleistocene extinctions. *In:* Martin, P.S. and Klein, R.G. eds. *Quaternary extinctions: a prehistoric revolution.* University of Arizona Press, Tucson, 223-249.

Greenwood, P.T. and Rittenhouse, L.R., 1997. Feeding area selection: the leader-follower phenomena. *Proceedings of the Western Section American Society of Animal Science,* 48, 267-269.

Grenfell, B.T., Wilson, K., Finkenstadt, B.F., et al., 1998. Noise and determinism in synchronized sheep dynamics. *Nature,* 394 (6694), 674-677.

Grinnell, J., 1917. The niche-relationships of the California thrasher. *Auk,* 34, 427-433.

Groombridge, B., 1992. *Global biodiversity: status of the earth's living resources.* Chapman and Hall, London.

Groot-Bruinderink, G.W.T.A., Lammertsma, D.R., Kramer, K., et al., 1999. *Dynamische interacties tussen hoefdieren en vegetatie in de Oostvaardersplassen.* IBN-DLO, Wageningen. IBN Rapport no. 436.

Gross, J.E., Hobbs, N.T. and Wunder, B.A., 1993a. Independent variables for predicting intake rate of mammalian herbivores: biomass density, plant density, or bite size? *Oikos,* 68 (1), 75-81.

Gross, J.E., Shipley, L.A., Hobbs, N.T., et al., 1993b. Functional response of herbivores in food-concentrated patches: tests of a mechanistic model. *Ecology,* 74 (3), 778-791.

Gross, J.E., Wang, Z. and Wunder, B.A., 1985. Effects of food quality and energy needs: changes in gut morphology and capacity of *Microtus ochrogaster. Journal of Mammalogy,* 66 (6), 661-667.

Gross, J.E., Zank, C., Hobbs, N.T., et al., 1995. Movement rules for herbivores in spatially heterogeneous environments: responses to small scale pattern. *Landscape Ecology,* 10 (4), 209-217.

Grünbaum, D., 1998. Using spatially explicit models to characterize foraging performance in heterogeneous landscapes. *American Naturalist,* 151 (2), 97-115.

Grunblatt, J., Said, M.Y., Wargute, P., et al., 1995. *DRSRS aerial surveys database.* Department of Resource Surveys and Remote Sensing, Ministry of Planning and National Development, Nairobi.

Gustafsson, M., Jensen, P., De Jonge, F.H., et al., 1999. Domestication effects on foraging strategies in pigs (*Sus scrofa*). *Applied Animal Behaviour Science,* 62 (4), 305-317.

Guthrie, R.D., 1990. *Frozen fauna of the mammoth steppe: the story of Blue Babe.* University of Chicago Press, Chicago.

Guyot, G., Baret, F. and Major, D.J., 1988. High spectral resolution: determination of spectral shifts between the red and near infrared. *International Archives of Photogrammetry and Remote Sensing,* 27, 750-760.

Hagerman, A.E., Riedl, K.M., Jones, G.A., et al., 1998. High molecular weight plant polyphenolics (tannins) as biological antioxidants. *Journal of Agricultural and Food Chemistry,* 46 (5), 1887-1892.

Haltenorth, T. and Diller, H., 1994. *Mammals of Africa including Madagascar.* Harper Collins Publishers, London.

Hanski, I., 1998. Metapopulation dynamics. *Nature,* 396 (6706), 41-49.

Hanski, I., 1999. *Metapopulation ecology.* Oxford University Press, Oxford.

Hardin, G., 1968. The tragedy of the commons: the population problem has no technical solution; it requires a fundamental extension in morality. *Science,* 162 (859), 1243-1248.

Harrington, R., Owen-Smith, N., Viljoen, P.C., et al., 1999. Establishing the causes of the roan antelope decline in the Kruger National Park, South Africa. *Biological Conservation,* 90 (1), 69-78.

Harris, N.R., Johnson, D.E., George, M.R., et al., 2002. The effect of topography, vegetation, and weather on cattle distribution at the San Joaquin Experimental Range, California. *In:* Standiford, R.B., McCreary, D. and Purcell, K.L. eds. *Proceedings of the fifth symposium on oak woodlands: oaks in California's changing landscape. 2001 October 22-25; San Diego, CA.* USDA Forest Service, Albany, 53-63. General Technical Report no. PSW-GTR-184.

Hary, I., Schwartz, H.J., Pielert, V.H.C., et al., 1996. Land degradation in African pastoral systems and the destocking controversy. *Ecological Modelling,* 86 (2/3), 227-233.

Haskell, J.P., Ritchie, M.E. and Olff, H., 2002. Fractal geometry predicts varying body size scaling relationships for mammal and bird home ranges. *Nature,* 418 (6897), 527-530.

Haxeltine, A., 1996. *Modelling the vegetation of the earth.* PhD Thesis, Lund University, Department of Ecology, Lund.

Heasley, L. and Delehanty, J., 1996. The politics of manure: resource tenure and the agropastoral economy in Southwestern Niger. *Society and Natural Resources,* 9 (1), 31-46.

Heath, B., 2000. Ranching: an economic yardstick. *In:* Prins, H.H.T. and Grootenhuis, J.G. eds. *Wildlife conservation by sustainable use.* Kluwer Academic Publishers, Boston, 21-33.

Hebblewhite, M. and Pletscher, D.H., 2002. Effects of elk group size on predation by wolves. *Canadian Journal of Zoology,* 80 (5), 800-809.

Hebblewhite, M., Pletscher, D.H. and Paquet, P.C., 2002. Elk population dynamics in areas with and without predation by recolonizing wolves in Banff National Park, Alberta. *Canadian Journal of Zoology,* 80 (5), 789-799.

Herbel, C.H. and Nelson, A.B., 1966. Activities of Hereford and Santa Gertrudis cattle on a southern New Mexico range. *Journal of Range Management,* 19 (4), 173-176.

Hernes, P.J. and Hedges, J.I., 2000. Determination of condensed tannin monomers in environmental samples by capillary gas chromatography of acid depolymerization extracts. *Analytical Chemistry,* 72 (20), 5115-5124.

Hester, A.J., Gordon, I.J., Baillie, G.J., et al., 1999. Foraging behaviour of sheep and red deer within natural heather/grass mosaics. *Journal of Applied Ecology,* 36 (1), 133-146.

Hiernaux, P. and Diarra, L., 1983. Pâturages de la zone d'inondation du Niger. *In:* Wilson, R.T., De Leeuw, P.N. and De Haan, C. eds. *Recherches sur les systèmes des zones arides du Mali: résultats préliminaires.* International Livestock Center for Africa ILCA, Addis Abeba, 42-48. Rapport de Recherche no. 5.

Hik, D.S., 1995. Does risk of predation influence population dynamics? Evidence from the cyclic decline of snowshoe hares. *Wildlife Research,* 22 (1), 115-129.

Hirst, S.M., 1969. Populations in a Transvaal lowland nature reserve. *Zoologica Africana,* 4, 199-230.

Hobbs, N.T., 1990. Diet selection by generalist herbivores: a test of the linear programming model. *In:* Hughes, R.N. ed. *Behavioural mechanisms of food selection.* Springer, Berlin, 395-413.

Hobbs, N.T., Gross, J.E., Shipley, L.A., et al., 2003. Herbivore functional response in heterogeneous environments: a contest among models. *Ecology,* 84 (3), 666-681.

Hodgson, J., Cosgrove, G.P. and Woodward, S.J.R., 1997. Research on foraging strategy: progress and priorities. *In:* Buchanan-Smith, J.G., Bailey, L.D. and McCaughey, P. eds. *Proceedings of the XVIII international grassland congress, Saskatoon, Saskatchewan, Canada, 8-13 June 1997.* Association Management Centre, Calgary, 109-118.

Hodgson, J., Niezen, J.H., Montossi, F., et al., 1996. Comparative studies on animal performance and parasite infestation in sheep grazing Yorkshire fog, perennial ryegrass, and tall fescue pastures. *Proceedings of the New Zealand Grassland Association,* 57, 89-93.

Hofmann, R.R., 1973. *The ruminant stomach: stomach structure and feeding habits of East African game ruminants.* East African Literature Bureau, Nairobi. East African Monographs in Biology no. 2.

Hofmann, R.R., 1989. Evolutionary steps of ecophysiological adaptation and diversification of ruminants: a comparative view of their digestive system. *Oecologia,* 78 (4), 443-457.

Hofmann, R.R. and Stewart, D.R.M., 1972. Grazer or browser: a classification based on the stomach structure and feeding habits of East African ruminants. *Mammalia,* 36 (2), 226-240.

Holdridge, L.R., 1947. Determination of world plant formations from simple climatic data. *Science,* 105 (2727), 367-368.

Holechek, J.L., Pieper, R.D. and Herbel, C.H., 2001. *Range management: principles and practices.* 4th edn. Prentice Hall, Englewood Cliffs.

Holling, C.S., 1959. Some characteristics of simple types of predation and parasitism. *Canadian Entomologist,* 91 (7), 385-398.

Homewood, K.M. and Rodgers, W.A., 1991. *Maasailand ecology: pastoralist development and wildlife conservation in Ngorongoro, Tanzania*. Cambridge Univ. Press, Cambridge.

Hooper, J.F., Workman, J.P., Grumbles, J.B., et al., 1969. Improved livestock distribution with fertilizer: a preliminary economic evaluation. *Journal of Range Management*, 22, 108-110.

Houseal, G.A. and Olson, B.E., 1995. Cattle use of microclimates on a northern latitude winter range. *Canadian Journal of Animal Science*, 75, 501-507.

Houston, A. and McNamara, J.M., 1999. *Models of adaptive behaviour: an approach based on state*. Cambridge University Press, Cambridge.

Howell, P., Lock, M. and Cobb, S., 1988. *The Jonglei Canal: impact and opportunity*. Cambridge University Press, Cambridge.

Howery, L.D., Bailey, D.W., Ruyle, G.B., et al., 2000. Cattle use visual cues to track food locations. *Applied Animal Behaviour Science*, 67 (1/2), 1-14.

Howery, L.D., Provenza, F.D., Banner, R.E., et al., 1996. Differences in home range and habitat use among individuals in a cattle herd. *Applied Animal Behaviour Science*, 49 (3), 305-320.

Howery, L.D., Provenza, F.D., Banner, R.E., et al., 1998. Social and environmental factors influence cattle distribution on rangeland. *Applied Animal Behaviour Science*, 55 (3/4), 231-244.

Hubbell, S.P., 2001. *The unified neutral theory of biodiversity and biogeography*. Princeton University Press, Princeton. Monographs in Population Biology no. 32.

Hudson, R.J., 1985. Body size, energetics, and adaptive radiation. *In:* Hudson, R.J. and White, R.G. eds. *Bioenergetics of wild herbivores*. CRC Press, Boca Raton, 1-24.

Hudson, R.J. and Frank, S., 1987. Foraging ecology of bison in aspen boreal habitats. *Journal of Range Management*, 40 (1), 71-75.

Huggard, D.J., 1994. A linear programming model of herbivore foraging: imprecise, yet successful? *Oecologia*, 100 (4), 470-474.

Humphrey, C. and Sneath, D., 1999. *The end of nomadism? Society, state and the environment in inner Asia*. White Horse Press, Cambridge.

Hunter, R.F., 1962. Hill sheep and their pasture: a study of sheep-grazing in south-east Scotland. *The Journal of Ecology*, 50 (3), 651-680.

Huntingford, G.W.B., 1933. Some aspects of Nandi stock-raising. *Journal of the East Africa and Uganda Natural History Society*, 11, 250-262.

Huntingford, G.W.B., 1953a. *The northern Nilo-Hamites*. International African Institute, London.

Huntingford, G.W.B., 1953b. *The southern Nilo-Hamites*. International African Institute, London.

Hutchings, M.J., 1986. The structure of plant populations. *In:* Crawley, M.J. ed. *Plant ecology*. Blackwell, Oxford, 97-136.

Hutchins, M., Hancocks, D. and Crockett, C., 1984. Naturalistic solutions to the behavioral problems of captive animals. *Der Zoologische Garten*, 54, 28-42.

Hutchinson, G.E., 1957. Concluding remarks: Cold Spring Harbor symposium. *Quantitative Biology*, 22, 415-427.

Iason, G.R., Mantecon, A.R., Sim, D.A., et al., 1999. Can grazing sheep compensate for a daily foraging time constraint? *Journal of Animal Ecology*, 68 (1), 87-93.

Iason, G.R. and Van Wieren, S.E., 1999. Digestive and ingestive adaptations of mammalian herbivores to low quality forage. *In:* Olff, H., Brown, V.K. and Drent, R.H. eds. *Herbivores: between plants and predators : the 38th symposium of the British Ecological Society in cooperation with the Netherlands Ecological Society, held at Wageningen Agricultural University, The Netherlands, 1997*. Blackwell Scientific Publications, Oxford, 337-369.

ILCA, 1979. *Livestock production in the subhumid zone of West Africa: a regional review*. International Livestock Center for Africa ILCA, Addis Ababa. ILCA Systems Study no. 2.

Illius, A.W. and Gordon, I.J., 1987. The allometry of food intake in grazing ruminants. *Journal of Animal Ecology*, 56 (3), 989-999.

Illius, A.W. and Gordon, I.J., 1991. Prediction of intake and digestion in ruminants by a model of rumen kinetics integrating animal size and plant characteristics. *Journal of Agricultural Science*, 116 (1), 145-157.

Illius, A.W. and Gordon, I.J., 1992. Modelling the nutritional ecology of ungulate herbivores: evolution of body size and competitive interactions. *Oecologia*, 89 (3), 428-434.

Illius, A.W. and O'Connor, T.G., 1999. On the relevance of nonequilibrium concepts to arid and semiarid grazing systems. *Ecological Applications*, 9 (3), 798-813.

Illius, A.W. and O'Connor, T.G., 2000. Resource heterogeneity and ungulate population dynamics. *Oikos*, 89 (2), 283-294.

Janis, C.M., 1995. Correlation between craniodental morphology and feeding behavior in ungulates: reciprocal illumination between living and fossil taxa. *In:* Thomason, J.J. ed. *Functional morphology in vertebrate paleontology*. Cambridge University Press, Cambridge, 76-98.

Janis, C.M. and Ehrhardt, D., 1988. Correlation of relative muzzle width and relative incisor width with dietary preference in ungulates. *Zoological Journal of the Linnean Society,* 92 (3), 267-284.

Jarman, M.V. and Jarman, P.J., 1973. Daily activity of impala. *East African Wildlife Journal,* 11, 75-92.

Jarman, P.J., 1974. The social organization of antelope in relation to their ecology. *Behaviour,* 48 (2), 15-66.

Jarman, P.J. and Sinclair, A.R.E., 1979. Feeding strategy and pattern of resource partitioning in ungulates. *In:* Sinclair, A.R.E. and Norton-Griffiths, M. eds. *Serengeti: dynamics of an ecosystem.* University of Chicago Press, Chicago, 130-160.

Jensen, S.P. and Doncaster, C.P., 1999. Lethal toxins in non-preferred foods: how plant chemical defences can drive microtine cycles. *Journal of Theoretical Biology,* 199 (1), 63-85.

Jetz, W., Carbone, C., Fulford, J., et al., 2004. The scaling of animal space use. *Science,* 306 (5694), 266-268.

Jiang, Z. and Hudson, R.J., 1993. Optimal grazing of wapiti (*Cervus elaphus*) on grassland: patch and feeding station departure rules. *Evolutionary Ecology,* 7 (5), 488-498.

Jina, P.S., 1999. The Chang Pa of the Ladakh Himalayas: some observations on their economy and environment. *Nomadic Peoples,* 3 (1), 94-105.

Johnson, B.J., Fogel, M.L. and Miller, G.H., 1998. Stable isotopes in modern ostrich eggshell: a calibration for paleoenvironmental applications in semi-arid regions of southern Africa. *Geochimica et Cosmochimica Acta,* 62 (14), 2451-2461.

Johnson, C.J., Parker, K.L. and Heard, D.C., 2001. Foraging across a variable landscape: behavioral decisions made by woodland caribou at multiple spatial scales. *Oecologia,* 127 (4), 590-602.

Johnson, C.J., Parker, K.L., Heard, D.C., et al., 2002a. Movement parameters of ungulates and scale-specific responses to the environment. *Journal of Animal Ecology,* 71 (2), 225-235.

Johnson, C.J., Parker, K.L., Heard, D.C., et al., 2002b. A multiscale behavioral approach to understanding the movements of woodland caribou. *Ecological Applications,* 12 (6), 1840-1860.

Jones, R.J., Meyer, J.H.F., Bechaz, F.M., et al., 2001. Comparison of rumen fluid from South African game species and from sheep to digest tanniniferous browse. *Australian Journal of Agricultural Research,* 52 (4), 453-460.

Kalliola, R. and Syrjanen, K., 1991. To what extent are vegetation types visible in satellite imagery? *Annales Botanici Fennici,* 28 (1), 45-57.

Kaluzny, S.P., Vega, S.C., Cardoso, T.P., et al., 1998. *S+SpatialStats: user's manual for Windows and UNIX.* Springer, New York.

Kari, S. and Scholte, P., 2001. La réhabilitation pastorale de la plaine d'inondation Waza-Logone (Cameroun): comment consolider sa réussite écologique? *In:* Tielkes, E. and Schlecht, E. eds. *Elevage et gestion de parcours au Sahel, implications pour le développement: comptes-rendus d'un atelier régional ouest-africain sur le thème "la gestion des pâturages et les projets de développement: quelles perspectives?", tenu du 2 au 6 octobre 2000 à Niamey, Niger.* Grauer, Beuren, 315-316.

Kavoori, P.S., 1999. *Pastoralism in expansion: the transhuming herders of western Rajasthan.* Oxford University Press, Oxford.

Kelt, D.A. and Martínez, D.R., 1989. Notes on distribution and ecology of two marsupials endemic to the Valdivian forests of southern South America. *Journal of Mammalogy,* 70 (1), 220-224.

Kelt, D.A., Rogovin, K., Shenbrot, G., et al., 1999. Patterns in the structure of Asian and North American desert small mammal communities. *Journal of Biogeography,* 26 (4), 825-841.

Kennedy, M. and Gray, R.D., 1993. Can ecological theory predict the distribution of foraging animals? A critical analysis of experiments on the ideal free distribution. *Oikos,* 68 (1), 158-166.

Kerven, C., 2002. The loss of livestock mobility in the new market economies of Kazakstan and Turkmenistan: implications for range degradation and regeneration. *In:* Chuluun, T. and Ojima, D. eds. *Fundamental issues affecting sustainability of the Mongolian steppe.* International Institute for Study of Nomadic Civilizations, Ulaanbaatar, 34-51.

Kerven, C., Alimaev, I.I., Behnke, R., et al., 2004. Retraction and expansion of flock mobility in Central Asia: costs and consequences. *In:* Allsopp, N., Palmer, A.R., Milton, S.J., et al. eds. *Proceedings of the VII International Rangelands Congress, Durban South Africa, 26 July-1 August 2003.* 543-556.

Key, C. and MacIver, R.M., 1980. The effects of maternal influences on sheep: breed differences in grazing, resting and courtship behaviour. *Applied Animal Ethology,* 6, 33-48.

Khaemba, W.M. and Stein, A., 2000. Use of GIS for a spatial and temporal analysis of Kenyan wildlife with generalised linear modelling. *International Journal of Geographical Information Science,* 14 (8), 833-853.

Kijazi, A., Mkumbo, S. and Thompson, D.M., 1997. Human and livestock population trends. *In:* Thompson, D.M. ed. *Multiple land-use: the experience of the Ngorongoro conservation area, Tanzania.* IUCN, Gland, 169-180.

Kimani, K. and Pickard, J., 1998. Recent trends and implications of group ranch sub-division and fragmentation in Kajiado district, Kenya. *The Geographical Journal,* 164 (2), 202-213.

Kingdon, J., 1982. *East African mammals: an atlas of evolution in Africa. Vol. 3, pt. C: Bovids.* Academic Press, London.

Kingdon, J., 1997. *The Kingdon field guide to African mammals.* Academic Press, San Diego.

Klein, D.R. and Fairall, N., 1986. Comparative foraging behaviour and associated energetics of impala and blesbok. *Journal of Applied Ecology,* 23 (2), 489-502.

Knipling, E.B., 1970. Physical and physiological basis for the reflectance of visible and near-infrared radiation. *Remote Sensing of Environment,* 1, 155-159.

Kolasa, J. and Pickett, S.T.A. (eds.), 1991. *Ecological heterogeneity.* Springer, Berlin. Ecological Studies, Analysis and Synthesis no. 86.

Kondo, M., Kita, K. and Yokota, H.O., 2004. Feeding value to goats of whole-crop oat ensiled with green tea waste. *Animal Feed Science and Technology,* 113 (1/4), 71-81.

Köppen, W., 1931. *Grundriss der Klimakunde.* Willem de Gruyter, Berlin.

Kotler, B.P. and Brown, J.S., 1999. Mechanisms of coexistence of optimal foragers as determinants of local abundances and distributions of desert granivores. *Journal of Mammalogy,* 80 (2), 361-374.

Kotliar, N.B. and Wiens, J.A., 1990. Multiple scales of patchiness and patch structure: a hierarchical framework for the study of heterogeneity. *Oikos,* 59 (2), 253-260.

Kramer, K., Groot Bruinderink, G.W.T.A. and Prins, H.H.T., 2006. Spatial interactions between ungulate herbivory and forest management. *Forest Ecology and Management,* 226 (1/3), 238-247.

Kraus, T.E.C., Zasoski, R.J., Dahlgren, R.A., et al., 2004. Carbon and nitrogen dynamics in a forest soil amended with purified tannins from different plant species. *Soil Biology and Biochemistry,* 36 (2), 309-321.

Krause, J. and Ruxton, G.D., 2002. *Living in Groups.* Oxford University Press, Oxford.

Krebs, C.J., 1989. *Ecological methodology.* Harper Collins, New York.

Krebs, J.R. and Davies, N.B. (eds.), 1997. *Behavioural ecology: an evolutionary approach.* 2nd edn. Blackwell, Oxford.

Krebs, J.R., Ryan, J.C. and Charnov, E.L., 1974. Hunting by expectation or optimal foraging? A study of patch use by chickadees. *Animal Behaviour,* 22 (4), 953-964.

Kristjanson, P., Radeny, M., Nkedianye, D., et al., 2002. *Valuing alternative land-use options in the Kitengela wildlife dispersal area of Kenya.* International Livestock Research Institute ILRI, Nairobi. ILRI Impact Assessment Series no. 10.

Křivan, V. and Vrkoč, I., 2000. Patch choice under predation hazard. *Theoretical Population Biology,* 58 (4), 329-340.

Ksiksi, T. and Laca, E.A., 2000. Can social interactions affect food searching efficiency of cattle? *Rangeland Journal,* 22 (2), 235-242.

Kumar, L., Schmidt, K.S., Dury, S., et al., 2001. Imaging spectroscopy and vegetation science. *In:* Van der Meer, F.D. and De Jong, S.M. eds. *Imaging spectrometry: basic principles and prospective applications.* Kluwer Academic Publishers, Dordrecht, 111-155. Remote Sensing and Digital Image Processing no. 4.

Kumar, L. and Skidmore, A.K., 2000. Radiation-vegetation relationships in a Eucalyptus forest. *Photogrammetric Engineering and Remote Sensing,* 66 (2), 193-204.

Kunkel, K. and Pletscher, D.H., 2001. Winter hunting patterns of wolves in and near glacier National Park, Montana. *Journal of Wildlife Management,* 65 (3), 520-530.

Kyriazakis, I. and Oldham, J.D., 1993. Diet selection in sheep: the ability of growing lambs to select a diet that meets their crude protein (nitrogen x 6.25) requirements. *British Journal of Nutrition,* 69 (3), 617-629.

Kyriazakis, I. and Oldham, J.D., 1997. Food intake and diet selection in sheep: the effect of manipulating the rates of digestion of carbohydrates and protein of the foods offered as a choice. *British Journal of Nutrition,* 77 (2), 243-254.

Kyriazakis, I., Oldham, J.D., Coop, R.L., et al., 1994. The effect of subclinical intestinal nematode infection on the diet selection of growing sheep. *British Journal of Nutrition,* 72 (5), 665-677.

Laca, E.A., 1998. Spatial memory and food searching mechanisms of cattle. *Journal of Range Management,* 51 (4), 370-378.

Laca, E.A. and Demment, M.W., 1996. Foraging strategies of grazing animals. *In:* Hodgson, J. and Illius, A.W. eds. *The ecology and management of grazing systems*. CAB international, Wallingford, 137-157.

Laca, E.A., Distel, R.A., Griggs, T.C., et al., 1994a. Effects of canopy structure on patch depression by grazers. *Ecology,* 75 (3), 706-716.

Laca, E.A., Distel, R.A., Griggs, T.C., et al., 1993. Field test of optimal foraging with cattle: the marginal value theorem successfully predicts patch selection and utilization. *In:* Baker, M.J. ed. *Grasslands for our world: proceedings of the XVII international grassland congress, held in February 1993, in New Zealand and Queensland.* SIR, Wellington, 709-710.

Laca, E.A., Ungar, E.D. and Demment, M.W., 1994b. Mechanisms of handling time and intake rate of a large mammalian grazer. *Applied Animal Behaviour Science,* 39 (1), 3-19.

Laca, E.A., Ungar, E.D., Seligman, N., et al., 1992. Effects of sward height and bulk density on bite dimensions of cattle grazing homogeneous swards. *Grass and Forage Science,* 47 (1), 91-102.

Lamprey, H.F., 1963. Ecological separation of the large mammal species in the Tarangire Game Reserve, Tanganyika. *East African Wildlife Journal,* 1, 63-92.

Lane, C. and Moorehead, R., 1994. New directions in rangeland resource tenure and policy. *In:* Scoones, I. ed. *Living with uncertainty: new directions in pastoral development in Africa.* Intermediate Technology Publications, London, 96-133.

Langer, P., 1988. *Mammalian herbivore stomach: comparative anatomy, function and evolution.* Fischer, Stuttgart.

Langvatn, R. and Hanley, T.A., 1993. Feeding-patch choice by red deer in relation to foraging efficiency: an experiment. *Oecologia,* 95 (2), 164-170.

Lawler, I.R., Foley, W.J., Eschler, B.M., et al., 1998. Intraspecific variation in Eucalyptus secondary metabolites determines food intake by folivorous marsupials. *Oecologia,* 116 (1/2), 160-169.

Laws, R.M., 1970. Elephants as agents of habitat and landscape change in East Afica. *Oikos,* 21 (1), 1-15.

Lawson, A., Biggeri, A. and Boehning, D. , 1999. *Disease mapping and risk assessment for public health.* Wiley, Chichester.

Lazo, A., 1994. Social segregation and the maintenance of social stability in a feral cattle population. *Animal Behaviour,* 48 (5), 1133-1141.

Le Houerou, H.N., 1989. *The grazing land ecosystems of the African Sahel.* Springer, Berlin.

Ledger, H.P., 1968. Body composition as a basis for a comparative study of some East African mammals. *In:* Crawford, M.A. ed. *Comparative nutrition of wild animals: proceedings of a symposium held at the Zoological Society of London, on 10 and 11 November 1966.* Academic Press, London, 289-310.

LeDoux, J.E., 1992. Brain mechanisms of emotion and emotional learning. *Current Opinion in Neurobiology,* 2 (2), 191-197.

LeDoux, J.E., 1994. Emotion, memory and the brain. *Scientific American,* 270 (6), 50-57.

Legendre, P. and Fortin, M.J., 1989. Spatial pattern and ecological analysis. *Vegetatio,* 80 (2), 107-138.

Leuthold, B.M. and Leuthold, W., 1978. Ecology of the giraffe in Tsavo East National Park, Kenya. *East African Wildlife Journal,* 16 (1), 1-20.

Levin, S.A., 1992. The problem of pattern and scale in ecology: the Robert H. MacArthur Award Lecture. *Ecology,* 73 (6), 1943-1967.

Levins, R., 1968. *Evolution in changing environments: some theoretical explorations.* Princeton University Press, Princeton.

Lewis, J.E., Rowland, J. and Nadeau, A., 1998. Estimating maize production in Kenya using NDVI: some statistical considerations. *International Journal of Remote Sensing,* 19 (13), 2609-2617.

Lewis, J.G., 1977. Game domestication for animal production in Kenya: activity patterns of eland, oryx, buffalo and zebu cattle. *Journal of Agricultural Science, UK,* 89 (3), 551-563.

Lewis, M.M., 1994. Species composition related to spectral classification in an Australian spinifex hummock grassland. *International Journal of Remote Sensing,* 15 (16), 3223-3239.

Li, B.L., Gorshkov, V.G. and Makarieva, A.M., 2004. Energy partitioning between different-sized organisms and ecosystem stability. *Ecology,* 85 (7), 1811-1813.

Li, W., Wang, Z., Ma, Z., et al., 1997. A regression model for the spatial distribution of red-crown crane in Yancheng Biosphere Reserve, China. *Ecological Modelling,* 103 (2), 115-121.

Li, W.K., 2002. Macroecological patterns of phytoplankton in the northwestern North Atlantic Ocean. *Nature,* 419 (6903), 154-157.

Lima, S.L., 2002. Putting predators back into behavioral predator-prey interactions. *Trends in Ecology and Evolution,* 17 (2), 70-75.

Lincoln, D.E., Newton, T.S., Ehrlich, P.R., et al., 1982. Coevolution of the checkerspot butterfly *Euphydryas chalcedona* and its larval food plant *Diplacus aurantiacus*: larval response to protein and leaf resin. *Oecologia,* 52 (2), 216-223.

Lind, J. and Sheikh, N., 2001. Armaments, environments: small arms and control of natural resources. *Eco-Conflicts,* 1 (3), 1-4. [http://www.acts.or.ke/pubs/policybriefs/pubs/ecopolicy%20brief2.pdf]

Lindsay, R.W., Percival, D.B. and Rothrock, D.A., 1996. The discrete wavelet transform and the scale analysis of the surface properties of sea ice. *IEEE Transactions on Geoscience and Remote Sensing,* 34 (3), 771-787.

Liu, X., 2001. *Mapping and modelling the habitat of giant pandas in Foping Nature Reserve, China.* Wageningen University, Wageningen. PhD thesis

Liu, X., Skidmore, A.K., Wang, T., et al., 2002. Giant Panda movements in Foping Nature Reserve, China. *The Journal of Wildlife Management,* 66 (4), 1179-1188.

Lloyd, L.E., MacDonald, B.E. and Crampton, E.W., 1978. *Fundamentals of nutrition.* Freeman, San Francisco.

Loeb, S.C. and Hooper, R.G., 1997. An experimental test of interspecific competition for red-cockaded woodpecker cavities. *Journal of Wildlife Management,* 61 (4), 1268-1280.

Los, S.O., 1998. *Linkages between global vegetation and climate: an analysis based on NOAA-Advanced Very High Resolution Radiometer data.* NASA. NASA Report no. GSFC/CR-1998-206852.

Loth, P.E. and Prins, H.H.T., 1986. Spatial patterns of the landscape and vegetation of the Lake Manyara National Park. *ITC Journal,* 2, 115-130.

Low, W.A., Tweedie, R.L., Edwards, C.B.H., et al., 1981. The influence of environment on daily maintenance behaviour of free-ranging Shorthorn cows in central Australia. I. General introduction and descriptive analysis of day-long activities. *Applied Animal Ethology,* 7 (1), 11-26.

Ludwig, F., 2001. *Tree-grass interactions on an East African savanna: the effects of facilitation, competition, and hydraulic lift.* Wageningen University, Wageningen. PhD thesis

Ludwig, F., De Kroon, H., Prins, H.H.T., et al., 2001. Effects of nutrients and shade on tree-grass interactions in an East African savanna. *Journal of Vegetation Science,* 12 (4), 579-588.

M'Closkey, R.T., 1980. Spatial patterns in sizes of seeds collected by four species of heteromyid rodents. *Ecology,* 61 (3), 486-489.

Macandza, V.A., Owen-Smith, N. and Cross, P.C., 2004. Forage selection by African buffalo (*Syncerus caffer*) through the dry season in two landscapes of the Kruger National Park. *South African Journal of Wildlife Research,* 34, 113-121.

MacArthur, R.H., 1958. Population ecology of some warblers of northeastern coniferous forests. *Ecology,* 39 (4), 599-619.

MacFadden, B.J. and Shockey, B.J., 1997. Ancient feeding ecology and niche differentiation of Pleistocene mammalian herbivores from Tarija, Bolivia: morphological and isotopic evidence. *Paleobiology,* 23 (1), 77-100.

Macfadyan, W.A., 1950. Soil and vegetation in British Somaliland. *Nature,* 165 (4186), 121.

Makkar, H.P.S., 2003. Effects and fate of tannins in ruminant animals, adaptation to tannins, and strategies to overcome detrimental effects of feeding tannin-rich feeds. *Small Ruminant Research,* 49 (3), 241-256.

Mangan, J.L., 1988. Nutritional effects of tannins in animal feeds. *Nutrition Research Reviews,* 1 (1), 209-231.

Mangel, M. and Clark, C.W., 1988. *Dynamic modelling in behavioral ecology.* Princeton University Press, Princeton.

Manly, B.F.J., 1999. *Randomization, bootstrap and Monte Carlo methods in biology.* 2nd edn. Chapman & Hall, London.

Manseau, M., 1996. *Relation réciproque entre les caribous et la végétation des aires d'estivage: le cas du troupeau de caribous de la rivière George.* Universite Laval, Laval. PhD thesis

Marell, A., Ball, J.P. and Hofgaard, A., 2002. Foraging and movement paths of female reindeer: insights from fractal analysis, correlated random walks, and Levy flights. *Canadian Journal of Zoology,* 80 (5), 854-865.

Markus, S., Bailey, D.W., Jensen, D., et al., 1998. A preliminary evaluation of a fenceless livestock control system. *Journal of Animal Science,* 76 (Suppl. 1), 103.

Marotti, D.M., Cosgrove, G.P., Chapman, D.F., et al., 2002. Growing grass and clover separately allows animals to sustain a high nutrient intake. *Proceedings of the New Zealand Society of Animal Production,* 62, 163-166.

Marschner, H., 1995. *Mineral nutrition of higher plants.* 2nd edn. Academic Press, London.

Martin, M.E., Newman, S.D., Aber, J.D., et al., 1998. Determining forest species composition using high spectral resolution remote sensing data. *Remote Sensing of Environment,* 65 (3), 249-254.

Martin, P.S., 1982. The pattern and meaning of holarctic mammoth extincion. *In:* Hopkins, D.M., Matthews, J.V., Schweger, C.S., et al. eds. *The paleoecology of Beringia.* Academic Press, New York, 398-408.

Martin, P.S. and Wright, H.E., 1967. *Pleistocene extinctions: the search for a cause*. Yale University Press, New Haven.

Mason, S.J., Waylen, P.R., Mimmack, G.M., et al., 1999. Changes in extreme rainfall events in South Africa. *Climatic Change,* 41 (2), 249-257.

Matson, K.D., Millam, J.R. and Klasing, K.C., 2004. Cockatiels (*Nymphicus hollandicus*) reject very low levels of plant secondary compounds. *Applied Animal Behaviour Science,* 85 (1/2), 141-156.

Mautz, W.W., Kanter, J. and Pekins, P.J., 1992. Seasonal metabolic rhythms of captive female white-tailed deer: a reexamination. *The Journal of Wildlife Management,* 56 (4), 656-661.

Mbogo, S., 2003. Pastoral shoot-outs: breaking the link between small arms and environmental conflict. *Toward Freedom* (51). [http://towardfreedom.com/home/content/view/106/0/]

McCabe, J.T., 1994. Mobility and land use among African pastoralists: old conceptual problems and new interpretations. *In:* Fratkin, E., Galvin, K.A. and Roth, E.A. eds. *African pastoral systems: an integrated approach.* Lynne Riener Publishers, Boulder, 69-89.

McComb, K., Moss, C., Durant, S.M., et al., 2001. Matriarchs as repositories of social knowledge in African elephants. *Science* (292), 491-494.

McIlvain, E.H. and Shoop, M.C., 1971. Shade for improving cattle gains and rangeland use. *Journal of Range Management,* 24 (3), 181-184.

McNair, J.N., 1982. Optimal giving-up times and the marginal value theorem. *The American Naturalist,* 119 (4), 511-529.

McNamara, J.M., 1987. Starvation and predation as factors limiting population size. *Ecology,* 68 (5), 1515-1519.

McNamara, J.M. and Houston, A.I., 1987. Partial preferences and foraging. *Animal behaviour,* 35 (4), 1084-1099.

McNaughton, S.J., 1976. Serengeti migratory wildebeest: facilitation of energy flow by grazing. *Science, USA,* 191 (4222), 92-94.

McNaughton, S.J., 1984. Grazing lawns: animals in herds, plant form, and coevolution. *The American Naturalist,* 124 (6), 863-886.

McNaughton, S.J., 1985. Ecology of a grazing ecosystem: the Serengeti. *Ecological Monographs,* 55 (3), 259-294.

McNaughton, S.J., 1990. Mineral nutrition and seasonal movements of African migratory ungulates. *Nature London,* 345 (6276), 613-615.

McNaughton, S.J., 1995. Plant communities and herbivory. *In:* Sinclair, A.R.E. and Arcese, P. eds. *Serengeti 2: dynamics, management, and conservation of an ecosystem: based on papers at a workshop held in December 1991 at the Serengeti Research Institute, Tanzania.* University of Chicago Press, Chicago.

McSweeney, C.S., Palmer, B., McNeill, D.M., et al., 2001. Microbial interactions with tannins: nutritional consequences for ruminants. *Animal Feed Science and Technology,* 91 (1), 83-93.

Mearns, R. and Swift, J., 1996. Pasture tenure and management in the retreat from a centrally planned economy in Mongolia. *In:* West, N.E. ed. *Rangelands in a sustainable biosphere: proceedings of the fifth international rangeland congress, held in Salt Lake City, Utah, July 23-28, 1995.* Society for Range Management, Denver, 96-98.

Meissner, H.H. and Paulsmeier, D.V., 1995. Plant compositional constituents affecting between-plant and animal species prediction of forage intake. *Journal of Animal Science,* 73 (8), 2447-2457.

Mendl, M. and Held, S., 2001. Living in groups: an evolutionary approach. *In:* Keeling, L.J. and Gonyou, H.W. eds. *Social behaviour in farm animals.* Cabi, Wallingford, 7-31.

Mendoza, M., Janis, C.M. and Palmqvist, P., 2002. Characterizing complex craniodental patterns related to feeding behaviour in ungulates: a multivariate approach. *Journal of Zoology,* 258 (2), 223-246.

Mills, M.G.L., 1984. Prey selection and feeding habits of the large carnivores in the southern Kalahari. *Koedoe,* 27 (Suppl.), 281-294.

Milne, B.T., Turner, M.G., Wiens, J.A., et al., 1992. Interactions between the fractal geometry of landscapes and allometric herbivory. *Theoretical Population Biology,* 41 (3), 337-353.

Milne, J.A., Hodgson, J., Thompson, R., et al., 1982. The diet ingested by sheep grazing swards differing in white clover and perennial ryegrass content. *Grass and Forage Science,* 37 (3), 209-218.

Mishra, C., Prins, H.H.T. and Van Wieren, S.E., 2001. Overstocking in the trans-Himalayan rangelands of India. *Environmental Conservation,* 28 (3), 279-283.

Mishra, C., Prins, H.H.T. and Van Wieren, S.E., 2003. Diversity, risk mediation, and change in a trans-Himalayan agropastoral system. *Human Ecology,* 31 (4), 595-609.

Mishra, C., Van Wieren, S.E., Ketner, P., et al., 2004. Competition between domestic livestock and wild bharal *Pseudois nayaur* in the Indian Trans-Himalaya. *Journal of Applied Ecology,* 41 (2), 344-354.

Moen, A.N., 1973. *Wildlife ecology: an analytical approach.* Freeman, San Francisco.

Moen, R., 1998. Linking moose population and plant growth models with a moose energetics model. *Ecosystems,* 1 (1), 52-63.

Moen, R., Pastor, J. and Cohen, Y., 1997. A spatially explicit model of moose foraging and energetics. *Ecology,* 78 (2), 505-521.

Mole, S., 1989. Polyphenolics and the nutritional ecology of herbivores. *In:* Cheeke, P.R. ed. *Toxicants of plant origin. IV, Phenolics.* CRC Press, Boca Raton, 191–221.

Monteith, J.L., 1972. Solar radiation and productivity in tropical ecosystems. *Journal of Applied Ecology,* 9 (3), 747-766.

Monteith, J.L., 1977. Climate and the efficiency of crop production in Britain. *Philosophical Transactions of the Royal Society of London, B,* 281, 277-294.

Mooney, H.A., 1986. Photosynthesis. *In:* Crawley, M.J. ed. *Plant ecology.* Blackwell, Oxford, 345-373.

Moore, D.S., 2002. *The dependent gene: the fallacy of nature vs. nurture.* Henry Holt and Company, New York.

Moreau, S., Bosseno, R., Gu, X.F., et al., 2003. Assessing the biomass dynamics of Andean bofedal and totora high-protein wetland grasses from NOAA/AVHRR. *Remote Sensing of Environment,* 85 (4), 516-529.

Morin, P.J., 2000. Biodiversity's ups and downs. *Nature,* 406 (6795), 508-512.

Moritz, M., Scholte, P. and Kari, S., 2002. The demise of the nomadic contract: arrangements and rangelands under pressure in the far north of Cameroon (1). *Nomadic Peoples,* 6 (1), 124-144.

Morris, D.W., Fox, B.J., Luo, J., et al., 2000. Habitat-dependent competition and the coexistence of Australian heathland rodents. *Oikos,* 91 (2), 294-306.

Morrison, M.L. and Hall, L.S., 2002. Standard terminology: toward a common language to advance ecological understanding and application. *In:* Scott, J.M. ed. *Predicting species occurrences: issues of accuracy and scale.* Island Press, Washington.

Morse, D.R., Lawton, J.H., Dodson, M.M., et al., 1985. Fractal dimension of vegetation and the distribution of arthropod body lengths. *Nature, UK,* 314 (6013), 731-733.

Mott MacDonald, 1999. *Logone Floodplain model study report.* Mott MacDonald Group, Cambridge.

Mueggler, W.F., 1965. Cattle distribution on steep slopes. *Journal of Range Management,* 18 (5), 255-257.

Murray, M.G., 1995. Specific nutrient requirements and migration of wildebeest. *In:* Sinclair, A.R.E. and Arcese, P. eds. *Serengeti 2: dynamics, management, and conservation of an ecosystem: based on papers at a workshop held in December 1991 at the Serengeti Research Institute, Tanzania.* University of Chicago Press, Chicago, 231-256.

Murray, M.G., 2003. Overkill and sustainable use. *Science,* 299 (5614), 1851 - 1853.

Murray, M.G. and Brown, D., 1993. Niche separation of grazing ungulates in the Serengeti: an experimental test. *The Journal of Animal Ecology,* 62 (2), 380-389.

Murwira, A., 2003. *Scale matters!: a new approach to quantify spatial heterogeneity for predicting the distribution of wildlife.* Wageningen University, Wageningen. PhD thesis [http://www.library.wur.nl/wda/dissertations/dis3506.pdf]

Mutanga, O. and Skidmore, A.K., 2004a. Integrating imaging spectroscopy and neural networks to map grass quality in the Kruger National Park, South Africa. *Remote Sensing of Environment,* 90 (1), 104-115.

Mutanga, O. and Skidmore, A.K., 2004b. Narrow band vegetation indices overcome the saturation problem in biomass estimation. *International Journal of Remote Sensing,* 25 (19), 3999-4014.

Mutanga, O., Skidmore, A.K. and Prins, H.H.T., 2004a. Discriminating sodium concentration in a mixed grass species environment of the Kruger National Park using field spectrometry. *International Journal of Remote Sensing,* 25 (20), 4191-4201.

Mutanga, O., Skidmore, A.K. and Prins, H.H.T., 2004b. Predicting in situ pasture quality in the Kruger National Park, South Africa, using continuum-removed absorption features. *Remote Sensing of Environment,* 89 (3), 393-408.

Mwasi, S.M., 2002. *Compressed nature: co-existing grazers in a small reserve in Kenya.* Wageningen University, Wageningen. PhD thesis

Mysterud, A., Lian, L.B. and Hjermann, D.O., 1999. Scale-dependent trade-offs in foraging by European roe deer (*Capreolus capreolus*) during winter. *Canadian Journal of Zoology,* 77 (9), 1486-1493.

Mysterud, A., Yoccoz, N.G., Stenseth, N.C., et al., 2001. Effects of age, sex and density on body weight of Norwegian red deer: evidence of density-dependent senescence. *Proceedings of the Royal Society of London Series B, Biological Sciences,* 268 (1470), 911-919.

Naah, E., 1992. *Hydrologie du Grand Yaéré du nord-Cameroun.* Université de Yaoundé, Yaoundé. PhD thesis

Nagy, K.A., Girard, I.A. and Brown, T.K., 1999. Energetics of free-ranging mammals, reptiles, and birds. *Annual Review of Nutrition,* 19, 247-277.

NCAA, 2000. *1998 aerial boma count, 1999 people and livestock census, and human population trend between 1954 and 1999 in the NCA.* Ngorongoro Conservation Area Authority, Ngorongoro Crater.

Nellemann, C., 1997. Grazing strategies of muskoxen (*Ovibos moschatus*) during winter in Angujaartorfiup Nunaa in western Greenland. *Canadian Journal of Zoology,* 75 (7), 1129-1134.

Newman, J.A., Parsons, A.J. and Harvey, A., 1992. Not all sheep prefer clover: diet selection revisited. *Journal of Agricultural Science,* 119 (2), 275-283.

Newman, J.A., Parsons, A.J., Thornley, J.H.M., et al., 1995. Optimal diet selection by a generalist grazing herbivore. *Functional Ecology,* 9 (2), 255-268.

Niamir-Fuller, M., 1999. *Managing mobility in African Rangelands: the legitimization of transhumance.* Intermediate Technology Publications, London.

Niamir-Fuller, M. and Turner, M.D., 1999. A review of recent literature on pastoralism and transhumance in Africa. *In:* Niamir-Fuller, M. ed. *Managing mobility in African Rangelands: the legitimization of transhumance.* Intermediate Technology Publications, London, 18-46.

Northup, R.R., Yu, Z.S., Dahlgren, R.A., et al., 1995. Polyphenol control of nitrogen release from pine litter. *Nature London,* 377 (6546), 227-229.

Norton-Griffiths, M., 1975. *Counting animals.* Serengeti Ecological Monitoring Programme, Nairobi.

Novellie, P.A., 1978. Comparison of the foraging strategies of blesbok and springbok on the Transvaal highveld. *South African Journal of Wildlife Research,* 8, 137–144.

Novellie, P.A., Manson, J. and Bigalke, R.C., 1984. Behavioural ecology and communication in the Cape grysbok. *South African Journal of Zoology,* 19 (1), 22-30.

O'Neill, R.V., Gardner, R.H., Milne, B.T., et al., 1991. Heterogeneity and spatial hierarchies. *In:* Kolasa, J., Pickett, S.T.A., Allen, T.F.H., et al. eds. *Ecological heterogeneity.* Springer, New York, 85-96.

O'Reagain, P.J., Goetsch, B.C. and Owen-Smith, R.N., 1996. Effect of species composition and sward structure on the ingestive behaviour of cattle and sheep grazing South African sourveld. *Journal of Agricultural Science,* 127 (2), 271-280.

Oates, J.F., Waterman, P.G. and Choo, G.M., 1980. Food selection by the South Indian leaf-monkey, *Presbytis johnii,* in relation to leaf chemistry. *Oecologia,* 45 (1), 45-56.

Oesterheld, M., DiBella, C.M. and Kerdiles, H., 1998. Relation between NOAA-AVHRR satellite data and stocking rate of rangelands. *Ecological Applications,* 8 (1), 207-212.

Oesterheld, M., Sala, O.E. and McNaughton, S.J., 1992. Effect of animal husbandry on herbivore-carrying capacity at a regional scale. *Nature,* 356 (6366), 234-236.

Oftedal, O.T., 1984. Milk composition, milk yield and energy output at peak lactation: a comparative review. *Symposia of the Zoological Society of London,* 51 (Physiological strategies in lactation), 33-85.

Oftedal, O.T., 1985. Pregnancy and lactation. *In:* Hudson, R.J. and White, R.G. eds. *Bioenergetics of wild herbivores.* CRC, Boca Raton, 215-238.

Ogden, R.T., 1997. *Essential wavelets for statistical applications and data analysis.* Birkhaeuser, Boston.

Ogura, S., Hasegawa, H. and Hirata, M., 2002. Effects of herbage mass and herbage quality on spatially heterogeneous grazing by cattle in a bahia grass (*Paspalum notatum*) pasture. *Tropical Grasslands,* 36 (3), 172-179.

Oindo, B.O., 2001. *Spatial patterns of species diversity in Kenya.* Wageningen University, Wageningen. PhD thesis

Oindo, B.O., 2002. Predicting mammal species richness and abundance using multi-temporal NDVI. *PEandRS, Photogrammetric Engineering and Remote Sensing,* 68 (6), 623-629.

Oindo, B.O., De By, R.A. and Skidmore, A.K., 2000. Interannual variability of NDVI and bird species diversity in Kenya. *International Journal of Applied Earth Observation and Geoinformation,* 2 (3), 172-180.

Oindo, B.O., Skidmore, A.K. and Prins, H.H.T., 2001. Body size and abundance relationship: an index of diversity for herbivores. *Biodiversity and Conservation,* 10 (11), 1923-1931.

Ole Katampoi, K., Genga, G.O., Mwangi, M., et al., 1990. *Kajiado District atlas.* Arid and Semi Arid Lands (ASAL) Programme, Kajiado.

Olff, H., De Leeuw, J., Bakker, J.P., et al., 1997. Vegetation succession and herbivory in a salt marsh: changes induced by sea level rise and silt deposition along an elevational gradient. *Journal of Ecology Oxford,* 85 (6), 799-814.

Olff, H., Ritchie, M.E. and Prins, H.H.T., 2002. Global environmental determinants of diversity in large herbivores. *Nature,* 415, 901-905.

Olivry, J.C., Chouret, A., Vuillaume, G., et al., 1996. *Hydrologie du Lac Tchad.* ORSTOM editions, Paris.

Olton, D.S., 1978. Characteristics of spatial memory. *In:* Hulse, S.H., Fowler, H. and Honig, W.K. eds. *Cognitive processes in animal behavior*. Erlbaum, Hillsdale, 341-373.

Orlove, B.S., 1977. Integration through production: the use of zonation in Espinar. *American Ethnologist,* 4 (1), 84-101.

Ottichilo, W.K., 2000. *Wildlife dynamics: an analysis of change in the Masai Mara ecosystem of Kenya.* Wageningen University, Wageningen. PhD thesis

Ottichilo, W.K., De Leeuw, J. and Prins, H.H.T., 2001. Population trends of resident wildebeest [*Connochaetes taurinus hecki* (Neumann)] and factors influencing them in the Masai Mara ecosystem, Kenya. *Biological Conservation,* 97 (3), 271-282.

Ottichilo, W.K., De Leeuw, J., Skidmore, A.K., et al., 2000a. Population trends of large non-migratory wild herbivores and livestock in the Masai Mara ecosystem, Kenya, between 1977 and 1997. *African Journal of Ecology,* 38 (3), 202-216.

Ottichilo, W.K., Grunblatt, J., Said, M.Y., et al., 2000b. Wildlife and livestock population trends in the Kenya rangeland. *In:* Prins, H.H.T., Grootenhuis, J.G. and Dolan, T.T. eds. *Wildlife conservation by sustainable use*. Kluwer Academic Publishers, Boston, 203-218.

Owen-Smith, N., 1975. The social ethology of the white rhinoceros. *Zeitschrift für Tierpsychologie,* 38, 337-384.

Owen-Smith, N., 1982. Factors influencing the consumption of plant products by large herbivores. *In:* Huntley, B.J. and Walker, B.H. eds. *The ecology of tropical savannas*. Springer-Verlag, Berlin, 359-404.

Owen-Smith, N., 1989. Morphological factors and their consequences for resource partitioning among African savanna ungulates: a simulation modelling approach. *In:* Morris, D.W., Abramsky, Z., Fox, B.J., et al. eds. *Patterns in the structure of mammalian communities*. Texas Tech University Press, Lubbock, 155-165.

Owen-Smith, N., 1993. Evaluating optimal diet models for an African browsing ruminant, the kudu: how constraining are the assumed constraints? *Evolutionary Ecology,* 7 (5), 499-524.

Owen-Smith, N., 1994. Foraging responses of kudus to seasonal changes in food resources: elasticity in constraints. *Ecology,* 75 (4), 1050-1062.

Owen-Smith, N., 1996. Circularity in linear programming models of optimal diet. *Oecologia,* 108 (2), 259-261.

Owen-Smith, N., 1998. How high ambient temperature affects the daily activity and foraging time of a subtropical ungulate, the greater kudu (*Tragelaphus strepsiceros*). *Journal of Zoology,* 246 (02), 183-192.

Owen-Smith, N., 2000. Modeling the population dynamics of a subtropical ungulate in a variable environment: rain, cold and predators. *Natural Resource Modeling,* 13, 57-87.

Owen-Smith, N., 2002a. *Adaptive herbivore ecology: from resources to populations in variable environments*. Cambridge University Press, Cambridge.

Owen-Smith, N., 2002b. Credible models for herbivore-vegetation systems: towards an ecology of equations. *South African Journal of Science,* 98 (9/10), 445-449.

Owen-Smith, N., 2002c. A metaphysiological modelling approach to stability in herbivore-vegetation systems. *Ecological Modelling,* 149 (1), 153-178.

Owen-Smith, N., 2004. Functional heterogeneity in resources within landscapes and herbivore population dynamics. *Landscape Ecology,* 19 (7), 761-771.

Owen-Smith, N. and Cooper, S.M., 1987. Palatability of woody plants to browsing ruminants in a South African savanna. *Ecology, USA,* 68 (2), 319-331.

Owen-Smith, N. and Cooper, S.M., 1989. Nutritional ecology of a browsing ruminant, the kudu (*Tragelaphus strepsiceros*), through the seasonal cycle. *Journal of Zoology,* 219 (1), 29-43.

Owen-Smith, N. and Cumming, D.H.M., 1993. Comparative foraging strategies of grazing ungulates in African savanna grasslands. *In:* Baker, M. ed. *Grasslands for our world: proceedings of the 17th International Grassland Congress* SIR Publishing, 691-697.

Owen-Smith, N. and Danckwerts, J.E., 1997. Herbivory. *In:* Cowling, R.M., Richardson, D.M. and Pierce, S.M. eds. *Vegetation of Southern Africa*. Cambridge University Press, Cambridge.

Owen-Smith, N. and Novellie, P., 1982. What should a clever ungulate eat? *The American Naturalist,* 119 (2), 151-178.

Owen-Smith, R.N., 1985. Niche separation among African ungulates. *In:* Vrba, E.S. ed. *Species and speciation*. Transvaal Museum, Pretoria, 161-171. Transvaal Museum Monograph No. 4.

Owen-Smith, R.N., 1988. *Megaherbivores: the influence of very large body size on ecology*. Cambridge University Press, Cambridge.

Owens, M.K., Launchbaugh, K.L. and Holloway, J.W., 1991. Pasture characteristics affecting spatial distribution of utilization by cattle in mixed brush communities. *Journal of Range Management,* 44 (2), 118-123.

Oxby, C., 1982. *Group ranches in Africa.* Overseas Development Institute, London. Pastoral Network Paper Agricultural Administration Unit, Overseas Development Institute.

Paine, R.T., 1969. A note on trophic complexity and community stability. *The American Naturalist,* 103 (929), 91-93.

Palmer, M.W., 1992. The coexistence of species in fractal landscapes. *The American Naturalist,* 139 (2), 375-397.

Parker, I.S.C., 1983. The Tsavo story: an ecological case history. *In:* Owen-Smith, R.N. ed. *Management of large mammals in African conservation areas: proceedings of a symposium, held in Pretoria, South Africa, 29-30 April 1982.* Haum, Pretoria, 37-50.

Parker, K.L., 1988. Effects of heat, cold, and rain on coastal black-tailed deer. *Canadian Journal of Zoology,* 66 (11), 2475-2483.

Parker, K.L., Gillingham, M.P., Hanley, T.A., et al., 1993. Seasonal patterns in body mass, body composition, and water transfer rates of free-ranging and captive black-tailed deer (*Odocoileus hemionus sitkensis*) in Alaska. *Canadian Journal of Zoology,* 71, 1397-1404.

Parker, K.L., Gillingham, M.P., Hanley, T.A., et al., 1999. *Energy and protein balance of free-ranging black-tailed deer in a natural forest environment.* Wildlife Monographs no. 143.

Parlar, M. and Weng, Z.K., 2003. Balancing desirable but conflicting objectives in the newsvendor problem. *IIE Transactions,* 35 (2), 131-142.

Parsons, A.J., Newman, J.A., Penning, P.D., et al., 1994a. Diet preference of sheep: effects of recent diet, physiological state and species abundance. *Journal of Animal Ecology,* 63 (2), 465-478.

Parsons, A.J., Thornley, J.H.M., Newman, J., et al., 1994b. A mechanistic model of some physical determinants of intake rate and diet selection in a two-species temperate grassland sward. *Functional Ecology,* 8 (2), 187-204.

Paruelo, J.M., Epstein, H.E., Lauenroth, W.K., et al., 1997. ANPP estimates from NDVI for the central grassland region of the United States. *Ecology,* 78 (3), 953-958.

Perez, C., Ackroff, K. and Sclafani, A., 1996. Carbohydrate-and protein-conditioned flavor preferences: effects of nutrient preloads. *Physiology and Behavior,* 59 (3), 467-474.

Peters, R.H., 1983. *The ecological implications of body size.* Cambridge University Press, Cambridge.

Petren, K. and Case, T.J., 1996. An experimental demonstration of exploitation competition in an ongoing invasion. *Ecology,* 77 (1), 118-132.

Petren, K. and Case, T.J., 1998. Habitat structure determines competition intensity and invasion success in gecko lizards. *Proceedings National Academy of Sciences USA,* 95 (20), 11739-11744.

Pfister, J.A., Provenza, F.D., Manners, G.D., et al., 1997. Tall larkspur ingestion: can cattle regulate intake below toxic levels? *Journal of Chemical Ecology,* 23 (3), 759-777.

Pickup, G. and Smith, D.M.S., 1993. Problems, prospects and procedures for assessing the sustainability of pastoral land management in arid Australia. *Journal of Biogeography,* 20 (5), 471-487.

Pielou, E.C., 1984. *The interpretation of ecological data: a primer on classification and ordination.* Wiley, New York.

Pinchak, W.E., Smith, M.A., Hart, R.H., et al., 1991. Beef cattle distribution patterns on foothill range. *Journal of Range Management,* 41 (3), 267-275.

Pitt, W.C. and Ritchie, M.E., 2002. Influence of prey distribution on the functional response of lizards. *Oikos,* 96 (1), 157-163.

Pollard, J.C. and Littlejohn, R.P., 1996. The effects of pen size on the behaviour of farmed red deer stags confined in yards. *Applied Animal Behaviour Science,* 47 (3/4), 247-253.

Price, E.O., 1978. Genotype versus experience effects on aggression in wild and domestic Norway rats. *Behaviour* 64 (3/4), 340-353.

Price, E.O., 1984. Behavioral aspects of animal domestication. *Quarterly Review of Biology,* 59 (1), 1-32.

Prins, H.H.T., 1987a. *The buffalo of Manyara: the individual in the context of herd life in a seasonal environment of East Africa.* University of Groningen, Groningen. PhD thesis

Prins, H.H.T., 1987b. Nature conservation as an integral part of optimal land use in East Africa: the case of the Masai Ecosystem of northern Tanzania. *Biological Conservation,* 40 (2), 141-161.

Prins, H.H.T., 1988. Plant phenology patterns in Lake Manyara National Park, Tanzania. *Journal of Biogeography,* 15 (3), 465-480.

Prins, H.H.T., 1989a. Buffalo herd structure and its repercussions for condition of individual African buffalo cows. *Ethology,* 81 (1), 47-71.

Prins, H.H.T., 1989b. Condition changes and choice of social environment in African buffalo bulls. *Behaviour,* 108 (3), 297-324.

Prins, H.H.T., 1992. The pastoral road to extinction: competition between wildlife and traditional pastoralism in East Africa. *Environmental Conservation,* 19 (2), 117-123.

Prins, H.H.T., 1996. *Ecology and behaviour of the African buffalo: social inequality and decision making.* Chapman & Hall, London.

Prins, H.H.T., 1998. The origins of grassland communities in northwestern Europe. *In:* Wallis De Vries, M.F., Bakker, J.P. and Van Wieren, S.E. eds. *Grazing and conservation management.* Kluwer, Dordrecht, 55-105.

Prins, H.H.T., 1999. Comment on Ruttan and Borgerhoff-Mulder. *Current Anthropology,* 40, 643–645.

Prins, H.H.T., 2000. Competition between wildlife and livestock. *In:* Prins, H.H.T., Grootenhuis, J.G. and Dolan, T.T. eds. *Wildlife conservation by sustainable use.* Kluwer Academic Publishers, Boston, 51-80.

Prins, H.H.T. and Beekman, J.H., 1989. A balanced diet as a goal for grazing: the food of the Manyara buffalo. *African Journal of Ecology,* 27 (3), 241-259.

Prins, H.H.T. and Douglas-Hamilton, I., 1990. Stability in a multi-species assemblage of large herbivores in East Africa. *Oecologia,* 83 (3), 392-400.

Prins, H.H.T., Grootenhuis, J.G. and Dolan, T.T., 2000. *Wildlife conservation by sustainable use.* Kluwer Academic Publishers, Boston.

Prins, H.H.T. and Iason, G.R., 1989. Dangerous lions and nonchalant buffalo. *Behaviour,* 108 (3), 262-293.

Prins, H.H.T. and Loth, P.E., 1988. Rainfall patterns as background to plant phenology in northern Tanzania. *Journal of Biogeography,* 15 (3), 451-463.

Prins, H.H.T. and Olff, H., 1998. Species-richness of African grazer assemblages: towards a functional explanation. *In:* Newbery, D.M., Prins, H.H.T. and Brown, N.D. eds. *Dynamics of tropical communities: the 37th symposium of the British Ecological Society, Cambridge University, 1996.* Blackwell Science, Oxford, 449-490.

Provenza, F.D., 1995. Postingestive feedback as an elementary determinant of food preference and intake in ruminants. *Journal of Range Management,* 48 (1), 2-17.

Provenza, F.D., 1996. Acquired aversions as the basis for varied diets of ruminants foraging on rangelands. *Journal of Animal Science,* 74 (8), 2010-2020.

Provenza, F.D., 2003. Twenty-five years of paradox in plant-herbivore interactions and "sustainable" grazing management. *Rangelands,* 25 (6), 4-15.

Provenza, F.D., Lynch, J.J. and Nolan, J.V., 1994. Food aversion conditioned in anesthetized sheep. *Physiology and Behavior,* 55 (3), 429-432.

Provenza, F.D. and Malechek, J.C., 1984. Diet selection by domestic goats in relation to blackbrush twig chemistry. *Journal of Applied Ecology,* 21, 831-841.

Provenza, F.D., Pfister, J.A. and Cheney, C.D., 1992. Mechanisms of learning in diet selection with reference to phytotoxicosis in herbivores. *Journal of Range Management,* 45 (1), 36-45.

Provenza, F.D., Scott, C.B., Phy, T.S., et al., 1996. Preference of sheep for foods varying in flavors and nutrients. *Journal of Animal Science,* 74, 2355-2361.

Provenza, F.D., Villalba, J.J., Dziba, L.E., et al., 2003. Linking herbivore experience, varied diets, and plant biochemical diversity. *Small Ruminant Research,* 49 (3), 257-274.

Ralphs, M.H. and Provenza, F.D., 1999. Conditioned food aversions: principles and practices, with special reference to social facilitation. *Proceedings of the Nutrition Society,* 58 (4), 813-820.

Ramirez, I., 1997. Intragastric carbohydrate exerts both intake-stimulating and intake-suppressing effects. *Behavioral Neuroscience,* 111 (3), 612-622.

Redbo, I., 1992. The influence of restraint on the occurrence of oral stereotypies in dairy cows. *Applied Animal Behaviour Science,* 35 (2), 115-123.

Redbo, I., Redbo-Torstensson, P., Odberg, F.O., et al., 1998. Factors affecting behavioural disturbances in race-horses. *Animal Science,* 66 (2), 475-481.

Redfern, J.V., Ryan, S.J. and Getz, W.M., 2006. Defining herbivore assemblages in the Kruger National Park: a correlative coherence approach. *Oecologia,* 146 (4), 632-640.

Regelin, W.L., Schwartz, C.C. and Franzmann, A.W., 1985. Seasonal energy metabolism of adult moose. *Journal of Wildlife Management,* 49 (2), 388-393.

Reid, R.L., Jung, G.A. and Thayne, W.V., 1988. Relationships between nutritive quality and fiber components of cool season and warm season forages: a retrospective study. *Journal of Animal Science,* 66 (5), 1275.

Reinhardt, V. and Reinhardt, A., 1981. Cohesive relationships in a cattle herd (*Bos indicus*). *Behaviour,* 77, 121-151.

Renecker, L.A. and Hudson, R.J., 1989. Seasonal activity budgets of moose in aspen-dominated boreal forests. *The Journal of Wildlife Management,* 53 (2), 296-302.

Richardson, A.J., Everitt, J.H. and Gausman, H.W., 1983. Radiometric estimation of biomass and nitrogen content of Alicia grass. *Remote Sensing of Environment,* 13 (2), 179-184.

Ricklefs, R.E. and Schluter, D., 1993. *Species diversity in ecological communities: historical and geographical perspectives.* University of Chicago Press, Chicago.

Ridout, M.S. and Robson, M.J., 1991. Diet composition of sheep grazing grass/white clover swards: a re-evaluation. *New Zealand Journal of Agricultural Research,* 34 (1), 89-93.

Rietkerk, M., Boerlijst, M., Van Langevelde, F., et al., 2002a. Self-organization of vegetation in arid ecosystems. *American Naturalist,* 160 (4), 524-530.

Rietkerk, M., Van de Koppel, J., Kumar, L., et al., 2002b. The ecology of scale. *Ecological Modelling,* 149 (1/2), 1-4.

Ripley, B.D., 1981. *Spatial statistics.* Wiley, New York.

Ripple, W.J. and Beschta, R.L., 2003. Wolf reintroduction, predation risk, and cottonwood recovery in Yellowstone National Park. *Forest Ecology and Management,* 184 (1/3), 299-313.

Ripple, W.J., Larsen, E.J., Renkin, R.A., et al., 2001. Trophic cascades among wolves, elk and aspen on Yellowstone National Park's northern range. *Biological Conservation,* 102 (3), 227-234.

Risenhoover, K.L., 1986. Winter activity patterns of moose in interior Alaska. *Journal of Wildlife Management,* 50 (4), 727-734.

Ritchie, M.E., 2002. Competition and coexistence in mobile animals. *In:* Sommer, U. and Worm, B. eds. *Competition and coexistence.* Springer, Berlin, 127-141.

Ritchie, M.E. and Olff, H., 1999. Spatial scaling laws yield a synthetic theory of biodiversity. *Nature,* 400 (6744), 557-560.

Ritchie, M.E. and Tilman, D., 1993. Predictions of species interactions from consumer-resource theory: experimental tests with grasshoppers and plants. *Oecologia,* 94 (4), 516-527.

Ritchie, M.E., Tilman, D. and Knops, J.M.H., 1998. Herbivore effects on plant and nitrogen dynamics in oak savanna. *Ecology,* 79 (1), 165-177.

Roath, L.R. and Krueger, W.C., 1982. Cattle grazing and behaviour on a forested range. *Journal of Range Management,* 35 (3), 332-338.

Robbins, C.T., 1993. *Wildlife feeding and nutrition.* Academic Press, San Diego.

Robbins, C.T., Mole, S., Hagerman, A.E., et al., 1987. Role of tannins in defending plants against ruminants: reduction in dry matter digestion? *Ecology, USA,* 68 (6), 1606-1615.

Robbins, C.T., Spalinger, D.E. and Van Hoven, W., 1995. Adaptation of ruminants to browse and grass diets: are anatomical-based browser-grazer interpretations valid? *Oecologia,* 103 (2), 208-213.

Roberts, S.C. and Dunbar, R.I.M., 1991. Climatic influences on the behavioural ecology of Chanler's mountain reedbuck in Kenya. *African Journal of Ecology,* 29 (4), 316-329.

Roberts, W.A. and Van Veldhuizen, N., 1985. Spatial memory in pigeons on the radial maze. *Journal of Experimental Psychology. Animal Behavior Processes* 11, 241-260.

Rosenzweig, M.L., 1968. Net primary productivity of terrestrial communities: prediction from climatological data. *The American Naturalist,* 102 (923), 67-74.

Rosenzweig, M.L. and Abramsky, Z., 1993. How are diversity and productivity related? *In:* Ricklefs, R.E. and Schluter, D. eds. *Species diversity in ecological communities: historical and geographical perspectives.* University of Chicago Press, Chicago.

Rosenzweig, M.L. and Abramsky, Z., 1997. Two gerbils of the Negev: a long-term investigation of optimal habitat selection and its consequences. *Evolutionary Ecology,* 11 (6), 733-756.

Ruel, J.J. and Ayres, M.P., 1999. Jensen's inequality predicts effects of environmental variation. *Trends in Ecology and Evolution,* 14 (9), 361-366.

Ruttan, L.M. and Borgerhoff-Mulder, M., 1999. Are East African pastoralists truly conservationists. *Current Anthropology,* 40 (5), 621-652.

Rutten, M.M.E.M., 1992. *Selling wealth to buy poverty: the process of the individualization of landownership among Maasai pastoralists of Kajiado District, Kenya, 1890 - 1990.* Breitenbach, Fort Lauderdale.

Ruyle, G.B., Hasson, O. and Rice, R.W., 1987. The influence of residual stems on biting rates of cattle grazing *Eragrostis lehmanniana* Nees. *Applied Animal Behaviour Science,* 19 (1/2), 11-17.

Said, M.Y., 2003. *Multiscale perspectives of species richness in East Africa.* PhD Thesis, Wageningen University, Wageningen. [http://www.library.wur.nl/wda/dissertations/dis3351.pdf]

Sandford, S., 1994. Improving the efficiency of opportunism: new directions for pastoral development. *In:* Scoones, I. ed. *Living with uncertainty: new directions in pastoral development in Africa.* Intermediate Technology Publications, London, 174-182.

Sankhyan, S.K., Bhatta, R., Shinde, A.K., et al., 2001. Seasonal changes in dietary composition and rumen fermentation pattern of sheep on semi-arid rangeland. *Indian Journal of Animal Sciences,* 71 (2), 162-164.

Savory, A., 1988. *Holistic resource management*. Island Press, Washington.

Schaefer, J.A. and Messier, F., 1995. Habitat selection as a hierarchy: the spatial scales of winter foraging by muskoxen. *Ecography*, 18 (4), 333-344.

Schaller, G.B. and Junrang, R., 1988. Effects of a snowstorm on Tibetan antelope. *Journal of Mammalogy*, 69 (3), 631-634.

Schareika, N., 2001. Environmental knowledge and pastoral migration among the Wodaabe of South-Eastern Niger (1). *Nomadic Peoples*, 5 (1), 65-89.

Schmidt, K.S. and Skidmore, A.K., 2001. Exploring spectral discrimination of grass species in African rangelands. *International Journal of Remote Sensing*, 22 (17), 3421-3434.

Schmidt, K.S. and Skidmore, A.K., 2003. Spectral discrimination of vegetation types in a coastal wetland. *Remote Sensing of Environment*, 85 (1), 92-108.

Schmidt, K.S., Skidmore, A.K., Kloosterman, E.H., et al., 2004. Mapping coastal vegetation using an expert system and hyperspectral imagery. *PEandRS, Photogrammetric Engineering and Remote Sensing*, 70 (6), 703-715.

Schmitz, O.J., 1991. Thermal constraints and optimization of winter feeding and habitat choice in white-tailed deer. *Holarctic Ecology*, 14 (2), 104-111.

Schmitz, O.J., Adler, F.R. and Agrawal, A.A., 2003. Linking individual-scale trait plasticity to community dynamics: trait plasticity and community dynamics. *Ecology*, 84 (5), 1081-1082.

Schmitz, O.J., Cohon, J.L., Rothley, K.D., et al., 1997. Reconciling variability and optimal behaviour using multiple criteria in optimization models. *Evolutionary Ecology*, 12 (1), 73-94.

Schoener, T.W., 1971. Theory of feeding strategies. *Annual Review of Ecology and Systematics*, 2, 369-404.

Schoener, T.W., 1973. Population growth regulated by intraspecific competition for energy or time: some simple representations. *Theoretical Population Biology*, 4 (1), 56-84.

Schoener, T.W., 1976. The species-area relation within archipelagoes: models and evidence from land birds. *In:* Frith, H.J. and Calaby, J.H. eds. *Proceedings of the 16th International Ornithological Congress*. Australian Academy of Sciences, Canberra, 629-642.

Schoener, T.W., 1983. Field experiments on interspecific competition. *The American Naturalist*, 122 (2), 240-285.

Scholes, R.J., 1997. Savanna. *In:* Cowling, R.M., Richardson, D.M. and Pierce, S.M. eds. *Vegetation of Southern Africa*. Cambridge University Press, Cambridge.

Scholes, R.J. and Walker, B.H., 1993. *An African savanna: synthesis of the Nylsvley study*. Cambridge University Press, Cambridge.

Scholte, P., 2005. *Floodplain rehabilitation and the future of conservation & development: adaptive management of success in Waza-Logone, Cameroon*. Wageningen University and Research Centre, Wageningen. Tropical Resource Management Papers no. 67. Ook verschenen als proefschrift Leiden Universiteit, 2005

Scholte, P., Kari, S., Moritz, M., et al., 2006. Pastoralist responses to floodplain rehabilitation in North Cameroon. *Human Ecology*, 34 (1), 27-51.

Scholte, P., Kirda, P., Adam, S., et al., 2000. Floodplain rehabilitation in north Cameroon: impact on vegetation dynamics. *Applied Vegetation Science*, 3 (1), 33-42.

Scholte, P., Pamo, E., Kari, S., et al., 1996. Floodplain rehabilitation in North Cameroon: expected impact on vegetation, pastoralists and wildlife. *In:* West, N.E. ed. *Rangelands in a sustainable biosphere: proceedings of the fifth international rangeland congress*. Society for Range Management, Denver, 492-493.

Scholte, P.T., 1992. Leaf litter and Acacia pods as feed for livestock during the dry season in Acacia-Commiphora bushland, Kenya. *Journal of Arid Environments*, 22 (3), 271-276.

Schrader, T., 1986. *Les Yaérés au Nord du Cameroun: pâturages de saison sèche ?* . CML/Ecole de Faune, Leiden.

Sclafani, A., 2000. Macronutrient-conditioned flavour preferences. *In:* Berthoud, H.J. and Seeley, R.J. eds. *Neural and metabolic control of macronutrient intake*. CRC Press, New York, 93-106.

Scoones, I., 1991. Wetlands in drylands: key resources for agricultural and pastoral production in Africa. *Ambio*, 20 (8), 366-371.

Scoones, I., 1992. Coping with drought: responses of herders and livestock in contrasting Savanna environments in southern Zimbabwe. *Human Ecology*, 20 (3), 293-314.

Scoones, I., 1994. *Living with uncertainty: new directions in pastoral development in Africa*. Intermediate Technology Publications, London.

Scoones, I., 1995. Exploiting heterogeneity: habitat use by cattle in dryland Zimbabwe. *Journal of Arid Environments*, 29 (2), 221-237.

Scott, C.B., Provenza, F.D. and Banner, R.E., 1995. Dietary habits and social interactions affect choice of feeding location by sheep. *Applied Animal Behaviour Science,* 45 (3/4), 225-237.

Scott, L.L. and Provenza, F.D., 1998. Variety of foods and flavors affects selection of foraging location by sheep. *Applied Animal Behaviour Science,* 61 (2), 113-122.

Scott, L.L. and Provenza, F.D., 1999. Variation in food selection among lambs: effects of basal diet and foods offered in a meal. *Journal of Animal Science,* 77 (9), 2391-2397.

Scott, L.L. and Provenza, F.D., 2000. Lambs fed protein or energy imbalanced diets forage in locations and on foods that rectify imbalances. *Applied Animal Behaviour Science,* 68 (4), 293-305.

Sellers, P.J., 1985. Canopy reflectance, photosynthesis and transpiration. *International Journal of Remote Sensing,* 6 (8), 1335-1372.

Sellers, P.J., Los, S.O., Tucker, C.J., et al., 1996. A revised land surface parameterization (SiB2) for atmospheric GCMS. Part II: The generation of global fields of terrestrial biophysical parameters from satellite data. *Journal of Climate,* 9 (4), 706-737.

Sellers, P.J., Tucker, C.J., Collatz, G.J., et al., 1994. A global 1° by 1° NDVI data set for climate studies. Part 2: The generation of global fields of terrestrial biophysical parameters from the NDVI. *International Journal of Remote Sensing,* 15 (17), 3519-3545.

Senft, R.L., 1989. Hierarchical foraging models: effects of stocking and landscape composition on simulated resource use by cattle. *Ecological Modelling,* 46 (3/4), 283-303.

Senft, R.L., Coughenour, M.B., Bailey, D.W., et al., 1987a. Large herbivore foraging and ecological hierarchies. *BioScience,* 37 (11), 789-799.

Senft, R.L., Rittenhouse, L.R. and Woodmansee, D.R.G., 1985a. Factors influencing patterns of cattle grazing behavior on shortgrass steppe. *Journal of Range Management,* 38 (1), 82-87.

Senft, R.L., Rittenhouse, L.R. and Woodmansee, R.G., 1985b. Factors influencing selection of resting sites by cattle on shortgrass steppe. *Journal of Range Management,* 38 (4), 295-299.

Senft, R.L., Stillwell, M.A. and Rittenhouse, L.R., 1987b. Nitrogen and energy budgets of free-roaming cattle. *Journal of Range Management,* 40, 421.

Shipley, L.A., 1999. Grazers and browsers: how digestive morphology affects diet selection. *In:* Launchbaugh, K.L., Sanders, K.D. and Mosley, J.C. eds. *Grazing behavior of livestock and wildlife.* University of Idaho, Moscow, 20–27. Idaho Forest, Wildlife and Range Experimental Station Bulletin no. 70.

Shipley, L.A., Gross, J.E., Spalinger, D.E., et al., 1994. The scaling of intake rate in mammalian herbivores. *American Naturalist,* 143 (6), 1055-1082.

Shipley, L.A. and Spalinger, D.E., 1992. Mechanics of browsing in dense food patches: effects of plant and animal morphology on intake rate. *Canadian Journal of Zoology,* 70, 1743-1752.

Shipley, L.A., Spalinger, D.E., Gross, J.E., et al., 1996. The dynamics and scaling of foraging velocity and encounter rate in mammalian herbivores. *Functional Ecology,* 10 (2), 234-244.

Shiyomi, M., 1987. Characteristics of herbage biomass estimated by the quadrat method. *Journal of Japanese Society of Grassland Science,* 32 (4), 307-312.

Shiyomi, M., Akiyama, T. and Takahashi, S., 1983. A spatial pattern model of plant biomass in grazing pasture. *Journal of Japanese Society of Grassland Science,* 28 (4), 373-382.

Shiyomi, M., Akiyama, T. and Takahashi, S., 1984. A spatial pattern model of plant biomass in grazing pasture. II. *Journal of Japanese Society of Grassland Science,* 30 (1), 40-48.

Shiyomi, M., Koyama, N. and Tsuiki, M., 1991. Method for predicting available herbage biomass on grazed pasture. *Journal of Japanese Society of Grassland Science,* 36 (4), 327-337.

Shiyomi, M., Okada, M., Takahashi, S., et al., 1998. Spatial pattern changes in aboveground plant biomass in a grazing pasture. *Ecological Research,* 13 (3), 313-322.

Short, J., 1985. The functional response of kangaroos, sheep and rabbits in an arid grazing system. *The Journal of Applied Ecology,* 22 (2), 435-447.

Sibly, R.M. and Smith, R.H., 1985. *Behavioural ecology: ecological consequences of adaptive behaviour: the 25th symposium of the British Ecological Society, Reading 1984.* Blackwell, Oxford.

Silver, H., Colovos, N.F., Holter, J.B., et al., 1969. Fasting metabolism of white-tailed deer. *The Journal of Wildlife Management,* 33 (3), 490-498.

Simpson, S.J. and Raubenheimer, D., 1993. A multi-level analysis of feeding behaviour: the geometry of nutritional decisions. *Philosophical Transactions: Biological Sciences,* 342 (1302), 381-402.

Simpson, S.J. and Raubenheimer, D., 1999. Assuaging nutritional complexity: a geometrical approach. *Proceedings of the Nutrition Society,* 58 (4), 779-789.

Simpson, S.J., Sibly, R.M., Lee, K.P., et al., 2004. Optimal foraging when regulating intake of multiple nutrients. *Animal Behaviour,* 68 (6), 1299-1311.

Sinclair, A.R., Mduma, S. and Brashares, J.S., 2003. Patterns of predation in a diverse predator-prey system. *Nature,* 425 (6955), 288-290.

Sinclair, A.R.E., 1977. *African buffalo: a study of resource limitation of populations*. University of Chicago Press, Chicago.
Sinclair, A.R.E., 1985. Does interspecific competition or predation shape the African ungulate community? *Journal of Animal Ecology*, 54 (3), 899-918.
Sinclair, A.R.E. and Arcese, P., 1995a. *Serengeti 2: dynamics, management, and conservation of an ecosystem [based on papers at a workshop held in December 1991 at the Serengeti Research Institute, Tanzania*. University of Chicago Press, Chicago.
Sinclair, A.R.E. and Arcese, P., 1995b. Population consequences of predation-sensitive foraging: the Serengeti wildebeest. *Ecology*, 76 (3), 882-891.
Sinclair, A.R.E. and Norton-Griffiths, M., 1982. Does competition or facilitation regulate migrant ungulate populations in the Serengeti? A test of hypotheses. *Oecologia*, 53 (3), 364-369.
Sinclair, A.R.E. and Smith, N.M., 1984. Do plant secondary compounds determine feeding preferences of snowshoe hares? *Oecologia*, 61 (3), 403-410.
Skidmore, A.K., 1989. An expert system classifies eucalypt forest types using Thematic Mapper data and a digital terrain model. *PEandRS, Photogrammetric Engineering and Remote Sensing*, 55 (10), 1449-1464.
Skidmore, A.K., Bijker, W., Schmidt, K., et al., 1997. Use of remote sensing and GIS for sustainable land management. *International Journal of Applied Earth Observation and Geoinformation* 3(4), 302-315.
Skidmore, A.K., Schmidt, K.S., Kloosterman, H., et al., 2001. *Hyperspectral imagery for coastal wetland vegetation mapping*. Netherlands Remote Sensing Board (BCRS) Programme Bureau Rijkswaterstaat Survey Department, Delft.
Skidmore, A.K., Watford, F., Luckananurug, P., et al., 1996. An operational GIS expert system for mapping forest soils. *Photogrammetric Engineering and Remote Sensing*, 62 (5), 501-511.
Smith, N.S., 1970. Appraisal of condition estimation models for East African ungulates. *East African Wildlife Journal* 8, 123-129.
Solanki, G.S., 1994. Feeding habits and grazing behavior of goats in a semi-arid region of India. *Small Ruminant Research*, 14 (1), 39-43.
Spalinger, D.E., Hanley, T.A. and Robbins, C.T., 1988. Analysis of the functional response in foraging in the Sitka black-tailed deer. *Ecology*, 69 (4), 1166-1175.
Spalinger, D.E. and Hobbs, N.T., 1992. Mechanisms of foraging in mammalian herbivores: new models of functional response. *The American Naturalist*, 140 (2), 325-348.
Sperling, L. and Galaty, J.G., 1990. Cattle, culture, and economy: dynamics in East African pastoralism. *In:* Galaty, J.G. and Johnson, D.L. eds. *World of pastoralism: herding systems in comparative perspective*. Guildford Press, New York, 69-98.
Stanley Price, M.R., 1974. *The feeding ecology of Coke's hartebeest, Alcelaphus buselaphus cokei Günther, in Kenya*. University of Oxford, Oxford. DPhil thesis
Stein, A., Van Lieshout, M.N.M. and Booltink, H.W.G., 2001. Spatial interaction of methylene blue stained soil pores. *Geoderma*, 102 (1/2), 101-121.
Stenning, D.J., 1957. Transhumance, migratory drift, migration: patterns of pastoral Fulani nomadism. *The Journal of the Royal Anthropological Institute of Great Britain and Ireland*, 87 (1), 57-73.
Stenning, D.J., 1959. *Savannah nomads: a study of the Wodaabe pastoral Fulani of Western Bornu province Northern region, Nigeria*. Oxford University Press, Oxford.
Stephens, D.W. and Krebs, J.R., 1986. *Foraging theory*. Princeton University Press, Princeton.
Stobbs, T.H., 1973. The effect of plant structure on the intake of tropical pastures. I. Variation in the bite size of grazing cattle. *Australian Journal of Agricultural Research*, 24 (6), 809-819.
Stone, L., Dayan, T. and Simberloff, D., 2000. On desert rodents, favored states, and unresolved issues: scaling up and down regional assemblages and local communities. *The American Naturalist*, 156 (3), 322-328.
Stoyan, D., Kendall, W.S. and Mecke, J., 1995. *Stochastic geometry and its applications*. 2nd edn. Wiley, Chichester.
Stricklin, W.R., 2001. The evolution and domestication of social behaviour. *In:* Keeling, L.J. and Gonyou, H.W. eds. *Social behaviour in farm animals*. Cabi, Wallingford, 83-112. [http://www.cabi.org/bk_BookDisplay.asp?PID=1487]
Sullivan, S. and Rohde, R., 2002. On non-equilibrium in arid and semi-arid grazing systems. *Journal of Biogeography*, 29 (12), 1595-1618.
Suttie, J.M., 2000. *Country pasture/forage resource profiles: Mongolia*. Food and Agriculture Organization of the United Nations, Rome. [http://www.fao.org/ag/AGP/AGPC/doc/Counprof/mongol1.htm]
Swain, T., 1979. Tanins and lignins. *In:* Rosenthal, G.A. and Janzen, D.H. eds. *Herbivores: their interaction with secondary plant metabolites*. Academic Press, New York, 657-683.

Swift, J., 1977. Sahelian pastoralists: underdevelopment, desertification, and famine. *Annual Review of Anthropology,* 6 (1), 457-478.

Swift, J., 1986. The economics of production and exchange in West African pastoral societies. *In:* Adamu, M. and Kirk-Greene, A.H.M. eds. *The pastoralists of the West African savanna.* Manchester University Press, Manchester, 175-189.

Sylla, D., 1994. Pastoral organizations for uncertain environments. *In:* Scoones, I. ed. *Living with uncertainty: new directions in pastoral development in Africa.* Intermediate Technology Publications, London, 134-152.

Taylor, C.R., 1969. The eland and the oryx. *Scientific American,* 220 (1), 88-95.

Taylor, R.J., 1984. Foraging in the eastern grey kangaroo and the wallaroo. *The Journal of Animal Ecology,* 53 (1), 65-74.

Thornton, P.K., Galvin, K.A. and Boone, R.B., 2003. An agro-pastoral household model for the rangelands of East Africa. *Agricultural Systems,* 76 (2), 601-622.

Tieleman, B.I. and Williams, J.B., 2000. The adjustment of avian metabolic rates and water fluxes to desert environments. *Physiological and Biochemical Zoology,* 73 (4), 461-479.

Tilman, D., 1982. *Resource competition and community structure.* Princeton University Press, Princeton.

Tilman, D., 1988. *Plant strategies and the dynamics and structure of plant communities.* Princeton University Press, Princeton.

Tilman, D. and Kareiva, P., 1997. *Spatial ecology: the role of space in population dynamics and interspecific interactions.* Princeton University Press, Princeton.

Trudell, J. and White, R.G., 1981. The effect of forage structure and availability on food intake, biting rate, bite size and daily eating time of reindeer. *Journal of Applied Ecology,* 18 (1), 63-81.

Tucker, C.J., 1977. Asymptotic nature of grass canopy spectral reflectance. *Applied Optics,* 16 (5), 1151-1156.

Tucker, C.J., Vanpraet, C.L., Sharman, M.J., et al., 1985. Satellite remote sensing of total herbaceous biomass production in the Senegalese Sahel: 1980-1984. *Remote Sensing of Environment,* 17, 233-249.

Turchin, P., 2003. *Complex population dynamics: a theoretical/empirical synthesis.* Princeton University Press, Princeton.

Turner, M.D., 1998. Long-term effects of daily grazing orbits on nutrient availability in Sahelian West Africa: I. Gradients in the chemical composition of rangeland soils and vegetation. *Journal of Biogeography,* 25 (4), 669-682.

Turner, M.D., 1999a. Merging local and regional analyses of land-use change: the case of livestock in the Sahel. *Annals of the Association of American Geographers,* 89 (2), 191-219.

Turner, M.D., 1999b. The role of social networks, indefinite boundaries and political bargaining in maintaining the ecological and economic resilience of the transhumance systems of Sudano-Sahelian West Africa. *In:* Niamir-Fuller, M. ed. *Managing mobility in African rangelands: the legitimization of transhumance.* Intermediate Technology Publications, London, 97-123.

Turner, M.D., 2003. Multiple holders of multiple stakes: the multilayered politics of agropastoral resource management in semi-arid Africa. *In:* Allsopp N, P.A., Milton SJ, Kirkman KP, Kerley GIH, Hurt CR, Brown CJ ed. *Proceedings of the VIIth International Rangeland Congress, July 26-August 1, 2003, Durban, South Africa.* 1715-1725.

Turner, M.G., Wu, Y., Romme, W.H., et al., 1993. A landscape simulation model of winter foraging by large ungulates. *Ecological Modelling,* 69 (3/4), 163-184.

Tyler, N.J.C., 1987. Body composition and energy balance of pregnant and non-pregnant Svalbard reindeer during winter. *Symposia of the Zoological Society of London,* 57, 203–229.

Underwood, R., 1983. The feeding behaviour of grazing African ungulates. *Behaviour,* 84 (3/4), 195-243.

UNEP, 1992. *World atlas of desertification.* Arnold, London.

Ungar, E.D., 1996. Ingestive behaviour. *In:* Hodgson, J. and Illius, A.W. eds. *The ecology and management of grazing systems.* CAB international, Wallingford, 185-218.

Ungar, E.D., Genizi, A. and Demment, M.W., 1991. Bite dimensions and herbage intake by cattle grazing short hand-constructed swards. *Agronomy Journal,* 83 (6), 973-978.

Valentine, K.A., 1947. Distance from water as a factor in grazing capacity of rangeland. *Journal of Forestry,* 45, 749-754.

Vallentine, J.F., 2001. *Grazing management.* 2nd edn. Academic Press, San Diego.

Van de Koppel, J., 1997. *Trophic interactions in gradients of primary productivity.* University of Groningen, Groningen. PhD thesis

Van de Koppel, J., Huisman, J., Van der Wal, R., et al., 1996. Patterns of herbivory along a productivity gradient: an empirical and theoretical investigation. *Ecology,* 77 (3), 736-745.

Van de Koppel, J., Rietkerk, M., Van Langevelde, F., et al., 2002. Spatial heterogeneity and irreversible vegetation change in semiarid grazing systems. *American Naturalist,* 159 (2), 209-218.
Van de Vijver, C., Poot, P. and Prins, H.H.T., 1999. Causes of increased nutrient concentrations in post-fire regrowth in an East African savanna. *Plant and Soil,* 214 (1), 173-185.
Van de Vijver, C.A.D.M., 1999. *Fire and life in Tarangire: effects of burning and herbivory on an East African Savanna system.* Wageningen University, Wageningen. PhD thesis
Van der Jeugd, H.P. and Prins, H.H.T., 2000. Movements and group structure of giraffe (*Giraffa camelopardalis*) in Lake Manyara National Park, Tanzania. *Journal of Zoology,* 251 (01), 15-21.
Van der Wal, R., Madan, N., Van Lieshout, S., et al., 2000. Trading forage quality for quantity? Plant phenology and patch choice by Svalbard reindeer. *Oecologia,* 123 (1), 108-115.
Van Hoven, W. and Frustenburg, D., 1992. The use of purified condensed tannin as a reference in determining its influence on rumen fermentation. *Comparative Biochemistry and Physiology A, Comparative Physiology,* 101 (2), 381-385.
Van Keulen, H. and Breman, H., 1990. Agricultural development in the West African Sahelian region: a cure against land hunger? *Agriculture, Ecosystems and Environment,* 32 (3/4), 177-197.
Van Langevelde, F., Van de Vijver, C.A.D.M., Lalit, K., et al., 2003. Effects of fire and herbivory on the stability of savanna ecosystems. *Ecology,* 84 (2), 337-350.
Van Lieshout, M.N.M. and Baddeley, A.J., 1999. Indices of dependence between types in multivariate point patterns. *Scandinavian Journal of Statistics,* 26 (4), 511-532.
Van Soest, P.J., 1982. *Nutritional ecology of the ruminant: ruminant metabolism, nutritional strategies, the cellulolytic fermentation and the chemistry of forages and plant fibers.* O & B Books, Corvallis.
Van Soest, P.J., 1994. *Nutritional ecology of the ruminant.* 2nd edn. Comstock, Ithaca.
Van Wieren, S.E., 1992. Factors limiting food intake in ruminants and non-ruminants in the temperate zone. *In:* Spitz, F., Janeau, G., Gonzalez, G., et al. eds. *Ongules - ungulates: proceedings of the international symposium "Ongules - ungulates", Toulouse, France, September 2-6, 1991.* S.F.E.P.M., Paris, 139-145.
Van Wieren, S.E., 1996a. *Digestive strategies in ruminants and nonruminants.* Wageningen University, Wageningen. PhD thesis
Van Wieren, S.E., 1996b. Do large herbivores select a diet that maximizes short-term energy intake rate? *Forest Ecology and Management,* 88 (1/2), 149-156.
Vavra, M. and Ganskopp, D., 1998. Grazing behavior in ungulates: current concepts and future challenges. *Annals of Arid Zones,* 37 (3), 319-335.
Verheyden-Tixier, H. and Duncan, P., 2000. Selection for small amounts of hydrolysable tannins by a concentrate-selecting mammalian herbivore. *Journal of Chemical Ecology,* 26 (2), 351-358.
Verweij, R.J.T., Verrelst, J., Loth, P.E., et al., 2006. Grazing lawns contribute to the subsistence of mesoherbivores on dystrophic savannas. *Oikos,* 114 (1), 108-116.
Vesey-FitzGerald, D.F., 1974. Utilisation of grazing resources by buffalo in the Arusha National Park, Tanzania. *East African Wildlife Journal* 12, 107-134.
Villalba, J.J. and Provenza, F.D., 1996. Preference for flavored wheat straw by lambs conditioned with intraruminal administrations of sodium propionate. *Journal of Animal Science,* 74 (10), 2362-2368.
Villalba, J.J. and Provenza, F.D., 1997a. Preference for flavored wheat straw by lambs conditioned with intraruminal infusions of acetate and propionate. *Journal of Animal Science,* 75 (11), 2905-2914.
Villalba, J.J. and Provenza, F.D., 1997b. Preference for flavoured foods by lambs conditioned with intraruminal administration of nitrogen. *British Journal of Nutrition,* 78 (4), 545-561.
Villalba, J.J. and Provenza, F.D., 1997c. Preference for wheat straw by lambs conditioned with intraruminal infusions of starch. *British Journal of Nutrition,* 77 (2), 287-297.
Villalba, J.J. and Provenza, F.D., 1999. Nutrient-specific preferences by lambs conditioned with intraruminal infusions of starch, casein, and water. *Journal of Animal Science,* 77 (2), 378-387.
Villalba, J.J. and Provenza, F.D., 2002. Polyethylene glycol influences selection of foraging location by sheep consuming quebracho tannin 1. *Journal of Animal Science,* 80 (7), 1846-1851.
VITO, 2003. *VEGETATION.* VITO. Available: [http://www.vgt.vito.be/indexstart.htm].
Voeten, M.M., 1999. *Living with wildlife: coexistence of wildlife and livestock in an East African Savanna system.* Wageningen University, Wageningen. PhD thesis
Voeten, M.M. and Prins, H.H.T., 1999. Resource partitioning between sympatric wild and domestic herbivores in the Tarangire region of Tanzania. *Oecologia,* 120 (2), 287-294.
Vogelmann, J. and Moss, D., 1993. Spectral reflectance measurements in the genus Sphagnum. *Remote Sensing of Environment,* 45 (3), 273-279.
Vos, J. and Bom, M., 1993. Hand-held chlorophyll meter: a promising tool to assess the nitrogen status of potato foliage. *Potato Research,* 36 (4), 301-308.

Vucetich, J.A., Peterson, R.O. and Schaefer, C.L., 2002. The effect of prey and predator densities on wolf predation. *Ecology,* 83 (11), 3003-3013.

Wackernagel, H., 1995. *Multivariate geostatistics: an introduction with applications.* Springer, Berlin.

Wagenaar, K.T., Diallo, A. and Sayers, A.R., 1986. *Productivity of transhumant Fulani cattle in the inner Niger delta of Mali.* ILCA, Addis Ababa.

Walker, B.H., Emslie, R.H., Owen-Smith, R.N., et al., 1987. To cull or not to cull: lessons from a Southern African drought. *The Journal of Applied Ecology,* 24 (2), 381-401.

Walker, J., Jupp, D.L.B., Penridge, L.K., et al., 1986. Interpretation of vegetation structure in Landsat MSS imagery: a case study in disturbed semi-arid eucalypt woodlands. Part 1. Field data analysis. *Journal of Environmental Management,* 23 (1), 19-33.

Walker, P.A., 1990. Modelling wildlife distributions using a geographic information system: kangaroos in relation to climate. *Journal of Biogeography,* 17 (3), 279-289.

Wallace, J.S., Brouwer, J. and Allen, S. J., 1994. *Hapex-Sahel southern super-site report: an overview of the Hapex-Sahel southern super-site and the experimental programme during the intensive observation period in 1992.* Institute of Hydrology, Wallingford.

Wallis de Vries, M.F., 1996. Effects of resource distribution patterns on ungulate foraging behaviour: a modelling approach. *Forest Ecology and Management,* 88 (1/2), 167-177.

Wallis de Vries, M.F. and Daleboudt, C., 1994. Foraging strategy of cattle in patchy grassland. *Oecologia,* 100 (1/2), 98-106.

Wallis de Vries, M.F., Laca, E.A. and Demment, M.W., 1999. The importance of scale of patchiness for selectivity in grazing herbivores. *Oecologia,* 121 (3), 355-363.

Wallis de Vries, M.F. and Schippers, P., 1994. Foraging in a landscape mosaic: selection for energy and minerals in free-ranging cattle. *Oecologia,* 100 (1/2), 107-117.

Wallis, I.R., Green, B. and Newgrain, K., 1997. Seasonal field energetics and water fluxes of the long-nosed potoroo (*Potorous tridactylus*) in southern Victoria. *Australian Journal of Zoology,* 45 (1), 1-11.

Walter, H., 1971. *Ecology of tropical and subtropical vegetation.* Oliver and Boyd, Edinburgh.

Wang, J. and Provenza, F.D., 1996. Food deprivation affects preference of sheep for foods varying in nutrients and a toxin. *Journal of Chemical Ecology,* 22 (11), 2011-2021.

Ward, D., 1992. The role of satisficing in foraging theory. *Oikos,* 63 (2), 312-317.

Ward, D. and Saltz, D., 1994. Foraging at different spatial scales: dorcas gazelles foraging for lilies in the Negev desert. *Ecology,* 75 (1), 48-58.

Waser, P., 1975. Diurnal and nocturnal strategies of the bushbuck. *East African Wildlife Journal,* 13, 49-64.

Weiher, E. and Keddy, P.A., 1999. Relative abundance and evenness patterns along diversity and biomass gradients. *Oikos,* 87 (2), 355-361.

Weiner, J., 1977. Energy metabolism of the roe deer. *Acta Theriologica,* 22 (1/8), 3-24.

West, G.B., Savage, V.M., Gillooly, J., et al., 2003. Why does metabolic rate scale with body size? *Nature,* 421 (6924), 713.

West, G.B., Woodruff, W.H. and Brown, J.H., 2002. Allometric scaling of metabolic rate from molecules and mitochondria to cells and mammals. *PNAS,* 99 (Suppl. 1), 2473-2478.

Westoby, M., 1978. What are the biological bases of varied diets? *The American Naturalist,* 112 (985), 627-631.

Westoby, M., 1979. Elements of a theory of vegetation dynamics in arid rangelands. *Israel Journal of Botany,* 28 (3/4), 169-194.

Westoby, M., Walker, B. and Noy-Meir, I., 1989. Opportunistic management for rangelands not at equilibrium. *Journal of Range Management,* 42 (4), 266-274.

Wickstrom, M.L., Robbins, C.T., Hanley, T.A., et al., 1984. Food intake and foraging energetics of elk and mule deer. *Journal of Wildlife Management,* 48 (4), 1285-1301.

Wiens, J.A., 1992. *The ecology of bird communities.* Repr. edn. Cambridge University Press, Cambridge.

Williams, J.B., Anderson, M.D. and Richardson, P.R.K., 1997. Seasonal differences in field metabolism, water requirements, and foraging behavior of free-living aardwolves. *Ecology,* 78 (8), 2588-2602.

Williams, R.E., 1954. Modern methods of getting uniform use of ranges. *Journal of Range Management,* 7 (2), 77-81.

Wilmshurst, J.F. and Fryxell, J.M., 1995. Patch selection by red deer in relation to energy and protein intake: a re-evaluation of Langvatn and Hanley's (1993) results. *Oecologia,* 104 (3), 297-300.

Wilmshurst, J.F., Fryxell, J.M. and Bergman, C.M., 2000. The allometry of patch selection in ruminants. *Proceedings of the Royal Society of London Series B, Biological Sciences,* 267 (1441), 345-349.

Wilmshurst, J.F., Fryxell, J.M. and Colucci, P.E., 1999a. What constrains daily intake in Thomson's gazelles? *Ecology,* 80 (7), 2338-2347.

Wilmshurst, J.F., Fryxell, J.M., Farm, B.P., et al., 1999b. Spatial distribution of Serengeti wildebeest in relation to resources. *Canadian Journal of Zoology,* 77 (8), 1223-1232.
Wilmshurst, J.F., Fryxell, J.M. and Hudson, R.J., 1995. Forage quality and patch choice by wapiti (*Cervus elaphus*). *Behavioral Ecology,* 6 (2), 209-217.
Wilson, E.O. and Bossert, W.H., 1971. *A primer of population biology*. Sinauer, Sunderland.
Wilson, R.T., 1980. Fuelwood consumption in a central Malian town and its effects on browse availability. *In:* Le Houerou, H.N. ed. *Browse in Africa: the current state of knowledge*. International Livestock Centre for Africa, Addis Ababa, 473-476.
Wilson, S.L. and Kerley, G.I.H., 2003. Bite diameter selection by thicket browsers: the effect of body size and plant morphology on forage intake and quality. *Forest Ecology and Management,* 181 (1/2), 51-65.
Winston, W.L., 1994. *Operations research: applications and algorithms*. Duxbury, Belmont.
Wint, W., 1998. Rapid resource assessment and environmental monitoring using low level aerial surveys. *In:* Squires, V.R. and Sidahmed, A.E. eds. *Drylands: sustainable use of rangelands into the twenty-first century*. IFAD, Rome, 277-301.
Winterbach, H.E.K. and Bothma, J.D.P., 1998. Activity patterns of the Cape buffalo *Syncerus caffer caffer* in the Willem Pretorius Game Reserve, Free State. *South African Journal of Wildlife Research,* 28 (3), 73-81.
Wissinger, S.A., Sparks, G.B., Rouse, G.L., et al., 1996. Intraguild predation and cannibalism among larvae of detritivorous caddisflies in subalpine wetlands. *Ecology,* 77 (8), 2421-2430.
Woodward, F.I., 1987. *Climate and plant distribution*. Cambridge University Press, Cambridge.
Worden, J., Reid, R. and Gichohi, H., 2003. *Land-use impacts on large wildlife and livestock in the swamps of the greater Amboseli ecosystem, Kajiado District, Kenya*. International Livestock Research Institute Nairobi. Land Use Change Impacts and Dynamics (LUCID) Working Paper no. 27.
Wyant, J.G. and Ellis, J.E., 1990. Compositional patterns of riparian woodlands in the Rift Valley of northern Kenya. *Vegetatio,* 89 (1), 23-37.
Wynne-Edwards, V.C., 1962. *Animal dispersion in relation to social behaviour*. Oliver and Boyd, Edinburgh.
Ydenberg, R.C., Prins, H.H.T. and Van Dijk, J., 1983. Postroost gatherings of wintering Barnacle Geese: information centres. *Ardea,* 73, 125-132.
Yearsley, J., Hastings, I.M., Gordon, I.J., et al., 2002. A lifetime perspective on foraging and mortality. *Journal of Theoretical Biology,* 215 (4), 385-397.
Zervas, G., 1998. Quantifying and optimizing grazing regimes in Greek mountain systems. *Journal of Applied Ecology,* 35 (6), 983-986.
Zonneveld, I.S., 1988. Survey approaches. *In:* Küchler, A.W. and Zonneveld, I.S. eds. *Vegetation mapping*. Kluwer, Dordrecht.

LIST OF CONTRIBUTORS

Derek W. Bailey	Northern Agricultural Research Centre, Montana State University, Havre, USA
Randall B. Boone	Natural Resource Ecology Laboratory, Colorado State University, Fort Collins, USA
Joost Brouwer	Resource Ecology Group, Wageningen University, Wageningen, The Netherlands
Arend M.H. Brunsting	Resource Ecology Group, Wageningen University, Wageningen, The Netherlands
Shauna B. BurnSilver	Natural Resource Ecology Laboratory, Colorado State University, Fort Collins, USA
Willem F. de Boer	Resource Ecology Group, Wageningen University, Wageningen, The Netherlands
Michael Drescher	Ontario Forest Research Institute, Sault Ste. Marie, Canada
Jelle G. Ferwerda	International Institute for Geo-Information Science and Earth Observation (ITC), Enschede, The Netherlands
John M. Fryxell	Department of Zoology, University of Guelph, Guelph, Canada
Kathleen A. Galvin	Natural Resource Ecology Laboratory and Department of Anthropology, Colorado State University, Fort Collins, USA
Nicholas Georgiadis	Property and Environment Research Center, Bozeman, MT, USA
Iain J. Gordon	CSIRO, Davies Laboratory, Aitkenvale, Australia
Thomas A. Groen	Resource Ecology Group, Wageningen University, Wageningen, The Netherlands
Ignas M.A. Heitkönig	Resource Ecology Group, Wageningen University, Wageningen, The Netherlands
N. Thompson Hobbs	Natural Resource Ecology Laboratory, Colorado State University, Fort Collins, USA
Koen Kramer	Alterra, Wageningen University and Research Centre, Wageningen, The Netherlands
Emilio A. Laca	Department of Plant Sciences, University of California, Davis, USA
Norman Owen-Smith	Centre for African Ecology, School of Animal, Plant and Environmental Sciences, University of the Witwatersrand, Wits, South Africa

Herbert H.T. Prins	Resource Ecology Group, Wageningen University, Wageningen, The Netherlands
Frederick D. Provenza	Department of Forest, Range, and Wildlife Sciences, Utah State University, Logan, USA
Paul Scholte	NC-IUCN, Leiden University & Independent Consultant, Nieuwe Teertuinen 12C, 1013 LV Amsterdam, The Netherlands
Andrew K. Skidmore	International Institute for Geo-Information Science and Earth Observation (ITC), Enschede, The Netherlands
Alfred Stein	International Institute for Geo-Information Science and Earth Observation (ITC), Enschede, The Netherlands
Frank van Langevelde	Resource Ecology Group, Wageningen University, Wageningen, The Netherlands
Sipke E. van Wieren	Resource Ecology Group, Wageningen University, Wageningen, The Netherlands
Jeffrey S. Worden	Natural Resource Ecology Laboratory, Colorado State University, Fort Collins, USA

Wageningen UR Frontis Series

1. A.G.J. Velthuis, L.J. Unnevehr, H. Hogeveen and R.B.M. Huirne (eds.): *New Approaches to Food-Safety Economics*. 2003
ISBN 1-4020-1425-2; Pb: 1-4020-1426-0
2. W. Takken and T.W. Scott (eds.): *Ecological Aspects for Application of Genetically Modified Mosquitoes.* 2003
ISBN 1-4020-1584-4; Pb: 1-4020-1585-2
3. M.A.J.S. van Boekel, A. Stein and A.H.C. van Bruggen (eds.): *Proceedings of the Frontis workshop on Bayesian Statistics and quality modelling.* 2003
ISBN 1-4020-1916-5
4. R.H.G. Jongman (ed.): *The New Dimensions of the European Landscape.* 2004
ISBN 1-4020-2909-8; Pb: 1-4020-2910-1
5. M.J.J.A.A. Korthals and R.J.Bogers (eds.): *Ethics for Life Scientists.* 2004
ISBN 1-4020-3178-5; Pb: 1-4020-3179-3
6. R.A. Feddes, G.H.de Rooij and J.C. van Dam (eds.): *Unsaturated-zone modeling.* Progress, challenges and applications. 2004 ISBN 1-4020-2919-5
7. J.H.H. Wesseler (ed.): *Environmental Costs and Benefits of Transgenic Crops.* 2005 ISBN 1-4020-3247-1; Pb: 1-4020-3248-X
8. R.S. Schrijver and G. Koch (eds.): *Avian Influenza.* Prevention and Control. 2005
ISBN 1-4020-3439-3; Pb: 1-4020-3440-7
9. W. Takken, P. Martens and R.J. Bogers (eds.): *Environmental Change and Malaria Risk.* Global and Local Implications. 2005
ISBN 1-4020-3927-1; Pb: 1-4020-3928-X
10. L.J.E.J. Gilissen, H.J. Wichers, H.F.J. Savelkoul and R.J. Bogers, (eds.): *Allergy Matters.* New Approaches to Allergy Prevention and Management. 2005
ISBN 1-4020-3895-X; Pb: 1-4020-3896-8
11. B.G.J. Knols and C. Louis (eds.): *Briding Laboratory and Field Research for Genetic Control of Disease Vectors.* 2006
ISBN 1-4020-3800-3; Pb: 1-4020-3799-6
12. B. Tress, G. Tress, G. Fry and P. Opdam (eds.): *From Landscape Research to Landscape Planning.* Aspects of Integration, Education and Application. 2006
ISBN 1-4020-3979-4; Pb: 1-4020-3978-6
13. J. Hassink and M. van Dijk (eds.): *Farming for Health.* Green-Care Farming Across Europe and the United States of America. 2006
ISBN 1-4020-4540-9; Pb: 1-4020-4541-7
14. R. Ruben, M. Slingerland and H. Nijhoff (eds.): *The Agro-Food Chains and Networks for Development.* 2006 ISBN 1-4020-4592-1; Pb: 1-4020-4600-6
15. C.J.M. Ondersteijn, J.H.M. Wijnands, R.B.M. Huiren and O. van Kooten (eds.): *Quantifying the Agri-Food Supply Chain.*
ISBN 1-4020-4692-8; Pb: 1-4020-4693-6
16. M. Dicke and W. Takken (eds.): *Chemical Ecology.* From Gene to Ecosystem. 2006 ISBN 1-4020-4783-5; Pb: 1-4020-4792-4
17. R.J. Bogers, L.E. Craker and D. Lange (eds.): *Medicinal and Aromatic Plants.* Agricultural, Commercial, Ecological, Legal, Pharmacological, and Social Aspects. 2006 ISBN 1-4020-5447-5; Pb: 1-4020-5448-3

18. A. Elgersma, J. Dijkstra and S. Tamminga (eds.): *Fresh Herbage for Dairy Cattle.* The Key to a Sustainable Food Chain. 2006
ISBN 1-4020-5450-5; Pb: 1-4020-5451-3
19. N. Koning and P. Pinstrup-Andersen (eds.): *Agricultural Trade Liberalization and the Least Developed Countries*. 2007
ISBN 978-1-4020-6079-3; Pb: 978-1-4020-6085-4
20. A.G.J.M. Oude Lansink (ed.): *New Approaches to the Economics of Plant Health.* 2007 ISBN 1-4020-5825-X; Pb: 1-4020-5826-8
21. J.H.J. Spiertz, P.C. Struik and H.H. van Laar (eds.): *Scale and Complexity in Plant Systems Research.* Gene-Plant-Crop Relations. 2007
ISBN 1-4020-5904-3; Pb: 1-4020-5905-1
22. J. Vos, L.F.M. Marcelis, P.H.B. de Visser, P.C. Struik and J.B. Evers (eds.): *Functional-Structural Plant Modelling in Crop Production.* 2007
ISBN 978-1-4020-6032-8; Pb: 978-1-4020-6033-5
23. H.H.T. Prins and F. van Langevelde (eds.): *Resource Ecology*. Spatial and Temporal Dynamics of Foraging. 2008
ISBN 978-1-4020-6848-5; Pb: 978-1-4020-6849-2